Caper

The Genus *Capparis*

Traditional Herbal Medicines for Modern Times

Each volume in this series provides academia, health sciences, and the herbal medicines industry with in-depth coverage of the herbal remedies for infectious diseases, certain medical conditions, or the plant medicines of a particular country.

Series Editor: Dr. Roland Hardman

Volume 1
Shengmai San, edited by Kam-Ming Ko

Volume 2
Rasayana: Ayurvedic Herbs for Rejuvenation and Longevity, by H.S. Puri

Volume 3
Sho-Saiko-To: (Xiao-Chai-Hu-Tang) Scientific Evaluation and Clinical Applications, by Yukio Ogihara and Masaki Aburada

Volume 4
Traditional Medicinal Plants and Malaria, edited by Merlin Willcox, Gerard Bodeker, and Philippe Rasoanaivo

Volume 5
Juzen-taiho-to (Shi-Quan-Da-Bu-Tang): Scientific Evaluation and Clinical Applications, edited by Haruki Yamada and Ikuo Saiki

Volume 6
Traditional Medicines for Modern Times: Antidiabetic Plants, edited by Amala Soumyanath

Volume 7
Bupleurum *Species: Scientific Evaluation and Clinical Applications,* edited by Sheng-Li Pan

Volume 8
Herbal Principles in Cosmetics: Properties and Mechanisms of Action, by Bruno Burlando, Luisella Verotta, Laura Cornara, and Elisa Bottini-Massa

Volume 9
Figs: The Genus Ficus, by Ephraim Philip Lansky and Helena Maaria Paavilainen

Volume 10
Phyllanthus Species: Scientific Evaluation and Medicinal Applications edited by Ramadasan Kuttan and K. B. Harikumar

Volume 11
Honey in Traditional and Modern Medicine, edited by Laïd Boukraâ

Traditional Herbal Medicines for Modern Times

Caper

The Genus *Capparis*

Ephraim Philip Lansky
Helena Maaria Paavilainen
Shifra Lansky

CRC Press
Taylor & Francis Group
Boca Raton London New York

CRC Press is an imprint of the
Taylor & Francis Group, an **informa** business

Cover photo: Tsachi Klein, Cover watercolor: Zipora Lansky.

CRC Press
Taylor & Francis Group
6000 Broken Sound Parkway NW, Suite 300
Boca Raton, FL 33487-2742

First issued in paperback 2019

ISBN-13: 978-1-4398-6136-3 (hbk)
ISBN-13: 978-0-367-37920-9 (pbk)

Library of Congress Cataloging-in-Publication Data

Lansky, Ephraim P.
Caper : the genus Capparis / authors: Ephraim Philip Lansky, Helena Maaria Paavilainen, Shifra Lansky.
p. cm. -- (Traditional herbal medicines for modern times ; v. 12)
ISBN 978-1-4398-6136-3
1. Capparis. 2. Capparis--Therapeutic use. I. Paavilainen, Helena M. II. Lansky, Shifra, 1991- III. Title. IV. Series: Traditional herbal medicines for modern times ; v. 12.

QK495.C198L36 2013
583'.64--dc23 2013019519

Visit the Taylor & Francis Web site at
http://www.taylorandfrancis.com

and the CRC Press Web site at
http://www.crcpress.com

Dedication

for Yu

E.P.L.

for my family

S.L.

for Kaarina, the best of sisters

H.M.P.

Contents

SECTION II Medical Uses

SECTION III Miscellany

List of Tables

List of Illustrations

Acknowledgments

E.L. I am indebted to Aryeh Naftali, poet, musician, and gentleman farmer, and his family of Reb Shlomo Carlebach *z"l chevra* on Moshav Modi'in for first introducing me to the joy, mystery, and practical benefits of wildcrafting and pickling capers during my first year in Israel, 18 years ago. Shelly Shen-Aridor of the University of Haifa Library was a loyal and devoted research partner, miraculously turning up rare articles at all times of day and night, and was always ready to review photographs. Professor Eviatar Nevo, founder of the Institute of Evolution, International Graduate Center of Evolution, and prime mover of the Faculty of Sciences at the University of Haifa, first introduced to me the five species of *Capparis* in Israel. I hope our treatment here is worthy of his scholarship! Dr. Eli Harlev was instrumental in making contacts for safaris into the Negev Desert to locate the more obscure species. Family members Dr. Yu Shen, Zipora, Shifra, Aaron, Gail, and Yale Lansky and Rabbi Sue Ann Wasserman sampled *C. spinosa* buds and fruits preserved with balsamic vinegar and shared their experiences. Drs. Martin Philip Goldman and Daniel Rubin provided generous spiritual and material support. Finally, I wish to thank our editors at CRC: Barbara Norwitz and John Sulzycki, whose vision and patience provided the context once again for our project to occur, and Dr. Roland Hardman, who generously shared his lifetime of pharmacognostical knowledge and enthusiasm to help us traverse the roads less traveled. Thank you again for all.

H.P. My first expression of gratitude is due to Dr. Ephraim Lansky as the initiator and moving force of this project. His choice of *Capparis* as the focus of our work surprised me at first, but time—and capers themselves—have shown that his choice was a good one. Thank you for letting me participate in this. And, if the project would not have begun without Dr. Lansky, it would not have come to completion without Shifra Lansky and her grasp on the intricacies of chemistry. Thank you.

During the data-collecting stage and in particular when collecting the photographic material, I was again and again surprised by the generosity and kindness of people I met. Dr. Elaine Soloway from Kibbutz Ketura introduced me personally to three types of *Capparis* that she cultivates and studies in the surroundings of the kibbutz and hosted my photographing visits in the kibbutz. Dr. Hagar Leschner, director of the National Herbarium at the Hebrew University of Jerusalem, searched the best caper plants for me during our botanic trips. Ryan Brookes, Eli Harlev, William Hawthorne, Satish Nikam, C. E. Timothy Paine, Sébastien Sant, Abhijeet Shiral, Ming-I Weng, Bart T. Wursten, and the staff of SCCF Native Plant Nursery generously shared their photographs and thus enabled me to overcome geographic limitations. (My special thanks to the photographer who not only sent me the photos I asked for but also went to the botanical garden and took some extra photos of another *Capparis* species for the book.) You all have been an enormous help and encouragement.

I am also very grateful to the editorial team at CRC—Barbara Norwitz, John Sulzycki, Jill Jurgensen, Roland Hardman, and Marsha Hecht—for accepting the idea of this project and for their encouraging and supportive attitude and endless patience. It has been very good to work with you on this project. I hope the results are satisfying.

And last but not least, my gratitude goes to the three persons who have supported me most in my work: to Prof. Samuel Kottek, my guide and mentor, who has helped and advised me from the beginning of my historical-medical studies and still does, and to my mother and sister, Maija and Kaarina Paavilainen, for encouraging me in tired moments and simply listening. Without you, my academic path would have been much lonelier.

About the Authors

Ephraim Philip Lansky holds an interdisciplinary BA in psychology/biology from New College in Sarasota, Florida, USA. He completed his MD at the University of Pennsylvania and an internship in medicine and psychiatry at Pennsylvania Hospital, both in Philadelphia. He also achieved certification as an instructor of advanced trauma and cardiac life support and is experienced in basic emergency medicine. He received an MBA from the University of Bradford (United Kingdom) and a PhD from Leiden University (The Netherlands) for his thesis, "Anticancer Pharmacognosy of *Punica granatum*." Author or coauthor of 23 peer-reviewed publications, 5 patents, and 2 books (*Pomegranate: The Most Medicinal Fruit,* Basic Books, New York; and *Figs*: *the Genus Ficus,* CRC Press, Boca Raton, FL), he is also the founder of Rimonest Limited and Punisyn Pharmaceuticals Limited, companies devoted to the economic development of the pomegranate fruit for nutritional/cosmeceutical and medical applications, respectively. A second-degree black belt and teacher of Kokikai Aikido, Dr. Lansky also is an avid archer and downhill skier. He has over 25 years of clinical experience in complementary and holistic medicine, including acupuncture, herbology, homeopathy, and hypnosis. He makes his home in Haifa, Israel, where he is a member of the research faculty and director of the Laboratory of Applied Metabolomics and Pharmacognosy (LAMP) at the University of Haifa. His chief academic and clinical interest is anticancer pharmacognosy.

Shifra Lansky holds a BSc in chemistry from Hebrew University in Jerusalem where she is presently pursuing her graduate studies. Her focus is on characterizing the three-dimensional structures of naturally occuring proteins. Shifra enjoys playing the violin, painting, and skiing in her spare time.

Helena Paavilainen is a researcher at the Hadassah Medical School, Hebrew University of Jerusalem, Israel. Her main research interests are ethnomedicine, historical ethnopharmacology, and the history of pharmacology, especially the Hebrew, Arabic, and Latin traditions. She wrote her PhD thesis (published as "Medieval Pharmacotherapy: Continuity and Change; Case Studies from Ibn Sina and Some of His Late Medieval Commentators," Leiden: Brill 2009) on the development of medical drug therapy in medieval times and on the potential validity of medieval herbal treatments. She also coauthored with Dr. Lansky the monograph *Figs: The Genus Ficus* (Boca Raton, FL: CRC Press, 2010). She currently works as a freelance consultant bioprospecting ancient and medieval herbal texts for practical applications in medicine, functional nutrition, and agriculture.

1 Mythopoesis/Meditation

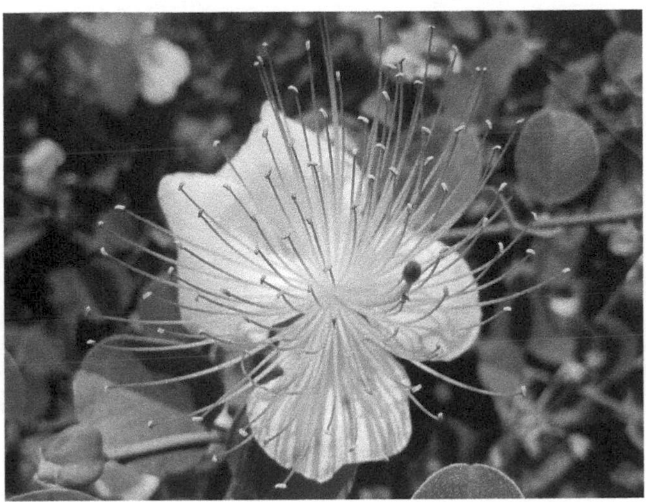

INTRODUCTION

The name of the genus, *Capparis*, and the common name for its edible parts, *caper*, both derive from the word for goat. C*aprification* (Lansky and Paavilainen 2011) is the fertilization of female fig trees by wasps from male trees carrying the latter's pollen. The word hints at the caprice of the goat, primordial symbol/spirit of the randy male.

Spreading of pollen resembles a casual event, and "caprice" conveys the lighthearted demeanor of Capricorn ("your hills skip like rams"). As this spirit of skipping lightly is sacrificed by the heavier elements of society, a need for expiation, כפרה (*kapara*, Hebrew) emerges, underlying the Day of Atonement, יום כפור (Yom Kippur). The goat is the universal symbol of internal forgiveness and "at-one-ment," the scapegoat *(se'ir le-azalel)* in Torah, and the sacrifice of the innocent and passive elements of the psyche and of their material "projections" in society and "reality," that is, the physical results of Mind.

The capture of the caper plant for this purpose is reflected in the "crown of thorns" from a plant, if not identical to *Capparis spinosa*, at least closely resembling it in terms of its prime prickly mechanism of defense. The plant conveys the "sacrificial attitude" required for the internality and passivity (parasympathetic arousal) needed for rest and repair, reduction of inflammation, and healing of many diseases. In that "like cures like," the pain of the caper's thorn ameliorates deeper and sharper pains.

Caper may be understood as nothing less than the spirit of life, and this is also reflected in the ability of *Capparis* to establish itself in the most hostile little niches, for example, growing directly out of a stone wall or in the fiercest and driest of deserts (Figures 1.1, 2.1). This attitude of adventure seeking and extreme adaptability is reflected in the slang of "caper," a kind of dangerous tale of adventure, often against logic or law. The exploratory urge in Capricorn and the caprice of goats is reflected in the capers of dalliance from the mean. Caper flowers, fruits, roots, and leaves may also provide solace and comfort.

FIGURE 1.1 The ability of caper bushes to grow on bare stone walls is astonishing: *Capparis spinosa* plants on the Western Wall. (7.4.2010, Western Wall, Old City of Jerusalem, Israel: by Helena Paavilainen.)

FIGURE 1.2 Spines of *Capparis spinosa*. A close look shows how sharp they really are, both in summer when the plant is fresh and in winter when it is dry. (7.7.2011 and 7.4.2010, respectively, Jerusalem, Israel: by Helena Paavilainen.)

A CAPER HAS BEEN PULLED OFF

So, now let us look deeply into the caper's heart and soul and see what it takes, and what it makes, in the way of phytoactive compounds that are able to benefit and tantalize *Homo sapiens* with health and pampering. The caper developed on a hill overlooking a valley to the sea off a Mediterranean coast who knows how many years ago, and now it has spread back to the hills. The evolution of the caper depended on multiple contingencies that, once converged, allowed the plot to be planted. Of note was its staying power even when nurture was most deeply hidden. And, what of the reason for the venture into new biospheres? Caprice? Capricorn, caper, *Capparis*, *Kaprisin* (Hebrew for "Cyprus"), from where it commenced one Cyprus morning?

Or, is the caper also a compensation to balance a biospheric need? Atonement for the human spirit: *kappara*, *kapooris*, *kippur*? (*Kappara*, *kapooris*, *kippur* are Hebrew words conveying different relations to the process of compensation on a spiritual phase and may also apply to an evolutionary phase, in-depth psychology, a sacrificial attitude, means of achieving at-one-ment, psychophysiological, somatopsychic integration.)

But, regarding the plant: First, consider the prickers (Figure 1.2). Before one can pull off the first caper, one must feel its sting. The prickers are sharp. At some point, you may evict droplets of blood, but the damage is not usually severe. Nevertheless, capricious would-be takers are discouraged.

REFERENCE

Lansky, E.P. and H.M. Paavilainen. 2011. *Figs: The Genus* Ficus. Boca Raton, FL: CRC Press.

2 Botany and Introduction

ECOLOGY OF *CAPPARIS* SPP.

Capparis plants are not the preferred food choice of goats (Garcia et al. 2008), and this owes to the presence of the secondary metabolites, which are more concentrated when the plants grow in harsh desert conditions (Figures 2.1 and 2.2), making such plants relied on in partially barren areas (e.g., those of northeast Brazil) undesirable from the standpoint of goat nutrition (Costa et al. 2011) and even potentially poisonous to the animal when, say, leaves from *C. tomentosa* are included as a forced part of a ruminant's regular diet, causing weakness of the hind limbs, staggering, and swaying (Ahmed et al. 1981). In Nubian goats, feeding of *C. tomentosa* leaves resulted in inappetence, locomotor disturbances, paresis (especially of the hind limbs), and recumbency. Associated lesions comprised perineuronal vacuolation in the gray matter of the spinal cord at the sacral region, centrilobular hepatocellular necrosis, degeneration of the renal proximal convoluted and collecting tubules, serous atrophy of the cardiac fat and renal pelvis, and straw-colored fluid in serious cavities. Later, anemia developed, and the results of kidney and liver function tests correlated with clinical abnormalities and pathologic changes (Ahmed et al. 1993).

Yet, fruits and buds of *Capparis* sp. *were* among the most preferred foods of six white-faced saki monkeys (*Pithecia pithecia*) on Round Island, Guri Lake, Venezuela, during a period of fruit abundance (Cunningham and Janson 2007). And, we contend here that these same potentially toxic secondary metabolites in the caper plant also infuse it with medicinal power worthy of our consideration. Secondary metabolites in caper plants, or "capers" (i.e., plants of the genus *Capparis* and their parts), their chemistry, and putative therapeutic functions are the subjects of this book (Figure 2.3).

The word *caper* usually refers to the flower bud of any member of the genus *Capparis*. Such capers particularly, plus the *Capparis* fruit, termed also caper berries, caper melons, and caper capsules, are preserved by pickling in brine or vinegar (Figure 2.4). In this way, they have provided a tasty addition to the Mediterranean diet for many centuries, mainly buds and fruits of *C. spinosa*, which to many Americans are known only as the centers of rolled anchovies. Other parts of *Capparis* plants, such as their leaves and roots, are also used in medicine, although usually not for food. However, recipes for brine-pickled *Capparis* leaves are extant. There is a long history of safe usage of these parts both in diet and as plant drugs throughout the world, and the details of this usage are summarized in this chapter, especially in tables.

Secondary metabolites in capers are, like secondary metabolites in all plants, significantly sensitive to environmental changes, stresses, biodiversity, and perturbations (Conforti et al. 2011a,b, Ozkur et al. 2009). This highly important principle that secondary metabolites can be manipulated by environmental factors leads to a science of plant manipulation independent of, or complementary to, genetic engineering per se.

Environmental stimuli push genetic changes, if not via mutation or selection, then at least epigenetically through methylations and acetylations. Thus, as always, environmental factors can be controlled and manipulated to achieve at least *epi*genetic changes. The important parameters for the scientist investigating such phenomena are light, temperature, moisture, and pressure (altitude). In *C. spinosa*, when water was harshly limited, the best photosynthetic performance occurred (Levizou et al. 2004). More is known regarding variation in mineral content of *C. spinosa* and *C. ovata* in response to harvest date and bud size (Özcan and Akgul 1998), N_2-fixing microorganisms from the *C. spinosa* rhizosphere (Andrade et al. 1997), and environmental stress: "Increased

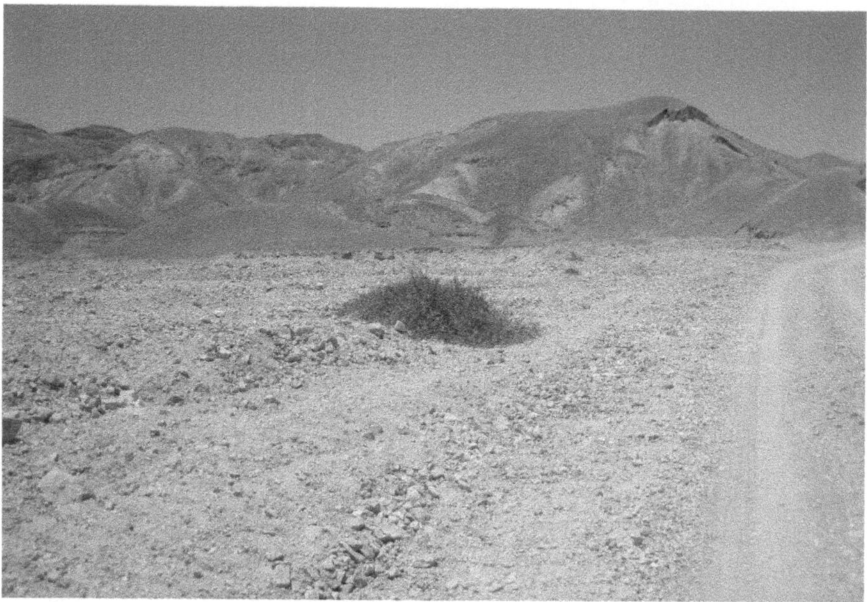

FIGURE 2.1 Caper bush (*Capparis aegyptia*) in the arid Arava Valley. (16.5.2011, near Kibbutz Ketura, Southern Arava, Israel: by Helena Paavilainen.)

FIGURE 2.2 As a response to the severe environmental stress in desert conditions, desert plants often contain higher concentrations of secondary metabolites. *Capparis cartilaginea* in the arid desert of Negev, near Kibbutz Ketura. (16.5.2011, Kibbutz Ketura, Southern Arava, Israel: by Helena Paavilainen.)

K and Ca uptake to maintain the hydric balance, thorny stems and a heavy investment in chemical defenses to prevent grazing, with a deep mycorrhizal root system allow *Capparis* to grow successfully in very infertile soils and to endure environmental stress" (Pugnaire and Esteban 1991). In greenhouse conditions, cold stratification treatments (i.e., layering of seeds between layers of moistened but not wet vermiculite, peat, or other absorbent material and storing at 1–5°C for

FIGURE 2.3 *Capparis* plants, such as this *C. aegyptia*, are an important source of nutrition also for different insects (cf. Peled 1997). (16.5.2011, near Kibbutz Ketura, Arava, Israel: by Helena Paavilainen.)

FIGURE 2.4 Home-pickled capers and caper fruit. (By Ephraim Lansky.)

60 days) was more effective than storing for 50, 40, 30, 20, or 10 days in *C. ovata* in promoting germination (Olmez et al. 2006). In the wild, growth of *C. decidua* was best in angular and blocky (loam) soils and poorest in granular and crumb (sandy) soils, while water-holding capacity and wilting coefficient associated with growth promotion were maximized when the pH of the subsurface soil was 6.7–7.6, and abundance was greatest when the pH was 7.2–8.2 (Qaiser and Qadir 1972).

Capparis leaves are tough, bright, and shiny and covered with epicuticular wax to aid in the conservation of moisture (Figures 2.5–2.9) (Oliveira et al. 2003). In the case of *C. spinosa*, a winter-deciduous perennial shrub and consistent floristic element of Mediterranean ecosystems growing from May to October (i.e., exclusively during the prolonged summer drought), the leaves are thick,

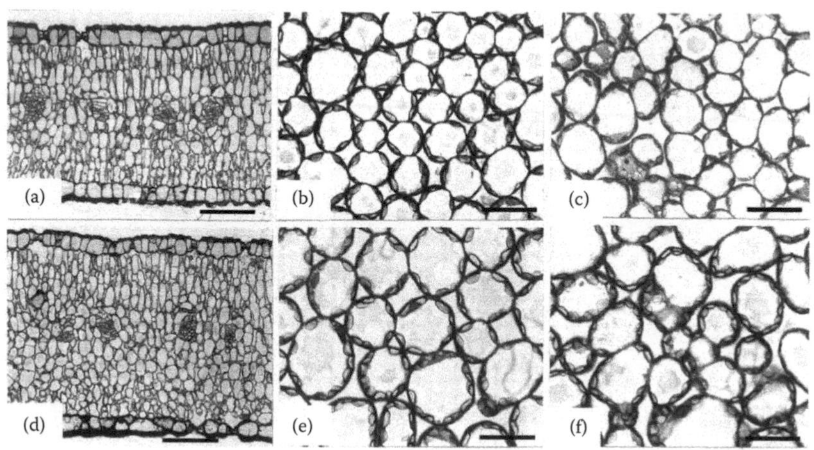

FIGURE 2.5 Anatomy of mature leaves of *C. spinosa* (L5 and L6): (a) and (d) cross-sectioned leaves; (b) and (e) paradermal sections through adaxial mesophyll of L5; (c) and (f) paradermal sections through abaxial mesophyll of L6. Cylindrical, densely packed mesophyll cells expose most of their surface to intercellular air space. Bars represent 50 μm (a and d) or 20 μm (all other micrographs). (From Rhizopoulou and Psaras 2003. *Ann Bot* 92(3): 377–83. Reprinted with permission.)

FIGURE 2.6 *Capparis spinosa* leaves are well suited to preserve the scanty moisture available. (7.4.2010, Jerusalem, Israel: by Helena Paavilainen.)

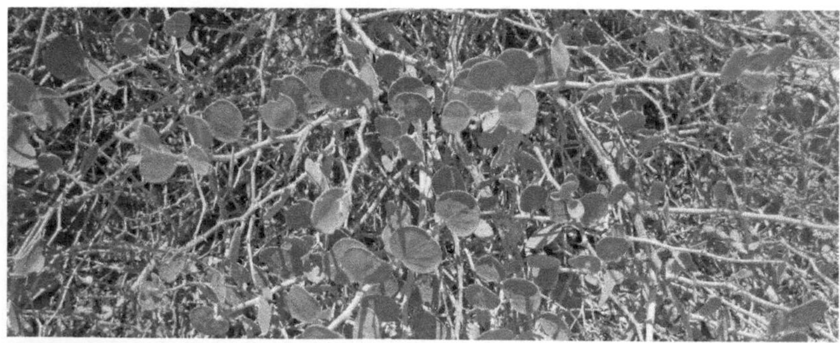

FIGURE 2.7 (See color insert.) The hard *Capparis aegyptia* leaves, as part of their adaptation system, have the bluish tinge typical of desert plants. (16.5.2011, near Kibbutz Ketura, Arava, Israel: by Helena Paavilainen.)

FIGURE 2.8 Detail of the leaf of *Capparis flexuosa* L. (15.6.2001, Grenada, Lesser Antilles: © William Hawthorne.)

FIGURE 2.9 Leaf of *Capparis cynophallophora*. (15.2.2013, Royal Botanic Garden, Edinburgh, Scotland: by C. E. Timothy Paine.)

amphistomatic, and homobaric with a multilayer mesophyll possessing an increased number of photosynthesizing cells per unit leaf surface and a large surface area of mesophyll cells facing intercellular spaces, presumably factors that facilitate transpiration and photosynthesis under field water shortage conditions (Rhizopoulou and Psaras 2003).

OVERVIEW OF GENUS *CAPPARIS*

Capparis is a large genus consisting of between 250 and 400 species, mostly in tropical and subtropical regions of both the Old and the New World, but some also are in temperate regions in the Mediterranean and Southwest Asian countries. It belongs to the family of Capparaceae, a family of some 45 genera and 675 species in the tropics and subtropics of both hemispheres, mainly in arid regions (Table 2.1) (Fahn et al. 1998, Fici et al. 1993, Jafri 1985, Oliver 1868, Pennington et al. 2004, Zhang and Tucker 2008, Zohary 1966–1986).

Concisely, the genus *Capparis* could be described as follows (see Figures 2.10–2.15) (Fici et al. 1993, Jafri 1985, Oliver 1868, Pennington et al. 2004, Sonder 1894, Wild 1960, Zhang and Tucker 2008, Zohary 1966–1986):

- Trees or shrubs, usually evergreen, often climbing or sometimes prostrate, unarmed or with short, often recurved, stipular spines
- Glabrescent or hairy (mostly the young twigs) with simple to stellate hairs or lepidote scales
- Leaves alternate, simple, entire or subentire, often leathery, with spiny or setaceous stipules; sometimes almost leafless
- Flowers hermaphrodite; solitary; racemose; corymbose or umbellate; axillary or terminal, rarely supraaxillary; bracts mostly present but early caduceus
- Sepals 4, free or fused at base, valvate or imbricate; the posterior sepal often larger and more concave than others
- Petals usually 4, imbricate, rather delicate, mostly obovate, usually caducous after anthesis; the 2 posterior ones coherent, forming a nectariferous fleshy protuberance or cavity at the thickened base
- Stamens 5, indefinite, radiating, exceeding the petals, glabrous, inserted on the torus at the base of the gynophore

TABLE 2.1
Classification of Genus *Capparis* L.

- Plantae
  - Viridaeplantae
    - Streptophyta
      - Tracheophyta
        - Spermatophytina
          - Angiospermae
            - Magnoliopsida
              - Rosanae
                - Brassicales
                  - Capparaceae
                    - *Capparis* L.

Source: Retrieved from the Integrated Taxonomic Information System (ITIS) (http://www.itis.gov).

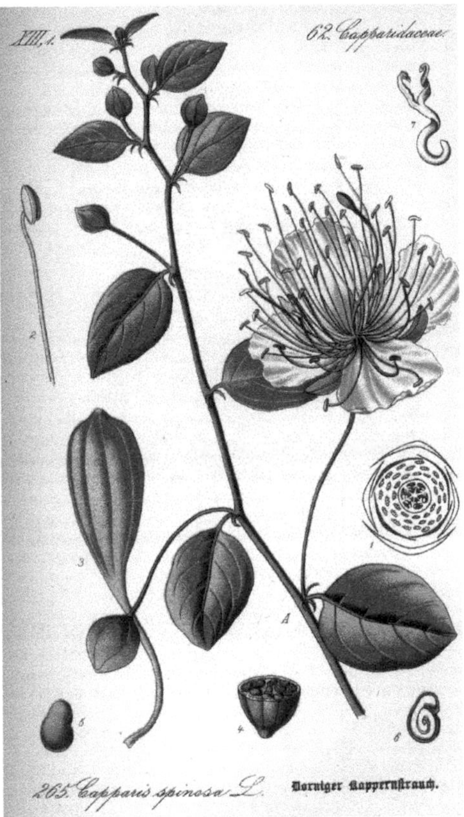

FIGURE 2.10 A morphological drawing of *Capparis spinosa* L. (Thomé, O.W. 1885–1905. *Flora von Deutschland, Österreich und der Schweiz in Wort und Bild fur Schule und Haus*. Gera-Untermhaus, Germany: Köhler. Via http://www.biolib.de. [Online library.] http://caliban.mpiz-koeln.mpg.de/thome/band2/tafel_108.jpg)

FIGURE 2.11 *Capparis aegyptia* flower. (15.5.2012, Negev, Israel: by Eli Harlev.)

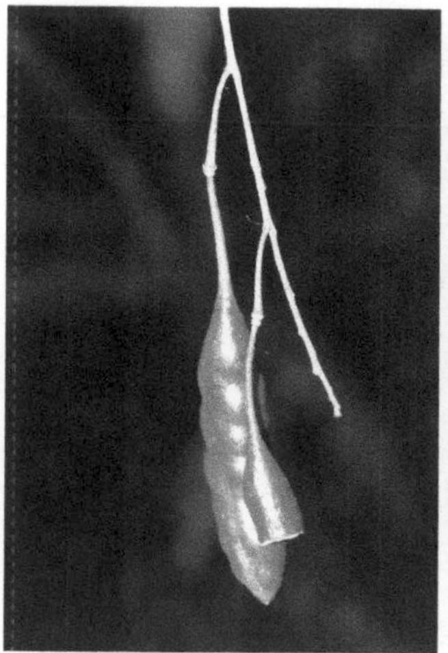
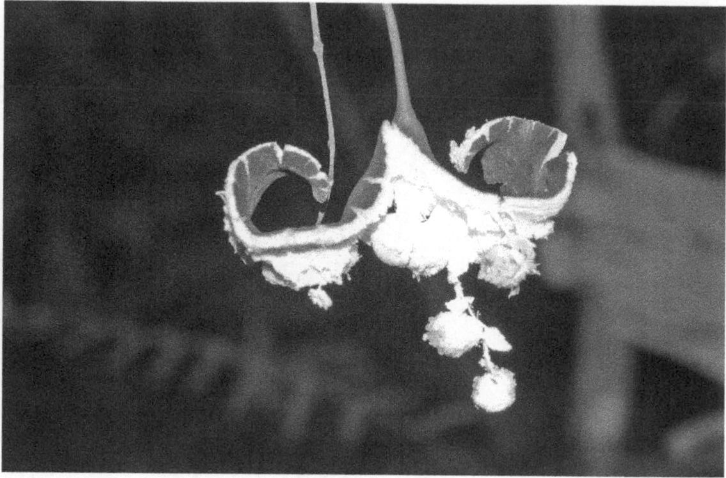

FIGURE 2.12 *Capparis sola*, fruit and dehisced fruit. (1.9.2007, Manu, Madre de Dios, Peru: by C. E. Timothy Paine.)

- Gynophore generally as long as the stamens, scarcely lengthening in fruit but often becoming thicker
- Ovary glabrous, cylindrical or ellipsoid, usually 1-locular with 2–8 (–10) placentas and few to many ovules
- Stigma sessile or subsessile
- Fruit globose; ovoid or elongated berry; usually with coriaceous, smooth, verruculose, or grooved pericarp and often with different color when mature or dry; indehiscent or tardily dehiscent and leaving a central membrane (Figure 2.12)
- Seeds 1 to numerous per fruit, reniform to nearly polygonal, embedded in pulp

In practice, however, this large genus shows much variation both in the physical appearance of the different species (Figures 2.16–2.23) and in the different ways that human cultures utilize them (Figure 2.24, Table 2.2). The genus *Capparis* in fact both shows great adaptability to the different natural surroundings in which it has found itself and serves as an example of the endless variety of human responses to an attracting and challenging plant. In the continuation of this chapter, we examine some of these responses through a short overview of the species of *Capparis* discussed in this book.

Capparis is a large genus and only partly studied in modern times. As this book was written and the scientific studies on the pharmacology of the genus delineated, different *Capparis* species were highlighted. Although representing only a small part of the total number of *Capparis* species, we felt that these species, selected for their scientific interest, constituted a reasonable sampling of the species comprising *Capparis*. Accordingly, we decided to use this sample as a means of discussing the range of species in the genus. The following discussion is therefore a species-by-species description encompassing the ethnography, distribution, and local uses, both common and medical, of each of the species in the following pages in the context of modern research (Tables 2.3–2.41, Figures 2.25–2.46). Thus, plants on which only ethnomedical data but no modern medical or pharmacological research exists were excluded. As a neutral and random means of arranging the species, alphabetical order is used. Description of the geographical distribution of the species is based on *World Geographical Scheme for Recording Plant Distributions* (2nd ed., Brummitt et al. 2001). The rest of the chapter focuses on modern ethnomedicinal uses of the plants, whereas the chemical composition and the medical effects of the various species of *Capparis* are discussed in detail in Chapters 3–25.

FIGURE 2.13 Morphology of different organs of *Capparis*. (Le Maout, E. and J. Decaisne. 1873. *A General System of Botany, Descriptive and Analytical* [English translation by Mrs. Hooker, with additional material by J.D. Hooker. Illustrations by L. Steinheil and A. Riocreux]. London: Longmans, Green. Via Watson, L. and M.J. Dallwitz, 1992 onward. *The Families of Flowering Plants: Descriptions, Illustrations, Identification, and Information Retrieval*. Version: 18 May 2012. http://delta-intkey.com/angio/images/cappa233.gif.)

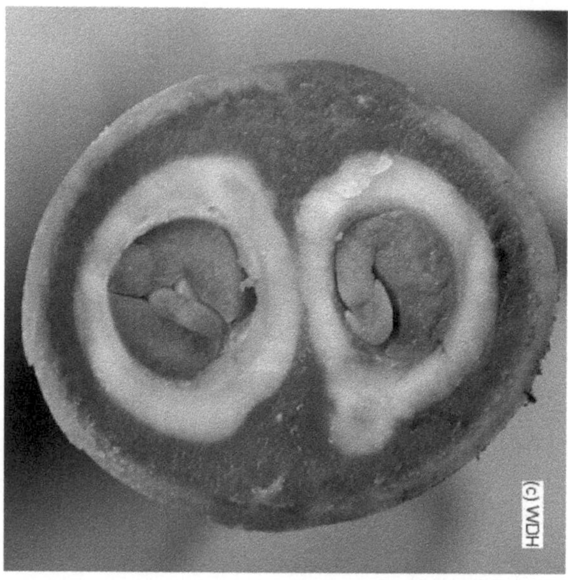

FIGURE 2.14 Fruit of *Capparis flexuosa* cut transversely, exposing the inner structure of the seeds. (15.6.2001, near River Sallee, Grenada, Lesser Antilles: © William Hawthorne). Cf. to Figure 2.13.

FIGURE 2.15 *Capparis spinosa*. Notice the relative positions of the bud stems and the leaves. (14.5.2010, Jerusalem, Israel: by Helena Paavilainen.)

(a)

FIGURE 2.16 (See color insert.) (a) Examples of the variability of flowers within genus *Capparis: Capparis spinosa* L. (23.5.2010, Hadassah Ein Kerem, Jerusalem, Israel: by Helena Paavilainen.)

(b)

FIGURE 2.16 (b) Examples of the variability of flowers within genus *Capparis: Capparis sepiaria* L. var. *subglabra* (Oliv.) DeWolf. (13.10.2007, Gorongosa National Park, Mozambique: by Bart T. Wursten. In Hyde, M.A., B.T. Wursten, P. Ballings, and S. Dondeyne. 2013b. Flora of Mozambique: Species Information: Individual Images: *Capparis sepiaria*. http://www.mozambiqueflora.com/speciesdata/image-display.php?species_id=124450&image_id=7, retrieved 20 February 2013.)

(c)

FIGURE 2.16 (c) Examples of the variability of flowers within genus *Capparis: Capparis flexuosa* L. (15.6.2001, near River Sallee, Grenada, Lesser Antilles: © William Hawthorne.)

(d)

FIGURE 2.16 (See color insert.) (d) Examples of the variability of flowers within genus *Capparis: Capparis decidua*. (15.3.2010, near Shirur, Pune, Maharashtra, India: by Abhijeet Shiral.)

(e)

FIGURE 2.16 (e) Examples of the variability of flowers within genus *Capparis: Capparis sola.* (2.4.2006, Manu, Madre de Dios, Peru: by C. E. Timothy Paine.)

(f)

FIGURE 2.16 (f) Examples of the variability of flowers within genus *Capparis: Capparis cartilaginea.* (16.5.2011, Kibbutz Ketura, Southern Arava, Israel: by Helena Paavilainen.)

(g)

FIGURE 2.16 (g) Examples of the variability of flowers within genus *Capparis: Capparis micracantha* var. *henryi.* (4.5.2010, Taiwan: by Ming-I Weng.)

(h)

FIGURE 2.16 (h) Examples of the variability of flowers within genus *Capparis: Capparis moonii*. (20.3.2011, Western Ghats, India: by Satish Nikam.)

FIGURE 2.17 (See color insert.) Considerable variation may also exist between the flowers of the same plant: *Capparis cartilaginea* with flowers of different age showing differing colors that range between white and red. (16.5.2011, Kibbutz Ketura, Southern Arava, Israel: by Helena Paavilainen.)

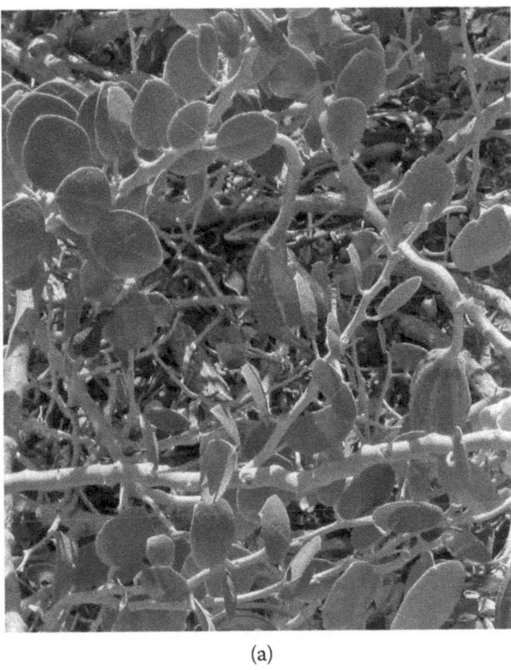

(a)

FIGURE 2.18 (See color insert.) (a) Variations in fruit within genus *Capparis: Capparis cartilaginea*, ripe red fruit. (8.8.2011, Kibbutz Ketura, Southern Arava, Israel: by Helena Paavilainen.)

(b)

FIGURE 2.18 (See color insert.) (b) Variations in fruit within genus *Capparis: Capparis spinosa*, unripe green fruit. (25.7.2010, Hadassah Ein Kerem, Jerusalem, Israel: by Helena Paavilainen.)

(c)

FIGURE 2.18 (c) Variations in fruit within genus *Capparis: Capparis decidua*, ripening red fruit. (Mauritania: by Sébastien Sant, in J.P. Peltier. 2006. Plant Biodiversity of South-Western Morocco. [Online website.] http://www.teline.fr, retrieved 13 March 2013.)

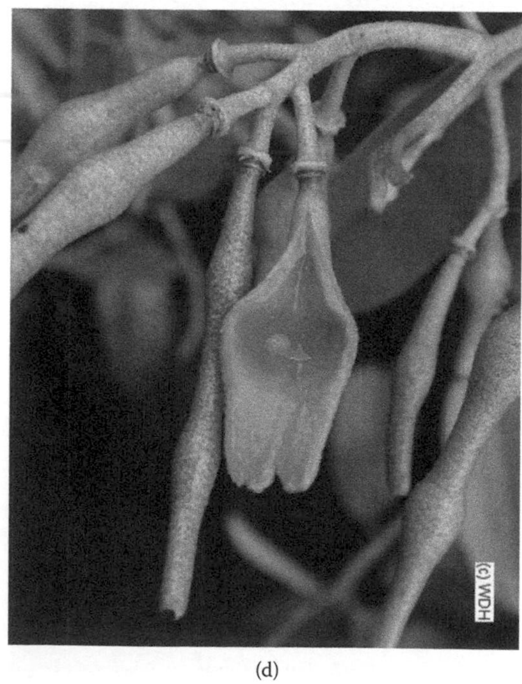

(d)

FIGURE 2.18 (d) Variations in fruit within genus *Capparis: Capparis odoratissima* Jacq., ripe red fruit, one of them cut vertically. (25.6.2001, Grenadan Dove Nature Reserve, Grenada, Lesser Antilles: © William Hawthorne.)

(e)

FIGURE 2.18 (e) Variations in fruit within genus *Capparis: Capparis tomentosa* Lam., unripe green fruit. (Gorongosa National Park, Mozambique, 9.11.2007: by Bart T. Wursten. In Hyde, M.A., B.T. Wursten, P. Ballings, and S. Dondeyne. 2013b. Flora of Mozambique: Species Information: Individual Images: *Capparis tomentosa*. http://www.mozambiqueflora.com/speciesdata/image-display.php?species_id=124460&image_id=7, retrieved 20 February 2013.)

(f)

FIGURE 2.18 (f) Variations in fruit within genus *Capparis: Capparis erythrocarpos* Isert var. *rosea* (Klotzsch) DeWolf, ripe, bright red fruit. (Gorongosa National Park, Mozambique, 26.1.2007: by Bart T. Wursten. In Hyde, M.A., B.T. Wursten, P. Ballings, and S. Dondeyne. 2013b. Flora of Mozambique: Species Information: Individual Images: *Capparis erythrocarpos*. http://www.mozambiqueflora.com/speciesdata/image-display.php?species_id=124430&image_id=4, retrieved 20 February 2013.)

(a)

FIGURE 2.19 (a) Different habits within *Capparis: Capparis spinosa* growing on the wall of a house. (July 2011, Haifa, Israel: by Ephraim Lansky.)

(b)

FIGURE 2.19 (b) Different habits within *Capparis: Capparis decidua.* (24.3.2010, Rajasthan, India: by Ryan Brookes.)

(c)

FIGURE 2.19 (c) Different habits within *Capparis: Capparis aegyptia*. (16.5.2011, near Kibbutz Ketura, Southern Arava, Israel: by Helena Paavilainen.)

(d)

FIGURE 2.19 (d) Different habits within *Capparis: Capparis sepiaria* var. *subglabra* (Oliv.) DeWolf. (21.10.2005, near Odzani River, northwest of Mutare, Zimbabwe: by Bart T. Wursten. In Hyde, M.A., B.T. Wursten, and P. Ballings. 2013a. Flora of Zimbabwe: Species Information: Individual Images: *Capparis sepiaria*. http://www.zimbabweflora.co.zw/speciesdata/image-display.php?species_id=124450&image_id=1, retrieved 20 February 2013.)

(e)

FIGURE 2.19 (e) Different habits within *Capparis: Capparis cartilaginea*. (16.5.2011, Kibbutz Ketura, Southern Arava, Israel: by Helena Paavilainen.)

(f)

FIGURE 2.19 (f) Different habits within *Capparis: Capparis moonii*. (20.3.2011, Western Ghats, India: by Satish Nikam.)

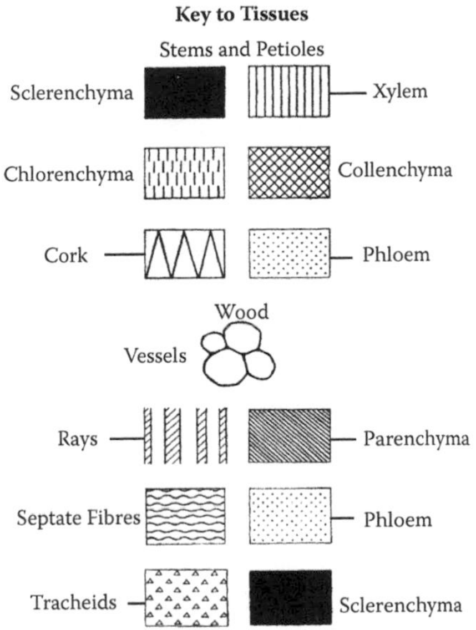

FIGURE 2.20 Morphology of *Capparis* spp. Key to the morphological drawings on the following pages. (From Metcalfe, C.R. and L. Chalk. 1957. *Anatomy of the Dicotyledons: Leaves, Stem, and Wood in Relation to Taxonomy with Notes on Economic Uses*. Vol. 1. Oxford, UK: Oxford University Press, inner cover page. Via Internet Archive, http://archive.org [Internet library.] http://archive.org/details/anatomyofthedico033552mbp)

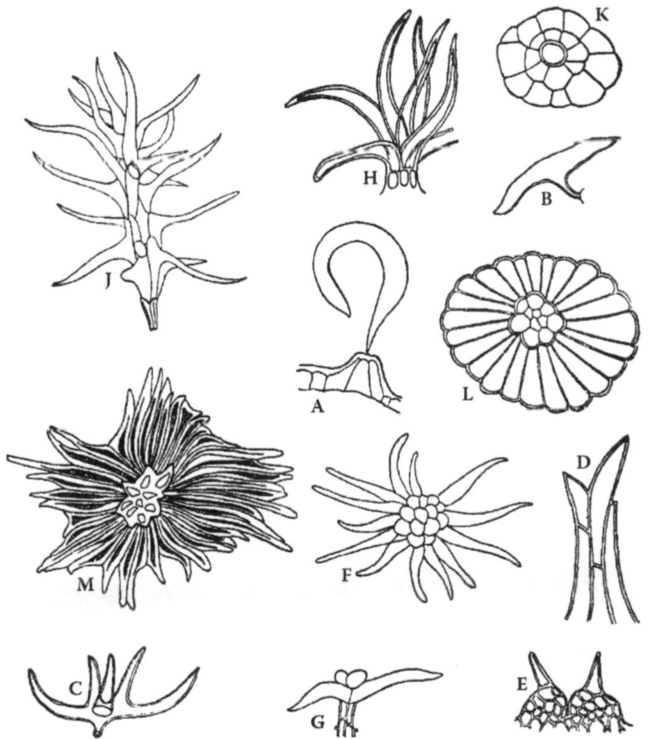

FIGURE 2.21 Morphology of *Capparis* spp. Structural differences in hairs of different *Capparis* spp. (Drawings A, B, C, E, H, J, and M from Metcalfe, C.R. and L. Chalk. 1957. *Anatomy of the Dicotyledons: Leaves, Stem, and Wood in Relation to Taxonomy with Notes on Economic Uses*. Vol. 1. Oxford, UK: Oxford University Press, 88. Via Internet Archive, http://archive.org. [Internet library.] http://archive.org/details/anatomyofthedico033552mb)

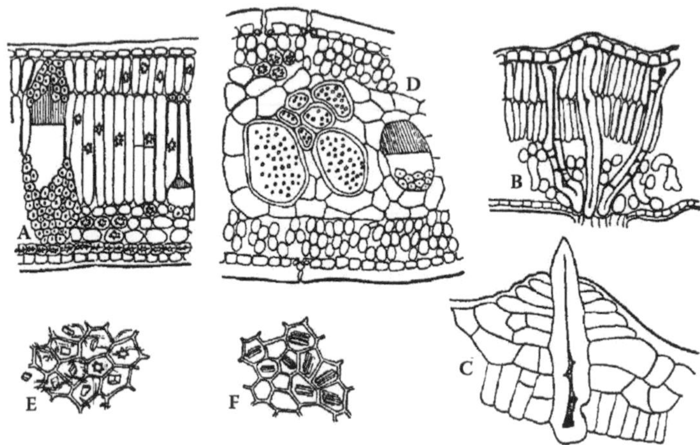

FIGURE 2.22 Morphology of *Capparis* spp. Differences in the leaf structure of various *Capparis* spp. (From Metcalfe, C.R. and L. Chalk. 1957. *Anatomy of the Dicotyledons: Leaves, Stem, and Wood in Relation to Taxonomy with Notes on Economic Uses.* Vol. 1. Oxford, UK: Oxford University Press, 90. Via Internet Archive, http://archive.org. [Internet library.] http://archive.org/details/anatomyofthedico033552mbp)

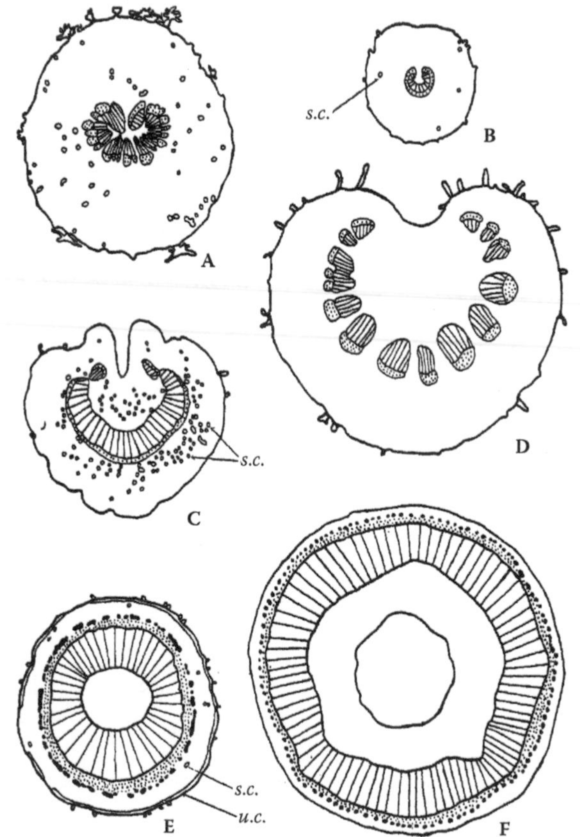

FIGURE 2.23 Morphology of *Capparis* spp. Structure of the stem of *Capparis* spp. (Drawings B and E from Metcalfe, C.R. and L. Chalk. 1957. *Anatomy of the Dicotyledons: Leaves, Stem, and Wood in Relation to Taxonomy with Notes on Economic Uses.* Vol. 1. Oxford, UK: Oxford University Press, 96. Via Internet Archive, http://archive.org. [Internet library.] http://archive.org/details/anatomyofthedico033552mbp)

TABLE 2.2
Overview of *Capparis* Species Covered in This Book

Capparis Species	Region	Traditional or Economic Uses	Research	References
acutifolia	China, India, Southwest Asia	Traditional medicine	Analgesic, anti-inflammatory	Deng 2010, *GRIN* 1994, Siu-cheong and Ning-hon 1978–1986, Zhang and Tucker 2008
aegyptia	Middle East, North Africa	Condiment; antitumor, vermifuge	Stachydrine	Hammouda et al. 1975, Manniche 1989, Rivera et al. 2003
amygdalina	Mexico, Caribbean, Venezuela	Neurotonic, for menstrual problems		Dragendorff 1898, Encyclopedia of Life 2007, Tropicos 2013
angulata	Ecuador, Peru	Food, fodder; antierosion; woodcrafting	Glucocapangulin	Kjaer et al. 1960, Rodríguez Rodríguez et al. 2007
aphylla	Northern and tropical Africa, Middle East	Food; preparation of salt; ant-resistant wood	Antioxidant, antidiabetic, hypotensive	Dangi and Mishra 2011a,b, GRIN 1994, Jabbar and Hassan 1993, Jafri 1985, McGuffin et al. 2000, Sturtevant and Hedrick 1919
assamica	Indian subcontinent, Indochina, China	Possibly antinociceptive		Encyclopedia of Life 2007, Maikhuri and Gangwar 1993, Tropicos 2013
baducca	Central and South America	Diuretic, sedative, antispasmodic	Hypotensive, anticonvulsant, anti-inflammatory	Duke 1993, GRIN 1994, Hashimoto 1967, IOPI 1991, IPNI 2012, Standley 1920–1926, USDA, NRCS 2012
bariensis	Indochina	For menstrual problems		IPNI 2012, Petelot 1952–1954
cartilaginea	North and East Africa, Middle East, India	Food, fodder; traditional medicine; veterinary medicine	Rutin; antiarthritic	Balar and Nakum 2010, Elffers et al. 1964, Fici et al. 1993, Hamed et al. 2007, Jafri 1985, Rivera et al. 2003
cordata	Ecuador, Peru	For neurological problems, rheumatism		Encyclopedia of Life 2007, IPNI 2012, Ramirez et al. 1988, Tropicos 2013
coriacea	South Africa		Polyphenols	IPNI 2012, Jankowski and Chojnacki 1991, Tropicos 2013
corymbifera	South Africa	Aphrodisiac		Bryant 1966, Encyclopedia of Life 2007, IPNI 2012
corymbosa	Central and Southern Africa	Food; hunting poison	Flavonoids, plant sterols	Burkill 1985, Hanelt et al. 2001, Sawadogo et al. 1981, Terra 1966
critifolia	South Africa	Emetic, for gall sickness		Simon and Lamla 1991
cynophallophora	Florida, Mexico, Central America, Caribbean, Brazil, Argentina	For menstrual problems		Encyclopedia of Life 2007, GRIN 1994, Roig y Mesa 1945, Tropicos 2013

(Continued)

TABLE 2.2 (*Continued*)
Overview of *Capparis* Species Covered in This Book

Capparis Species	Region	Traditional or Economic Uses	Research	References
decidua	North and tropical Africa, Middle East, India	Food; windbreaking hedges; anti-inflammatory, analgesic, veterinary medicine	Hepatoprotective, antibacterial	Ali et al. 2009, Boulos 1983, Duke 1993, Ganguly and Kaul 1969, GRIN 1994, Juneja et al. 1970
deserti	Middle East, North Africa	Food; antitumor, vermifuge	Rutin (antioxidant)	Hamed et al. 2007, *IOPI* 1991, Rivera et al. 2003
divaricata	India, Sri Lanka	Analgesic, diuretic	Palmitoleic acid	ENVIS 1999, Kittur et al. 1993, Patil et al. 2011
elaeagnoides	Tropical Africa	Analgesic, against sleeping sickness		Encyclopedia of Life 2007, Freiburghaus et al. 1996, IPNI 2012, Tropicos 2013, Watt and Breyer-Brandwijk 1962
erythrocarpos	West Africa, eastern tropical Africa	Galactagogue, anti-infective		Encyclopedia of Life 2007, GBIF 2001, IPNI 2012, Kisangau et al. 2009, Tropicos 2013, Vasileva 1969
fascicularis	Western and eastern Africa	Gastrointestinal problems		Encyclopedia of Life 2007, GBIF 2001, Johns et al. 1990, 1995
flavicans	Mainland Southeast Asia	Galactagogue	Antiestrogenic compounds	IPNI 2012, JSTOR Plant Science n.d., Luecha et al. 2009
flexuosa	Central and South America	Food; diuretic, sedative	Glucosinolates	Duke 1993, GRIN 1994, Kjaer and Schuster 1971
formosana	China, Japan, Vietnam		Plant sterols	Liu et al. 1977, Zhang and Tucker 2008
galeata	North and East Africa, Middle East to Pakistan	Food; analgesic, antipyretic		Al-Khalil 1995, Encyclopedia of Life 2007, Goodman and Hobbs 1988, Greuter et al. 1984/1986, IOPI 1991, Jafri 1985
grandiflora	India	Food		IPNI 2012, Ramachandran and Nair 1981
grandis	India, Southeast Asia	Timber	Glucosinolate	Brandis 1906, Gaind et al. 1975, IPNI 2012
gueinzii	South Africa, West Africa	For cough		Encyclopedia of Life 2007, Tropicos 2013, Watt and Breyer-Brandwijk 1962
hereroensis	South and Southwest Africa	Poison		IPNI 2012, Tropicos 2013
heyneana	India, Sri Lanka	Rheumatism, laxative	Anti-inflammatory, analgesic	Duke 1993, IPNI 2012, Vardhana 2008, Wang et al. 2009
himalayensis	Himalaya	Food; antirheumatic, analgesic	Alkaloids, flavonoids	IPNI 2012, Kirtikar et al. 1987, Li et al. 2008, Rivera et al. 2003
horrida	Tropical Asia	Food; antiperspirant, against gastric irritation, antisyphilitic	Alkaloids	Chakravarti and Venkatasubban 1932, Duke 1993, IPNI 2012, Quisumbing 1951, Sturtevant and Hedrick 1919
humilis	Argentina, Bolivia, Paraguay		Proteins	GBIF 2001, IPNI 2012, Jørgensen 2008
incanescens	Indochina, Australia, tropical Africa	Tonic, antipyretic	Glucosinolates	Duke 1993, Jafri 1985, *Wealth of India* 1948–1976
kirkii	Tropical East Africa	Disinfectant before nursing		Encyclopedia of Life 2007, GBIF 2001, IPNI 2012, Moller 1961

leucophylla	India, Iraq	Food; antirheumatic, antimalarial	Stachydrine	Hammouda et al. 1975, Hooper and Field 1937, IPNI 2012, Jafri 1985, Rivera et al. 2003
linearis	Colombia, Venezuela, Peru		Glucosinolates	GBIF 2001, IPNI 2012, JSTOR Plant Science n.d., Kjaer and Wagnieres 1965
longispina	Myanmar		Glucosinolates	IPNI 2012, Satyanarayana et al. 2008
masaikai	China		Sweetener; against dry mouth	Kitada et al. 2009, Liu et al. 1993, Zhang and Tucker 2008
micracantha	South Asia from India to Philippines and China	Antiasthmatic, antipyretic, diuretic		Burkill et al. 1966, Tropicos 2013, Wasuwat 1967
moonii	India, Sri Lanka	Antitubercular	Antidiabetic, tuberculostatic	Chopra et al. 1950, ENVIS 1999, IPNI 2012, Kanaujia et al. 2010, Mishra et al. 2007, Shah 1962, Tyagi 2005, 115
oleoides	South Africa	Food, fodder; antiepileptic		Encyclopedia of Life 2007, IPNI 2012, Tropicos 2013, Watt and Breyer-Brandwijk 1962
orientalis	Mediterranean	Food; ornamental	Sterols	Conforti et al. 2011b, GRIN 1994, Inocencio et al. 2000, Huxley et al. 1992, Rivera et al. 2003
ovalifolia	Panama, Ecuador, Peru	For rheumatism, neurological problems		IPNI 2012, Ramirez et al. 1988, Tropicos 2013
ovata	Mediterranean, Central Asia	Food; antinociceptive	Antinociceptive, anti-inflammatory	Arslan and Bektas 2010, Greuter et al. 1984/1986, *IPNI* 2012, Martínez-Lirola et al. 1996, Rivera et al. 2003, Villasenor 2007
retusa	Bolivia, Brazil, Argentina, Paraguay	Food; against chicken pox		Arenas 1987, Encyclopedia of Life 2007, Schmeda-Hirschmann 1994, Tropicos 2013
rotundifolia	India, Sri Lanka, Myanmar		Glucosinolates	Ravikumar et al. 2004, Satyanarayana et al. 2008
salicifolia	Argentina, Bolivia, Paraguay	Food; material for fire drill; analgesic	Glucocappasalin	Arenas and Suarez 2007, Arenas and Scarpa 2007, Filipoy 1994, GRIN 1994, Kjaer and Thomsen 1962, New World Fruits Database 2012, Schmeda-Hirschmann 1994
sepiaria	Tropical Africa, tropical Asia	Hedge plant; traditional medicine, veterinary medicine	Antidiabetic, anti-inflammatory, analgesic	Chaudhari et al. 2004, Duke 1993, GRIN 1994, Selvamani et al. 2008
sicula	Mediterranean	Food; anti-inflammatory, analgesic; veterinary medicine	Flavonoids, glucosinolates	Conforti et al. 2011a, Greuter et al. 1984/1986, IOPI 1991, Rivera et al. 2003
sikkimensis	Indochina, Himalaya		Anticancer	IPNI 2012, Taiwan Plant Names n.d., Wu et al. 2003, Zhang and Tucker 2008
sinaica	Middle East, East Africa, Southwest Asia	Food, fodder; stimulant, antirheumatic	Antiviral	Boulos 1983, Greuter et al. 1984/1986, IOPI 1991, MEDUSA n.d., Soltan and Zaki 2009, Tropicos 2013

(Continued)

TABLE 2.2 (*Continued*)
Overview of *Capparis* Species Covered in This Book

Capparis Species	Region	Traditional or Economic Uses	Research	References
sola	Northern South America	Analgesic, against skin problems	Alkaloids	Alexiades 1999, Folkers 1938, Friedman et al. 1993, IOPI 1991, IPNI 2012
speciosa	Mexico, Panama, Argentina	Removes thorns from skin		Arenas 1987, Encyclopedia of Life 2007, IPNI 2012, Tropicos 2013
spinosa	Mediterranean; temperate and tropical Asia	Food; cosmetic; eye problems, rheumatism, veterinary medicine	Anticancer, antileishmaniac	Bown 1995, Burkill 1985, Chiej 1984, GRIN 1994, Inocencio et al. 2000, Jacobson and Schlein 1999, Ji et al. 2008a,b, 2011
stylosa	India	Piscicidal	Antibacterial, piscicidal	Ambedkar and Muniyan 2009, Gamble 1915, 45, JSTOR Plant Science n.d., Mohan and Suganthi 1998
tenera	Africa, Southeast Asia	Hedge plant	Antioxidant	Brandis 1906, Encyclopedia of Life 2007, Su et al. 2007, Zhang and Tucker 2008
tomentosa	Tropical Africa, Saudi Arabia	Food, fodder; magic; traditional medicine, eye problems, veterinary medicine	Glucosinolates, stachydrine	Burkill 1985, Cornforth and Henry 1952, *GRIN* 1994, Schraudolf 1988
tweediana	Bolivia, Brazil, Argentina, Paraguay	Food; eye problems, antitussive		Encyclopedia of Life 2007, Filipoy 1994, Schmeda-Hirschmann 1994, Tropicos 2013
zeylanica	Tropical Asia	Anticholera, antipneumonia	Antioxidant, antipyretic	Duke 1993, Ghule et al. 2007, GRIN 1994, Satyanarayana et al. 2010

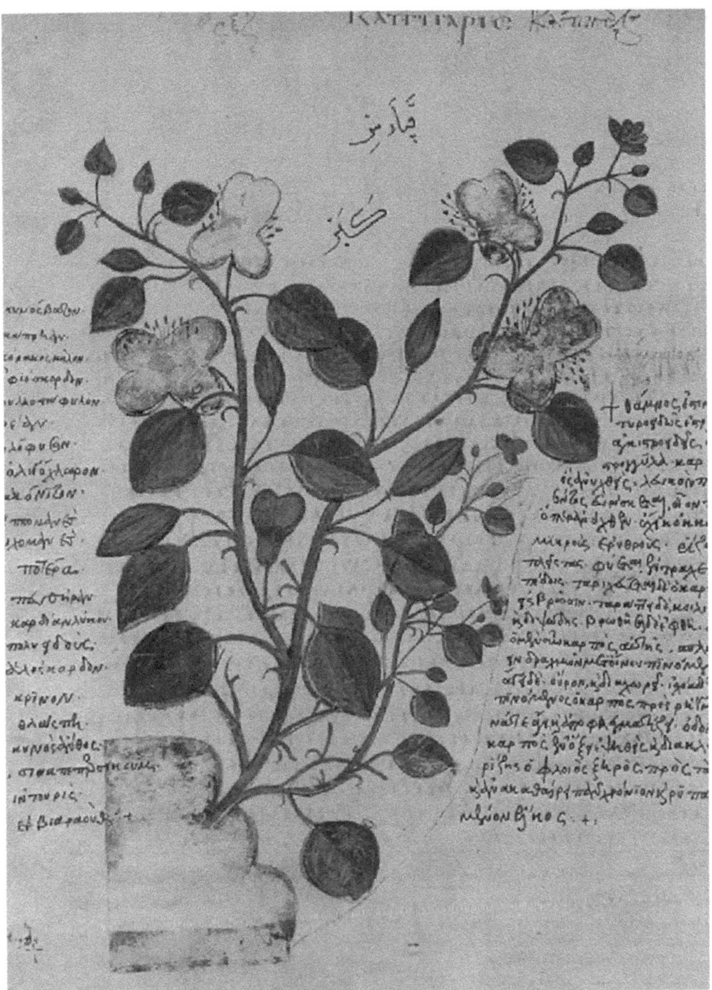

FIGURE 2.24 Dioscorides (2nd century CE) recommended capers, among other things, for spleen problems, sciatica, and toothache. (Dioscorides 1902, Book II, Chapter 204. The illustration is from *Codex Aniciae Julianae*, a manuscript of *De Materia Medica* written ca. 512 CE for the Byzantine princess Juliana Anicia. *kapparis*, v. 172: reproduced by Michel Chauvet, via Wikimedia, http://commons.wikimedia.org/wiki/File:Kapparis_172v_Dioscoride_Vienne.png, retrieved 15 July 2012.)

ETHNOGRAPHIC DATA ON VARIOUS *CAPPARIS* SPP.

TABLE 2.3
Capparis acutifolia

		References
Synonyms	*Capparis acutifolia* Sweet: *C. acuminata* Lindley *C. chinensis* G. Don *C. kikuchii* Hayata *C. leptophylla* Hayata *C. membranacea* Gardner & Champion *C. membranacea* var. *angustissima* Hemsley *C. membranacea* var. *puberula* B. S. Sun *C. tenuifolia* Hayata Subordinate taxa: **C. acutifolia* Sweet subsp. *acutifolia* = *C. membranacea* Gardner & Champ. **C. acutifolia* Sweet subsp. *bodinieri* (H. Lév.) M. Jacobs = *C. bodinieri* H. Lév. (basionym) = *C. subtenera* Craib & W. W. Sm. **C. acutifolia* Sweet subsp. *viminea* M. Jacobs = *C. membranifolia* Kurz	GRIN 1994, IOPI 1991, IPNI 2012, Zhang and Tucker 2008
Common names	Chinese: du xing qian li	Zhang and Tucker 2008
Habitat and distribution	Roadsides, stony mountains, thickets, forests; 300–1,100 m *China*: China *Eastern Asia*: Taiwan *Indian Subcontinent*: Bhutan, India *Indochina*: Cambodia, Laos, Myanmar, Thailand, Vietnam	Encyclopedia of Life 2007, GRIN 1994, IPNI 2012, Zhang and Tucker 2008
Economic importance	Leaves: food Part not specified: medicinal	GRIN 1994, Siu-cheong and Ning-hon 1978–1986, von Reis Altschul 1973
Local medicinal uses	Part not specified: medicinal	GRIN 1994, Siu-cheong and Ning-hon 1978–1986
Toxicity		

FIGURE 2.25 All parts of *Capparis*—buds, fruit, seeds, leaves, stems, bark, and root—are medicinal, although historically the most important part for medical use was the root bark. *Capparis aegyptia* in spring. (31.3.2012, Negev, Israel: by Eli Harlev.)

FIGURE 2.26 *Capparis aegyptia* in the beginning of summer. The fruit are still green. (16.5.2011, near Kibbutz Ketura, Southern Arava, Israel: by Helena Paavilainen.)

TABLE 2.4
Capparis aegyptia

		References
Synonyms	*Capparis aegyptia* Lam. (basionym): *C. spinosa* L. subsp. *aegyptia* (Lam.) Kit Tan & Runemark *C. spinosa* L. *C. spinosa* L. var. *deserti* Zohary	IPNI 2012, Hanelt et al. 2001, Rivera et al. 2003
Common names	Arabic: (Egypt) kabar, quabar, lassaf, shafellah; (Iraq) kabar, shafallah Coptic: kemegeoc, kemapic Hebrew: tsalaf mitsri	Blakelock and Townsend 1980, Danin 2006–, Hanelt et al. 2001, Rivera et al. 2003, Rolland 1967, Schweinfurth 1912
Habitat and distribution	Hard rock outcrops. Semisteppe shrub-lands, shrub steppes, deserts, and extreme deserts Northern Africa, Middle East *Northern Africa*: Morocco, Algeria, Tunisia, Libya, Egypt *Western Asia*: East Aegean Islands, Israel and Jordan, Sinai, Lebanon, and Syria	Danin 2006–, Greuter et al. 1984/1986, IOPI 1991, IPNI 2012, Rivera et al. 2003
Economic importance	Fruit: food (pickled) Seeds: condiment; added to wine to keep it sweet	Blakelock and Townsend 1980, Renfrew 1987, Rivera et al. 2003
Local medicinal uses	Root bark: wounds; skin diseases (historical); antitumoral (historical); vermifuge (historical); emmenagogue (historical)	Alpini and de Fenoyl 1980, Manniche 1989, Rivera et al. 2003
Toxicity		

TABLE 2.5
Capparis angulata

		References
Synonyms	*Capparis angulata* Ruiz & Pav.: *C. angulata* Ruiz & Pav. ex DC. *C. angulata* Ruiz, Pav., Ruiz & Pav. apud Lopez *C. scabrida* Kunth *Colicodendron scabridum* (Kunth) Seem.	Brako and Zarucchi 1993, GRIN 1994, IOPI 1991, IPNI 2012
Common names	Spanish: sapote/zapote, sapote/zapote de zorro, zapote de campo, zapote de perro	Rodríguez Rodríguez et al. 2007
Habitat and distribution	Altitude range: 0–2,500 m *Western South America:* Ecuador, Peru	IPNI 2012, Rodríguez Rodríguez et al. 2007
Economic importance	Flower: for apiculture (honey plant) Fruit: edible Fruit oil: food Leaf: fodder for sheep and goats, increases yield of milk in cows Plant: in stabilization of dunes, against erosion Wood: for woodcraft, furniture; firewood	Brack Egg 1999, Bussmann and Sharon 2007, Mostacero et al. 2002, Rodríguez Rodríguez et al. 2007
Local medicinal uses	Fruit: anti-inflammatory, hypotensive, against palpitation, refreshes liver, reduces anxiety Leaf: hypotensive Bark: antiallergic, against pulmonary hemorrhage, hypotensive	Brack Egg 1999, Bussmann and Sharon 2007, Mostacero et al. 2002, Rodríguez et al. 1996, Rodríguez Rodríguez et al. 2007
Toxicity		

FIGURE 2.27 *Capparis aphylla* is a useful medicinal plant, but also its wood has economic importance. (Brandis, D. 1874. *Illustrations of the Forest Flora of North-West and Central India*. London: [s.n.]. Via http://www.biolib.de. [Online library.] http://caliban.mpiz-koeln.mpg.de/brandis/screen/tafel_03_m.jpg)

TABLE 2.6
Capparis aphylla

		References
Synonyms	*Capparis aphylla* Roth: *C. aphylla* Hayne ex Roth *C. decidua* (Forssk.) Edgew. *C. sodada* R. Br. *Sodada decidua* Forssk. (basionym)	GRIN 1994, Hutchinson and Dalziel 1954, IPNI 2012, Jafri 1985, MMPND 1995
Common names	Arabic: khiran, khirar, tountoub (Chad) Chinese: wu ye shan gan English: karira; leafless caper bush Hindi: karer, kureel Sanskrit: karir, karira Tamil: sengam Telugu: kariramu	GRIN 1994, McGuffin et al. 2000, MMPND 1995
Habitat and distribution	Extending to eastern Sudan, Arabian peninsula and India; northern and tropical Africa. *Northern Africa*: Algeria, Libya, Egypt *Western Tropical Africa*: Mauritania, Niger, Nigeria, Senegal *Northeastern Tropical Africa*: Chad, Djibouti, Sudan, Ethiopia, Socotra *Western Asia*: Israel, Sinai, Iran *Arabian Peninsula*: Saudi Arabia, Oman, Yemen *Indian Subcontinent*: India, Pakistan *Northern South America*: Guyana (?)	GRIN 1994, Hutchinson and Dalziel 1954, IPNI 2012, Jafri 1985
Economic importance	Flower buds: food (pickled); potherb Fruits, young: food (pickled) Fruit, green/ripe: food Fruit, dried: food Root: burned yields salt Wood: for making knees of boats; resistant to white ants Part not specified: medicinal	GRIN 1994, Jafri 1985, McGuffin et al. 2000, Sturtevant and Hedrick 1919
Local medicinal uses	Part not specified: astringent; diaphoretic; skin problems: boils, eruptions; swelling; inflammation; arthritis, rheumatism; toothache; cough, asthma; biliousness; laxative; malaria; vermifuge; antidote	Broun and Massey 1929, Duke 1993, GRIN 1994, Jafri 1985, McGuffin et al. 2000, Sturtevant and Hedrick 1919
Toxicity		

TABLE 2.7
Capparis baducca

		References
Synonyms	*Capparis baducca* L.: A controversial name considered by some as lectotypified on an Old World plant (= *C. rheedei* DC.) and by others on an American plant (= *C. frondosa* Jacq.) *C. frondosa* Jacq. *C. baducca* auct. Amer.: *C. cynophallophora* var. *baducca* auct. *Capparidastrum frondosum* (Jacq.) Cornejo & Iltis	Brako and Zarucchi 1993, Encyclopedia of Life 2007, GRIN 1994, IOPI 1991, IPNI 2012, USDA, NRCS 2012
Common names	English: caper, church blossom Spanish: fruta de burro, naranjuelo, tinto (Colombia); palo de burro, sapo (Porto Rico); ajito (Venezuela)	Standley 1920–1926, USDA, NRCS 2012
Habitat and distribution	*Mexico:* Mexico *Central America*: Belize, Costa Rica, El Salvador, Guatemala, Honduras, Nicaragua, Panama *Caribbean*: Anguilla, Antigua and Barbuda, Barbados, Cuba, Dominica, Dominican Republic, Grenada, Guadeloupe, Haiti, Jamaica, Martinique, Montserrat, Netherlands Antilles, Puerto Rico, St. Kitts and Nevis, St. Lucia, St. Vincent and Grenadines, Virgin Islands (British), Virgin Islands (U.S.) *Northern South America:* French Guiana, Guyana, Suriname, Venezuela *Western South America:* Colombia, Ecuador, Peru *Brazil*: Brazil	GRIN 1994, IOPI 1991, IPNI 2012, USDA, NRCS 2012
Economic importance	Parts not specified: medicinal	Duke 1993
Local medicinal uses	Parts not specified: diuretic; sedative; skin problems; spasms; dropsy; emmenagogue; poison	Duke 1993, Standley 1920–1926, Standley and Steyermark 1952
Toxicity	Fruit: claimed to be poisonous	Standley 1920–1926

FIGURE 2.28 (See color insert.) The most important parts of *Capparis cartilaginea* for traditional use are the fruit (for food) and the leaves and stems (medicinally). The ripe fruit are red. (8.8.2011, Kibbutz Ketura, Southern Arava, Israel: by Helena Paavilainen.)

FIGURE 2.29 (See color insert.) The yellow inside of ripe split fruits of *Capparis cartilaginea* presented by Dr. Elaine Soloway. Note the seeds. (8.8.2011, Kibbutz Ketura, Southern Arava, Israel: by Helena Paavilainen.)

TABLE 2.8
Capparis cartilaginea

		References
Synonyms	*Capparis cartilaginea* Decne.: *C. galeata* Fres. *C. inermis* Forsskal *C. sinaica* Veillard *C. spinosa* var. *galeata* (Fres.) Hook. F. & Thorns. *C. uncinata* Edgew.	Elffers et al. 1964, Greuter et al. 1984/1986, IOPI 1991, IPNI 2012, Jafri 1985, JSTOR Plant Science n.d., Oliver 1868, Rivera et al. 2003
Common names	Arabic: felfel-jibbel, goah, goah-kulul, lassaf, kabar (Egypt); lassaf, lusef (plant), 'aslub, 'albelib (fruit), látssaf, nutssáf (Saudi Arabia); lúsfeh (plant), álbelib, áslub (fruit) (Dhofar); lattssaf, laşaf, nişaf (Yemen) Hebrew: tsalaf ṣhusi Jibbali: lósef, aselib (fruit) (Dhofar) In India: karat In Iran (Balouchistan): gorilimbuk blatter In Kenya: chepkogh, chepteretwa, gorra, ilngorochi, leachar, lokapilak, mtunguru, mbaruti, olatunde, qadhu In Somalia: goah, goah-kulul, goh, gombor	Beentje et al. 1994, Dale and Greenway 1961, Danin 2006–, Fici et al. 1993, Ghazanfar 1994, Lehmann 2001, Miller et al. 1988, Osborn 1968, Parsa 1951, Rivera et al. 2003, Schweinfurth 1912, Wood 1997
Habitat and distribution	Primarily a coastal species preferring rocks and hillocks as its habitat. Altitude range 0–1,800 m. Also semisteppe shrublands; deserts. From Yemen to the Middle East and India; northeastern Africa and tropical East Africa *Northern Africa*: Egypt *Northeastern Tropical Africa*: Djibouti, Eritrea, Ethiopia, Socotra, Somalia, Sudan *Eastern Tropical Africa*: Kenya, Tanzania *Western Asia*: Israel, Iraq, southern Iran *Arabian Peninsula*: Saudi Arabia, Yemen *Indian Subcontinent*: India, western Pakistan	Danin 2006–, Elffers et al. 1964, Fici et al. 1993, Jafri 1985, JSTOR Plant Science n.d., Seidemann 2005
Economic importance	Leaf: Important fodder for livestock, esp. for sickly camels and goats; medicinal, against cough Fruit: food; sweet pulp; high in vitamin C, significant amounts of protein and carbohydrate; medicinal Fruits, dried: main ingredient of a nutritious, spicy drink called *mariida*, which can be kept in a waterskin for 1 to 2 months	Fici et al. 1993, Goodman and Hobbs 1988, Miller et al. 1988, Rivera et al. 2003
Local medicinal uses	Fruit: rheumatism; veterinary (fevers) Leaf: skin inflammation; bruises; swellings; rheumatism, joint inflammation; knee problems; tendinitis; sprains; muscular contractions; paralysis of body members; headache; earache; for eye diseases (humans, cattle); cough; indigestion, colics; childbirth, after childbirth for pains and as an antiseptic; snakebites (pain, inflammation); veterinary medicine (parasites and ticks in livestock; cattle fevers; common cold in newborn goats; for sickly camels and goats: first causes severe diarrhea, then improvement of appetite, good condition, and milk increased in quantity and quality) Stems/Shoots: skin inflammation; bruises; swellings; rheumatism, joint inflammation, knee problems; tendinitis; sprains; muscular contractions; paralysis of body members; headache; earache; colics; childbirth; snakebites (pain, inflammation); veterinary medicine (parasites and ticks in livestock; cattle fevers) Root: dermatitis; skin ulcers; wounds	Fici et al. 1993, Ghazanfar 1994, Goodman and Hobbs 1988, Kokwaro 1987, Lehmann 2001, Lindsay and Hepper 1978, Miller et al. 1988, Osborn 1968, Prelude 1993, Rivera et al. 2003, Timberlake 1987
Toxicity		

TABLE 2.9
Capparis coriacea

		References
Synonyms	*Capparis coriacea* Burch. ex DC.: *C. clutifolia* Burch. ex DC. *C. oleoides* Burch. ex DC. *Boscia oleoides* (Burch. ex DC.) Toelken	Encyclopedia of Life 2007, JSTOR Plant Science n.d., Sonder 1894, Tropicos 2013
Common names		
Habitat and distribution	*Southern Africa*: South Africa	JSTOR Plant Science n.d., Sonder 1894
Economic importance		
Local medicinal uses		
Toxicity		

TABLE 2.10
Capparis corymbosa

		References
Synonyms	*Capparis corymbosa* Lam.: *C. cerasifera* Gilg *C. citrifolia*, sensu Arwidss. *C. djurica* Gilg & Bened. *C. fischeri* Pax *C. sepiaria* L. *C. sepiaria* var. *fischeri* (Pax) DeWolf *C. sepiaria* var. *fischeri* (Pax) Linn. *Capparis corymbosa* var. *sansibarensis* Pax: *C. citrifolia* Lam. *C. sansibarensis* Pax Gilg *Capparis corymbosa* var. *subglabra* Oliv.: *C. sepiaria* var. *subglabra* (Oliv.) DeWolf	Burkill 1985, Elffers et al. 1964, GRIN 1994, Hanelt et al. 2001, IPNI 2012, Wild 1960
Common names	Arabic ("Maure"): bauier, bulgui English: corymbose caper; hedge caper-bush Fula-Fulfulde (Nigeria): gorko nyangudoohi Fula-Pulaar (Mali): gumba, gumi, gumi balévi Fula-Pulaar (Senegal): gumba, gumi, gumi balévi Hausa (Nigeria): haujari, haujarin mutane Manding-Bambara (Mali): tabuti Manding-Bambara (Senegal): M-bukari, N-bukari, tabuti Manding-Bambara (Upper Volta): N-bukari, tabuti Moore (Upper Volta): gaongo Songhai (Mali): cobigna	Burkill 1985, Encyclopedia of Life 2007, Hyde et al. 2013a,b, USDA, NRCS 2012

TABLE 2.10 (*Continued*)
Capparis corymbosa

		References
Habitat and distribution	Savannah; altitude range 180–1,600 m Sahel savanna from Senegal to Niger and northern Nigeria and extending across Africa to Sudan. East Africa and southern Africa. *Western Tropical Africa*: Mali, Niger, northern Nigeria, Senegal *Northeastern Tropical Africa*: Sudan, Ethiopia, Somalia *Eastern Tropical Africa*: Kenya, Tanzania *Southern Tropical Africa*: Angola, Mozambique, Zambia *Southern Africa*: South Africa *Northern South America*: Guyana (?)	Burkill 1985, Elffers et al. 1964, GRIN 1994, Hanelt et al. 2001, Hutchinson and Dalziel 1954, IPNI 2012
Economic importance	Fruit: food; aphrodisiac Leaves: food Root: repellent, poison Root flour: hunting poison	Burkill 1985, Hanelt et al. 2001, Terra 1966
Local medicinal uses	Fruit: genital stimulants/depressants Bark: abscess Root: mouth disease; child's enuresis; veterinary (external parasites)	Ake-Assi 1992, Burkill 1985, Sita 1978, Thoen and Thiam 1990
Toxicity	Plant: graziers consider the plant toxic to stock, which will not graze it Root: said to be poisonous Root flour: hunting poison	Burkill 1985

FIGURE 2.30 Red flowers of *Capparis decidua* in the desert during the hot season. (24.3.2010, Rajasthan, India: by Ryan Brookes.)

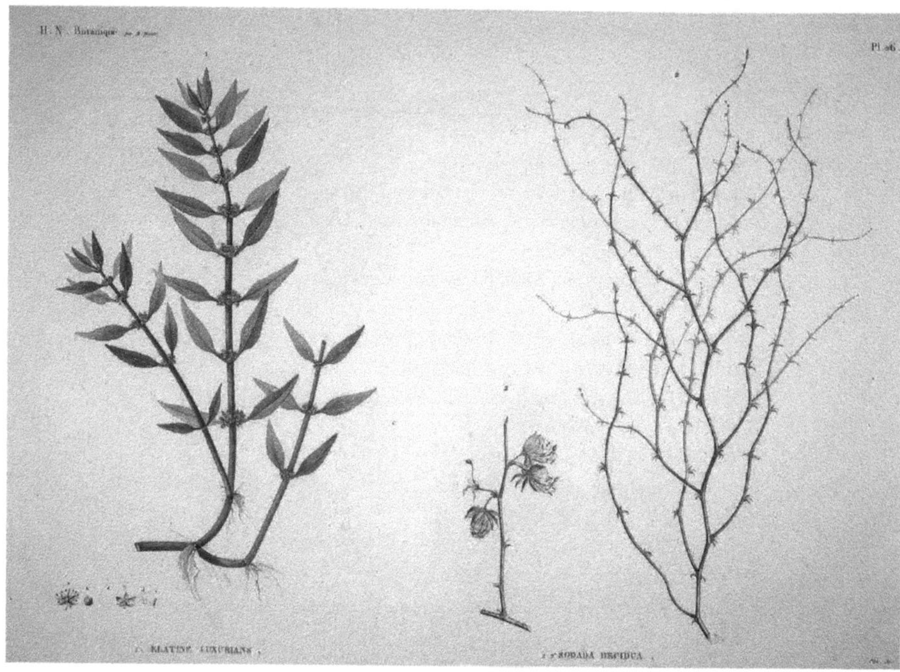

FIGURE 2.31 All the parts of *Capparis decidua* are traditionally considered medicinal. *Capparis decidua* (Forssk.) Edgew. (as *Sodada decidua* Forssk). (Raffeneau-Delile, A. 1813. *Flore d'Egypte: Explanation des Planches.* Plates, t. 26. Paris: Imprimerie Impériale. Via http://www.plantillustrations.org. 2009. [Online collection of illustrations]. http://www.plantgenera.org/illustration.php?id_illustration=48881 accessed 5 February 2013.)

TABLE 2.11
Capparis decidua

		References
Synonyms	*Capparis decidua* (Forssk.) Edgew. *Sodada decidua* Forssk. (basionym) *C. aphylla* Hayne ex Roth *C. aphylla* Roth *C. sodada* R. Br.	GRIN 1994, Hanelt et al. 2001, Hutchinson and Dalziel 1954, IPNI 2012, Jafri 1985, MMPND 1995
Common names	Arabic: khiran, khirar, habriga, sudad, sarwab, tundub, tountoub (Chad) Bengali: karil Berber: koussoms Chinese: wu ye shan gan English: karira, kureel, leafless caper bush French: câprier sans feuilles Gujarati: kerdo, kair, kera, kerda Hebrew: tsalaf ratmi Hindi: kair, karer, ker, delha, kurrel Sanskrit: karir, karira Tamil: sengam Telugu: kariramu	Bibliotheca Alexandrina n.d., Boulos 1983, Danin 2006–, Encyclopedia of Life 2007, GRIN 1994, Hanelt et al. 2001, Kress 1995, McGuffin et al. 2000, MMPND 1995, Seidemann 2005, 79

TABLE 2.11 (*Continued*)
Capparis decidua

		References
Habitat and distribution	On arid plains, flowering abundantly during the hot weather; deserts Northern and tropical Africa to eastern Sudan, Arabia, eastward to India *Northern Africa*: Morocco, Algeria, Libya, Egypt *Western Tropical Africa*: Guinea, Mauritania, Mali, Niger, Nigeria, Senegal *Northeastern Tropical Africa*: Chad, Djibouti, Ethiopia, Socotra, Sudan *Western Asia*: Israel, Sinai, Jordan, southern Iran *Arabian Peninsula*: Saudi Arabia, Oman, Yemen *Indian Subcontinent*: northwestern India, Pakistan *Northern South America*: Guyana (?)	Danin 2006–, Greuter et al. 1984/1986, GRIN 1994, Hanelt et al. 2001, Hutchinson and Dalziel 1954, IOPI 1991, Jafri 1985, Seidemann 2005, 79
Economic importance	Flower buds: food (pickled); potherb Fruits, young: food (pickled) Fruit: food (cultivated for fruit) Wood: for making knees of boats; resistant to white ants Plant: grown for hedges (windbreak) and the edible fruits Part not specified: medicinal	Ganguly and Kaul 1969, GRIN 1994, Hanelt et al. 2001, Jafri 1985, McGuffin et al. 2000, Seidemann 2005, 79
Local medicinal uses	Leaf: venereal disease Stem: headache; liver problems Twigs: astringent; diaphoretic; boils; swellings; inflammation; rheumatism, rheumatic arthritis; toothache; cough, asthma; cardiac problems; laxative; anthelmintic; fever Root: eye problems; internal parasites; venereal disease Bark: urinary problems; blennorragia; malaria Bark, ash of: astringent; hemostatic; disinfectant for wounds and sores Plant, ash of: veterinary (camels: hemostatic) Part not specified: astringent; diaphoretic; skin problems: boils, eruptions; swelling; inflammation; joint problems, arthritis, rheumatism; toothache; cough, asthma; biliousness, stomach disorders; laxative; malaria; vermifuge; antidote; veterinary (camels: scabies)	Bebawi and Neugebohrn 1991, Boulos 1983, Boury 1962, Broun and Massey 1929, Duke 1993, El Ghazali 1986, El Ghazali et al. 1994, 1997, GRIN 1994, Jafri 1985, Kerharo and Adam 1964a,b, 1974, McGuffin et al. 2000, Monteil and Sauvage 1953, Sturtevant and Hedrick 1919, Thoen and Thiam 1990
Toxicity	Part not specified: injurious to skin	FDA Poisonous Plant Database n.d., Mitchell and Rook 1979

TABLE 2.12
Capparis deserti

		References
Synonyms	*Capparis deserti* (Zohary) Täckholm & Boulos: *C. aegyptia* Lam. *C. spinosa* var. *deserti* Zohary (basionym)	Greuter et al. 1984/1986, IOPI 1991, IPNI 2012, Rivera et al. 2003, The Plant List 2010
Common names	Arabic: kabar, quabar, lassaf, shafellah (Egypt); kabar, shafallah (Iraq) Coptic: kemegeoc, kemapic Hebrew: tsalaf mitsri	Blakelock and Townsend 1980, Danin 2006–, Rolland 1967, Schweinfurth 1912
Habitat and distribution	Hard rock outcrops. Semisteppe shrublands, shrub steppes, deserts and extreme deserts Northern Africa; Middle East *Northern Africa*: Algeria, Morocco, Tunisia, Libya, Egypt *Western Asia*: eastern Aegean Islands, Lebanon and Syria, Israel and Jordan, Sinai	Danin 2006–, Greuter et al. 1984/1986, IOPI 1991, IPNI 2012, Rivera et al. 2003
Economic importance	Fruit: food (pickled) Seeds: condiment; added to wine to keep it sweet	Blakelock and Townsend 1980, Renfrew 1987, Rivera et al. 2003
Local medicinal uses	Leaf: against arteriosclerosis Root bark: wounds; skin diseases (historical); antitumoral (historical); vermifugue (historical); emmenagogue (historical)	Alpini and de Fenoyl 1980, Al-Said 1993, Manniche 1989, Rivera et al. 2003
Toxicity		

TABLE 2.13
Capparis divaricata

		References
Synonyms	*Capparis divaricata* Lam.: *C. horrida* Banks ex Wight & Arn. *C. stylosa* DC.	IPNI 2012; The Plant List 2010
Common names	English: spreading caper Kannada: bhandero, revadi, totla, totte, tottulla Marathi: pachunda Sanskrit: pakhoda, pakhauda Tamil: turatti Telugu: ambaram valli, ambaramvalli, badaraeni, boodari, budaroni, remidi, remmani	ENVIS 1999; efloraofindia 2007
Habitat and distribution	Tropics *Indian Subcontinent*: India, Sri Lanka	ENVIS 1999
Economic importance		
Local medicinal uses	Part not specified: analgesic; diuretic; antiulcer; aphrodisiac	Patil et al. 2011
Toxicity		

TABLE 2.14
Capparis flavicans

		References
Synonyms	*Capparis flavicans* Kurz: *C. cambodiana* Pierre ex Gagnep. *C. flavicans* Wall. *C. flavicans* Wall. ex Hook.f. & Thomson	Craib 1912, IPNI 2012, The Plant List 2010
Common names		
Habitat and distribution	*Indian Subcontinent*: India *Indochina*: Cambodia, Myanmar, Thailand	Craib 1912, IPNI 2012, JSTOR Plant Science n.d.
Economic importance		
Local medicinal uses	Part not specified: galactagogue	Luecha et al. 2009
Toxicity		

FIGURE 2.32 *Capparis flexuosa* L. plant growing beside a rock. (Near Grand Ton, Guadeloupe, Lesser Antilles: © William Hawthorne.)

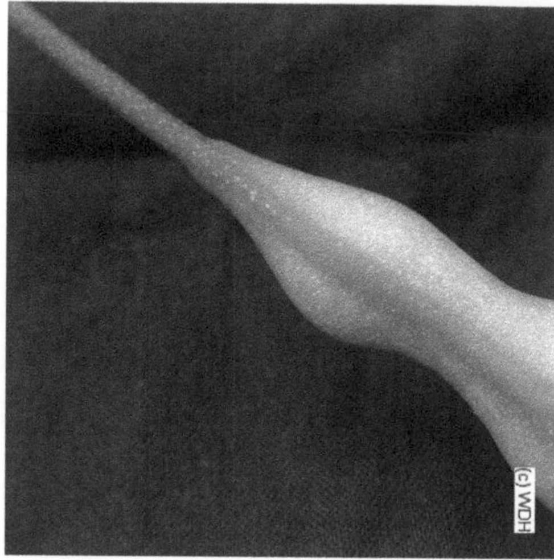

FIGURE 2.33 *Capparis flexuosa* is cultivated for its horseradish-tasting fruit. (15.6.2001, near River Sallee, Grenada, Lesser Antilles: © William Hawthorne.)

TABLE 2.15
Capparis flexuosa

		References
Synonyms	*Capparis flexuosa* L.: *C. flexuosa* (L.) L. *C. cynophallophora* L. *Cynophalla flexuosa* (L.) J. Presl. *Morisonia flexuosa* L. (basionym)	Brako and Zarucchi 1993, GRIN 1994, Hanelt et al. 2001, IOPI 1991, IPNI 2012, Jørgensen 2008, MMPND 1995
Common names	English: bay-leaved caper bush; bay-leaved capertree; capertree; falseteeth; limber caper German: jamaika kaper In Mexico: xpayumak, pan y aqua, burro, palo de burro Unidentified: fructo de burro	Encyclopedia of Life 2007, FDA Poisonous Plant Database n.d., GRIN 1994, Hanelt et al. 2001, Howard 1974–1989, MMPND 1995, Peckholt 1898, Seidemann 2005, 79, USDA, NRCS 2012
Habitat and distribution	Disturbed areas, forests. 0–1,500 m Western India, Mexico to southern USA and to Paraguay *Southeastern USA*: Florida *Mexico*: Mexico *Central America*: Belize, Costa Rica, El Salvador, Guatemala, Honduras, Nicaragua, Panama *Caribbean*: Anguilla, Antigua and Barbuda, Bahamas, Barbados, Cayman Islands, Cuba, Dominica, Dominican Republic, Grenada, Guadeloupe, Haiti, Jamaica, Martinique, Montserrat, Puerto Rico, St. Kitts and Nevis, St. Lucia, St. Vincent and Grenadines, Virgin Islands (British), Virgin Islands (U.S.) *Northern South America*: Suriname, Venezuela *Western South America*: Bolivia, Colombia, Ecuador, Peru *Brazil*: Brazil *Southern South America*: Paraguay	Brako and Zarucchi 1993, GRIN 1994, Hanelt et al. 2001, IOPI 1991, Jørgensen 2008, Seidemann 2005, 79, USDA, NRCS 2012
Economic importance	Flower bud: condiment Fruit: food (cultivated for the fruit that tastes like horseradish)	Hanelt et al. 2001, Sánchez-Monge y Parellada 1981, Seidemann 2005, 79

TABLE 2.15 (*Continued*)
Capparis flexuosa

		References
Local medicinal uses	Part not specified: sedative; vesicant; for skin problems; spasm; tooth hygiene; diuretic; dropsy; emmenagogue	Duke 1993, Liogier 1974, Martínez 1933, Pittier 1926, Standley 1920–1926
Toxicity	Part not specified: poisonous	Bernhard-Smith 1923, FDA Poisonous Plant Database n.d.

TABLE 2.16
Capparis formosana

		References
Synonyms	*Capparis formosana* Hemsl. (basionym): *C. kanehirae* Hayata ex Kanehira *C. sikkimensis* Kurz subsp. *formosana* (Hemsl.) M. Jacobs	IPNI 2012, Zhang and Tucker 2008
Common names	Chinese: tai wan shan gan	Zhang and Tucker 2008
Habitat and distribution	Dense montane forests, altitude range above 700 m *China*: China (Guangdong, Hainan) *Eastern Asia*: Taiwan, Japan (Ryukyu Islands) *Indochina*: Vietnam	IPNI 2012, Zhang and Tucker 2008
Economic importance		
Local medicinal uses		
Toxicity		

TABLE 2.17
Capparis grandis

		References
Synonyms	*Capparis grandis* L.f.: *C. auricans* (Kurz) Craib *C. bisperma* Roxb. *C. disperma* Walp. *C. maxima* Roth. *C. obovata* Buch.-Ham. ex DC. *C. racemifera* DC.	IPNI 2012, The Plant List 2010
Common names	Kannada: Torate Marathi: Pachunda, Katarni Tamili: Mudkondai Telugu: Guli, Ragota, Nallupi	Brandis 1906, Gamble 1915, 46
Habitat and distribution	Altitude range: 275–425 m *Indian Subcontinent*: India, Sri Lanka *Indochina*: Myanmar, Thailand	Brandis 1906, IPNI 2012, Prasad et al. 2013
Economic importance	Plant: timber	Prasad et al. 2013
Local medicinal uses	Leaf: swelling; eruption Bark: swelling; eruption	Chopra et al. 1950, Duke 1993, Mishra et al. 2007, *Wealth of India* 1948–1976
Toxicity		

TABLE 2.18
Capparis heyneana

		References
Synonyms	*Capparis heyneana* Wall. ex Wight & Arn.: *C. baducca* L.	IPNI 2012, The Plant List 2010
Common names		
Habitat and distribution	*Indian Subcontinent*: India, Sri Lanka	IPNI 2012, Vardhana 2008
Economic importance		
Local medicinal uses	Leaf: rheumatism Part not specified: laxative	Chopra et al. 1950, Duke 1993, Mishra et al. 2007, *Wealth of India* 1948–1976
Toxicity		

TABLE 2.19
Capparis himalayensis

		References
Synonyms	*Capparis himalayensis* Jafri: *C. leucophylla* Collett *C. napaulensis* DC. *C. spinosa* L. var. *himalayensis* (Jafri) Jacobs	IPNI 2012, Jafri 1985, Rivera et al. 2003, Zhang and Tucker 2008
Common names	Chinese: zhua jia shan gan German: Nepal-Kaper; Himalaya-Kaper Hindi/Hindustani (Pakistan): kabra Punjabi (Pakistan): kakri, kander, kabra Sanskrit (Pakistan): kakadani In India: karil, kabra	Baquar 1989, Collet 1902, Nadkarni 1976, Rivera et al. 2003, Seidemann 2005, 80, Sundara 1993, Zhang and Tucker 2008
Habitat and distribution	Plains, desert flats, open sunny areas; below 1,100 m Endemic to northwestern Himalayan areas of India and western Pakistan *Middle Asia*: Tajikistan *Caucasus*: Georgia *China*: China, Tibet *Indian Subcontinent*: northeastern Pakistan, northwestern India, Nepal	IPNI 2012, Jafri 1985, Zhang and Tucker 2008
Economic importance	Flower buds: condiment Fruit: condiment Root bark: medicinal	Rivera et al. 2003
Local medicinal uses	Leaf: antirheumatic Fruits: rheumatism; earache; snakebites Root: against sores Root bark: bitter, hot and dry; tonic; analgesic; diuretic; rheumatism; paralysis, kills worms in the ear; toothache; tubercular lymphadenitis; expectorant; splenomegaly; laxative; antihelminthic; emmenagogue Part not specified: rheumatism, gout; palsy; dropsy	Kakrani and Saluja 1994, Kirtikar et al. 1987, Nadkarni 1976, Rivera et al. 2003
Toxicity		

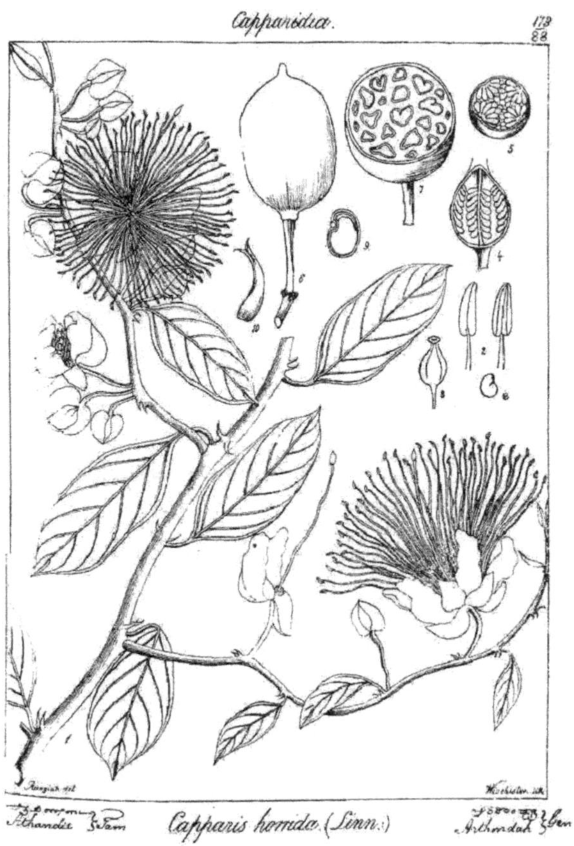

FIGURE 2.34 *Capparis horrida*. (Wight, R. 1846. *Icones Plantarum Indiae Orientalis*. Vol. 1, t. 173. Madras: published by J.B. Pharoah for the author, 1840–1853. Via http://www.plantillustrations.org. 2009. [Online collection of illustrations.] http://www.plantgenera.org/illustration.php?id_illustration=161692 accessed 5 February 2013.)

TABLE 2.20
Capparis horrida

		References
Synonyms	*Capparis horrida* L. f.: *C. brevispina* DC. *C. hastigera* Hance *C. hastigera* var. *obcordata* Merrill & F. P. Metcalf *C. swinhoei* Hance. *C. zeylanica* L.	IPNI 2012, Jafri 1985, MMPND 1995, Zhang and Tucker 2008
Common names	Chinese: niu yan jing English: Ceylon caper, Indian caper Hindi: ardanda, hinsa, kalhins Oriya: oserwa Sanskrit: vyakhra nakhi, hankaru Tagalog: dauag, halubágat-báging Tamil: adondai, adhandai, alanday, alavirukkam, tondai Telugu: adonda, adondathivva, adontha, aradoonda, aridonda, arthondah, arudonda, doddi	ENVIS 1999, MMPND 1995, Quisumbing 1951, 338, Zhang and Tucker 2008

(*Continued*)

TABLE 2.20 (*Continued*)
Capparis horrida

		References
Habitat and distribution	Forest margins, thickets, limestone slopes or sandy soil, scattered grasslands; below 700 m Tropical Asia; from India eastward to Malesia and Indochina *China*: China *Indian Subcontinent*: Pakistan, India, Sri Lanka, Nepal *Indochina*: Myanmar, Thailand, Vietnam *Malesia*: Indonesia, Malaya, Philippines Indian Ocean islands	IPNI 2012, Jafri 1985, Zhang and Tucker 2008
Economic importance	Fruit: food (pickled)	Sturtevant and Hedrick 1919
Local medicinal uses	Leaf: counterirritant; for boils, swellings; stomachic; improves appetite; piles; syphilis Root bark: antiperspirant; sedative; stomachic; cholera cure Part not specified: venereal diseases	Ashton et al. 1997, Batugal et al. 2004, Chakravarti and Venkatasubban 1932, Duke 1993, Quisumbing 1951, 338–339, Uphof 1968
Toxicity		

TABLE 2.21
Capparis humilis

		References
Synonyms	*Capparis humilis* Hassl. (basionym): *C. brasiliana* var. *puberula* Hieron. *Capparidastrum humilis* (Hassl.) Cornejo & Iltis	IPNI 2012, Jørgensen 2008
Common names		
Habitat and distribution	*Western South America*: Bolivia *Southern South America*: Argentina, Paraguay	GBIF 2001, IPNI 2012, Jørgensen 2008
Economic importance		
Local medicinal uses		
Toxicity		

TABLE 2.22
Capparis incanescens

		References
Synonyms	*Capparis incanescens* DC.: *C. sepiaria* L.	IPNI 2012, Jafri 1985
Common names		
Habitat and distribution	Tropical Africa and *China*: China *Indian Subcontinent*: Pakistan, India, Sri Lanka *Indochina*: Indochina *Malesia*: Malesia *Australia*: Australia	Jafri 1985
Economic importance		
Local medicinal uses	Part not specified: alterative; tonic; skin problems; fever	Duke 1993, *Wealth of India* 1948–1976
Toxicity		

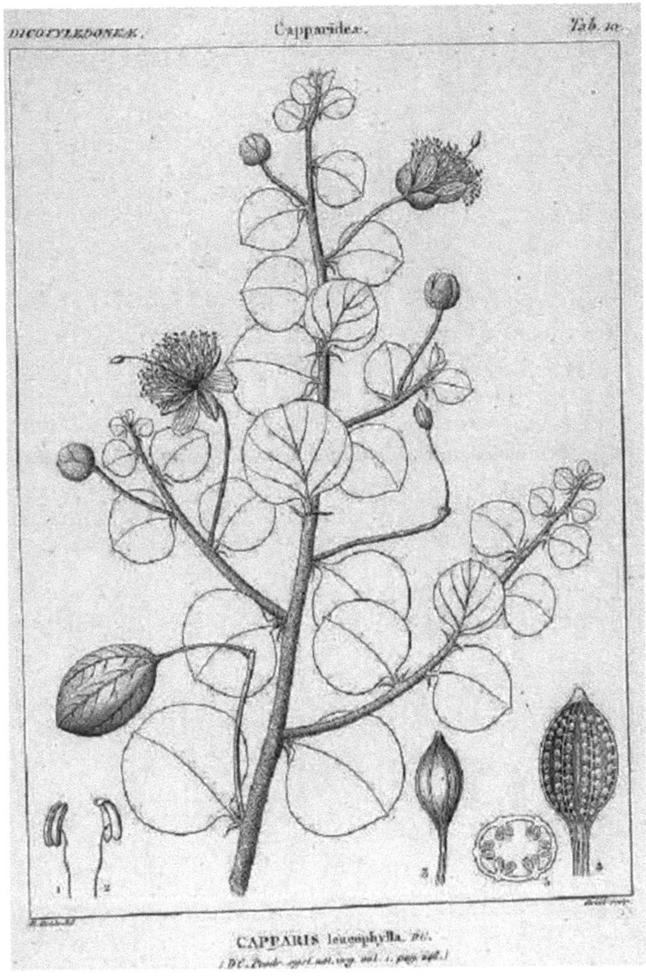

FIGURE 2.35 *Capparis leucophylla*. (Delessert, J.B. and A.P. de Candolle. 1837. *Icones Selectae Plantarum*. Vol. 3, t. 10. Via http://www.plantillustrations.org. 2009. [Online collection of illustrations.] http://www.plantgenera.org/illustration.php?id_illustration=79369 accessed 5 February 2013.)

TABLE 2.23
Capparis leucophylla

		References
Synonyms	*Capparis leucophylla* DC. (basionym): *C. sicula* Duhamel subsp. *leucophylla* (DC.) Inocencio, D. Rivera, Obón & Alcaraz *C. spinosa* L. var. *pubescens* Zohary	Greuter et al. 1984/1986, IOPI 1991, IPNI 2012, Jafri 1985, Rivera et al. 2003
Common names	Arabic: shafallah, kabar, kifri (Iraq) Persian: kabar (bark; Iraq); keverkai, mar gir, mar gaz (= "snakebite"), rishah-i-kabar (Iran)	Blakelock and Townsend 1980, Hooper and Field 1937, Rivera et al. 2003
Habitat and distribution	*Western Asia*: Iraq, Iran *Arabian Peninsula*: Bahrain	IPNI 2012, Jafri 1985, Rivera et al. 2003

(*Continued*)

TABLE 2.23 (*Continued*)
Capparis leucophylla

		References
Economic importance	Flower buds: food (pickled) Fruit, young: food (pickled) Fruit: food Root: medicinal Root bark: medicinal Plant: food for camels Part not specified: medicinal	Blakelock and Townsend 1980, Hooper and Field 1937, Mahasneh et al. 1996, Rivera et al. 2003
Local medicinal uses	Root: bitter; rheumatism; intermittent fever Root bark: bitter; medicine (Capparis Cortex Radicis of the old Persian Pharmacopoeia); rheumatism; intermittent fever Part not specified: tonic; expectorant; antidote for snakebite	Blakelock and Townsend 1980, Hooper and Field 1937, Mahasneh et al. 1996, Rivera et al. 2003
Toxicity		

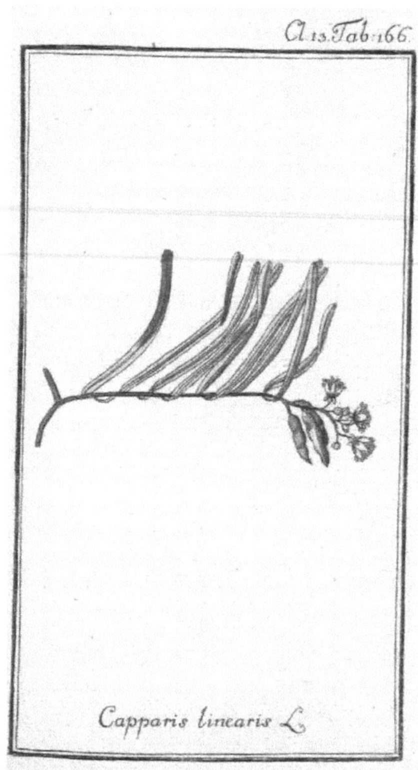

FIGURE 2.36 *Capparis linearis* L. (Zorn, J. and N.J.F. von Jacquin. 1786–1787. *Dreyhundert auserlesene amerikanische Gewächse*. Vol. 2, t. 166. Nuremberg, Germany: s.n. Via http://www.plantillustrations.org. 2009. [Online collection of illustrations.] http://www.plantgenera.org/illustration.php?id_illustration=72611&height=1080 accessed 5 February 2013.)

TABLE 2.24
Capparis linearis

		References
Synonyms	*Capparis linearis* Jacq.: *C. linearifolia* Linden ex Turcz. [illegitimate] *Cynophalla linearis* (Jacq.) J. Presl. *Pleuteron linearis* (Jacq.) Raf. *Uterveria linearis* (Jacq.) Bertol.	IPNI 2012, The Plant List 2010
Common names		
Habitat and distribution	*Northern South America*: Venezuela *Western South America*: Colombia, Peru	GBIF 2001, IPNI 2012, JSTOR Plant Science n.d.
Economic importance		
Local medicinal uses		
Toxicity		

TABLE 2.25
Capparis longispina

		References
Synonyms	*Capparis longispina* Hook. f. & Thomson (basionym): *C. rotundifolia* Rottler var. *longispina* (Hook. f. & Thomson) M. R. Almeida	IPNI 2012
Common names		
Habitat and distribution	*Indochina*: Myanmar	Kress and Lace 2003
Economic importance		
Local medicinal uses		
Toxicity		

TABLE 2.26
Capparis masaikai

		References
Synonyms	*Capparis masakai* H. Lév. (basionym): *C. sikkimensis* Kurz subsp. *masaikai* (H. Lév.) M. Jacobs	IPNI 2012, Zhang and Tucker 2008
Common names	Chinese: ma bing lang	Zhang and Tucker 2008
Habitat and distribution	Valleys, dense forests, slopes, limestone areas; below 1,600 m *China*: China	Zhang and Tucker 2008
Economic importance	Seeds: contain sweet protein mabilin, which is a powerful sweetener	Liu et al. 1993
Local medicinal uses		
Toxicity		

FIGURE 2.37 *Capparis moonii* buds and white flowers. (26.3.2011, Western Ghats, India: by Satish Nikam.)

TABLE 2.27
Capparis moonii

		References
Synonyms	*Capparis moonii* Wight	IPNI 2012, MMPND 1995
Common names	English: Indian caper	ENVIS 1999, MMPND 1995
	Kannada: bandiraroveldi, mullu kathari, mullu katthari balli, mullukarti, mullukathari, mullukattari, tatla, totte, tottulla	
	Marathi: rudrvanti, vaghati	
	Oriya: udipi	
	Sanskrit: rudanti	
	Telugu: aadsenda, adonda	
Habitat and distribution	Only in *Indian Subcontinent*: southern India, Sri Lanka	IPNI 2012, ENVIS 1999
Economic importance		
Local medicinal uses	Fruit: cough, spitting and weakness in tuberculosis Seed: cough, spitting	Chopra et al. 1950, Mishra et al. 2007, Tyagi 2005, 115
Toxicity		

FIGURE 2.38 *Capparis orientalis* Veill. (as *C. rupestris* Sm.). (Sibthrop, J. and J.E. Smith. 1825. *Flora Graeca* (drawings). Vol. 5, t. 87. [s.n.]: [s.n.] Via http://www.plantillustrations.org. 2009. [Online collection of illustrations.] http://www.plantgenera.org/illustration.php?id_illustration=141606 accessed 5 February 2013.)

TABLE 2.28
Capparis orientalis

		References
Synonyms	*Capparis orientalis* Veill.: *C. inermis* Turra *C. orientalis* Veill. in Duhamel *C. rupestris* Sm. (basionym) *C. spinosa* L. subsp. *inermis* A. Bolós & O. Bolós *C. spinosa* L. subsp. *orientalis* (Veillard) Jafri *C. spinosa* L. subsp. *rupestris* (Sm.) Nyman	Greuter et al. 1984/1986, GRIN 1994, IOPI 1991, IPNI 2012, Rivera et al. 2003
Common names	Arabic: cabbar (Libya) Berber: tilut (Libya) English: eastern caper, oriental caper French: câprier German: orientalische Kaper Greek: kappari Italian: caparen, caper, cappari, capparo, capperese, capperi, cappero (plants); cetriolini (fruits); chiappara, zucchette Portugese: alcaparras, alcaparreira Spanish: alcaparra; gorrinets (fruits); tapara, taparera (plant); táperes (flower buds)	Barbera and Di Lorenzo 1982, Franzan 2001, Guerau et al. 1981, Jafri 1977, Seidemann 2005, 80, Ungarelli 1985
Habitat and distribution	Mediterranean; southern Europe *Southwestern Europe*: southern France, Portugal, Spain (incl. Baleares) *Southeastern Europe*: Albania, Croatia, Slovenia, Yugoslavia, Greece (incl. Crete), Italy (incl. Sardinia, Sicily), Malta *Northern Africa*: Morocco, Algeria, Tunisia, Libya, Egypt *Western Asia*: Turkey *Northern South America*: Guyana (?)	Greuter et al. 1984/1986, GRIN 1994, IOPI 1991, IPNI 2012, Seidemann 2005, 80

(*Continued*)

TABLE 2.28 (*Continued*)
Capparis orientalis

		References
Economic importance	Flower bud: food (pickled) Fruit: food (pickled) Leaf: food (pickled) Plant: ornamental; environmental boundary/barrier/ support Part not specified: food flavoring; medicinal	Barbera and Di Lorenzo 1982, Duke et al. 2002, GRIN 1994, Guerau et al. 1981, Huxley et al. 1992, Inocencio et al. 2000, Jafri 1977, Komarov et al. 1934–1964, McGuffin et al. 2000, Rivera et al. 2003, Townsend and Guest 1966–1980, Tutin et al. 1993
Local medicinal uses	Plant: stomach ailments Part not specified: medicinal; tumors	Duke 1993, El-Gadi and Bshana 1986, Hartwell 1967–1971, Trotter 1915
Toxicity		

TABLE 2.29
***Capparis ovata* Desf.**
***C. ovata* Desf. var. *canescens* (Coss.) Heywood**
C. ovata* var. *palaestina

		References
Synonyms	*Capparis ovata* Desf.: *C. fontanesii* DC. *C. spinosa* L. *Capparis ovata* var. *canescens* (Cosson) Heywood: *C. spinosa* var. *canescens* Cosson *Capparis ovata* Desf. var. *palaestina* Zohary pro parte: *C. sicula* Veill. in Duham. subsp. *mesopotamica* Inocencio, D. Rivera, Obón and Alcaraz	Czerepanov 1996, Hanelt et al. 2001, Heywood 1996, IOPI 1991, IPNI 2012, MMPND 1995, Rivera et al. 2003
Common names	*C. ovata* Desf.: Arabic: khabbar, soukoum, felfel-el-gebel (Algeria); el kabbar (Morocco) Berber: teililout, teiloulout (Morocco) Berber Temacheck: talulut, telulut, teloûloût, touloulout (Algeria) Toubou: gozui, gozou, kozohou (Chad) *C. ovata* Desf. var. *palaestina* Zohary pro parte: Arabic: shafallah, shefellah, kabar Bakhtiari: lagajee Persian: guh-i-kamar (= "flower of the rocks"), alaf-i-mar (= "snake plant"), margaz (= "snakebite")	Benchelah et al. 2000, Blakelock and Townsend 1980, Chevalier 1938, Mandaville 1990, Nègre 1961, Rivera et al. 2003, Schweinfurth 1912, Seidemann 2005, 79

TABLE 2.29 (*Continued*)
***Capparis* ovata Desf.**
***C. ovata* Desf. var. *canescens* (Coss.) Heywood**
***C. ovata* var. palaestina**

		References
Habitat and distribution	*C. ovata* Desf.: *Southwestern Europe*: Spain (?) *Southeastern Europe*: Sicily (?) *Northern Africa*: Morocco, Algeria *Northeast Tropical Africa*: Chad "This species [= *Capparis ovata* Desf. = *C. fontanesii* DC.] is an African endemic which is often erroneously cited from Italy and Spain (Barbera and Di Lorenzo 1984), because of confusion with *C. sicula* Veill." (Rivera et al. 2003) *C. ovata* Desf. var. *herbacea* (Willd.) Zohary: *Middle Asia*: Turkmenistan *Caucasus*: Armenia, Azerbaijan, Georgia *Western Asia*: Turkey	Barbera and Di Lorenzo 1984, Greuter et al. 1984/1986, Heywood 1996, IOPI 1991, IPNI 2012, Rivera et al. 2003, Seidemann 2005, 79–80
Economic importance	*C. ovata* Desf.: Flower buds: food (as capers); condiment Fruit: food (as gruel) Part not specified: fodder for gazelle *C. ovata* Desf. var. *palaestina* Zohary pro parte: Flower buds: for the Turkish food industry Fruit, unripe: food	Benchelah et al. 2000, Chevalier 1938, Gast 2000, Rivera et al. 2003
Local medicinal uses	*C. ovata* Desf: Fruits: anti-inflammatory; common cold; stomach ailments Leaf: anti-inflammatory; lumbago; veterinary (camels: scabies) Root: blennorrhagia Part not specified: anti-inflammatory; headache; stomachache *C. ovata* Desf. var. *palaestina* Zohary pro parte: Root: palliative for rheumatism Part not specified: medicinal; antidote (snakebites)	Bellakhdar 1997, Benchelah et al. 2000, Blakelock and Townsend 1980, Gast 2000, Maire 1933, Rivera et al. 2003
Toxicity		

TABLE 2.30
Capparis rotundifolia

		References
Synonyms	*Capparis rotundifolia* Rottler: *C. longispina* Hook.f. & Thomson (basionym) *C. orbiculata* Wall. ex Hook.f. & Thomson *C. pedunculosa* Wall. ex Wight & Arn.	IPNI 2012, The Plant List 2010
Common names		
Habitat and distribution	Coastal plant *Indian Subcontinent*: India	JSTOR Plant Science n.d., Gamble 1915, 46
Economic importance		
Local medicinal uses		
Toxicity		

TABLE 2.31
Capparis salicifolia

		References
Synonyms	*Capparis salicifolia* Griseb. (basionym): *Colicodendron salicifolium* (Griseb.) *Sarcotoxicum salicifolium* (Griseb.) Cornejo & H. H. Iltis	GRIN 1994, IPNI 2012, Jørgensen 2008
Common names	Chorote: ójnak, nénuk In Argentine: sacha sandía, sandía hedionda, palo verde, sandía de cabra Unknown: maaning, sacha sandia	Barboza et al. 2009, New World Fruits Database 2012, Scarpa 2009
Habitat and distribution	*Western South America*: Bolivia *Southern South America*: Argentina, Paraguay	Cornejo and Iltis 2008b, Encyclopedia of Life 2007, GRIN 1994, IPNI 2012, Jørgensen 2008, New World Fruits Database 2012
Economic importance	Fruit: food Wood: for fire drill	Arenas and Scarpa 2007, Arenas and Suárez 2007, New World Fruits Database 2012
Local medicinal uses	Root: analgesic on toothache, earache; antisyphilitic Whole plant: purgative	Barboza et al. 2009, Scarpa 2009
Toxicity	Fruit, unripe: highly toxic if not boiled long (water changed at least five times) Part not specified: Vertebrate poison (mammals)	Arenas and Scarpa 2007, Cornejo and Iltis 2008b, GRIN 1994

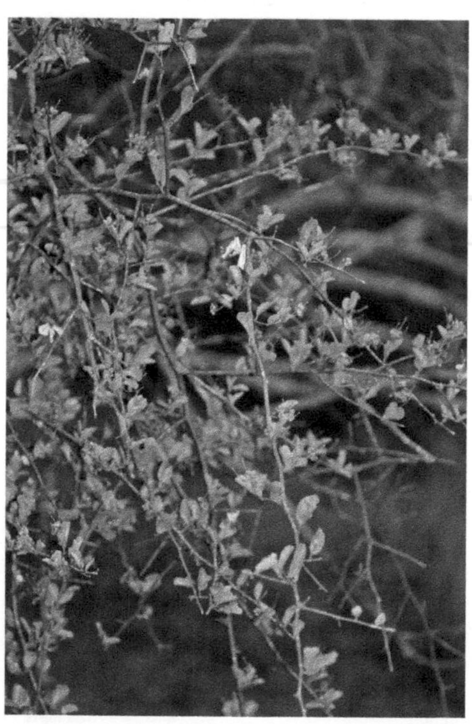

FIGURE 2.39 *Capparis sepiaria* is used as a hedge plant, but it has also recognized medicinal value (see Chapter 17, "Diabetes Mellitus") (*Capparis sepiaria* L. var. *subglabra* (Oliv.) DeWolf. 11.11.2006, Gorongosa National Park, Mozambique: by Bart T. Wursten. In Hyde, M.A., B.T. Wursten, P. Ballings, S. Dondeyne. 2013b. Flora of Mozambique: Species Information: Individual Images: *Capparis sepiaria*. http://www.mozambiqueflora.com/speciesdata/image-display.php?species_id=124450&image_id=5, retrieved 21 February 2013.)

TABLE 2.32
Capparis sepiaria

		References
Synonyms	*Capparis sepiaria* L.: *C. citrifolia*, sensu Arwidss. *C. corymbosa*, sensu Oliv. *C. flexicaulis* Hance *C. glauca* Wallich ex J. D. Hooker & Thomson. *C. incanescens* DC.	Burkill 1985, Elffers et al. 1964, Fici et al. 1993, GRIN 1994, IPNI 2012, Jafri 1985, JSTOR Plant Science n.d., MMPND 1995, Roux 2003, Wild 1960, Zhang and Tucker 2008
Common names	Chinese: qing pi ci English: Indian caper Hindi: aundi, chaili, chhail, hainsa, hium gama, hium garna, hium-garna, hiun, hiun garna, jal, kanthar, kanthari, kanthor, katan Kannada: baasingi, kaadu katthari, kaadukathari, kadukattari, kanthaari, kathari gida, kathiramullu, katthari gida, katthari mullu, katukatiri, musthodi, nibate, niputthige, olle uppina gida, thottilu mullu Marathi: kanthar, kantharvela, kantharyel Nepalese: junge laharo Oriya: solorakoli Sanskrit: ahimsra, ahinsra, amlaphala Tagalog: tarabtáb Tamil: ayinciram, cekkavuricceti, cenkaddari Telugu: nalla uppi, nallapuee, nallapuyyi, nallaupli, nallauppi, nallavuppi, nalluppi, puyyi, upp	Burkill 1985, ENVIS 1999, JSTOR Plant Science n.d., MMPND 1995, Quisumbing 1951, 340
Habitat and distribution	In low altitudes (0–300 m); in drier types of woodland and bushland or a climber in low-altitude riverine fringes. Seasides, slopes, thickets or scattered forests. Widespread in Africa with mainly an eastern African distribution, extending into India, China, and Malaysia *Western Tropical Africa*: Ghana, Mali, Niger, Nigeria, Senegal *Western-Central Tropical Africa*: Burundi, eastern Congo, Rwanda *Northeastern Tropical Africa*: Ethiopia, Somalia, Sudan, Chad *Eastern Tropical Africa*: Kenya, Tanzania, Uganda *Southern Tropical Africa*: Angola, Malawi, Mozambique, Zambia, Zimbabwe *Southern Africa*: South Africa, Swaziland *Western Indian Ocean*: Madagascar *China*: China *Indian Subcontinent*: Pakistan, India, Sri Lanka, Bangladesh, Maldives, Nepal *Indochina*: Cambodia, Laos, Myanmar, Thailand, Vietnam *Malesia*: Indonesia, Malaysia, Philippines *Papuasia*: Papua New Guinea *Australia*: Australia *Northern South America*: Guyana (?)	Elffers et al. 1964, GRIN 1994, IPNI 2012, Jafri 1985, Madagascar Catalogue 2012, Roux 2003, Wild 1960, Zhang and Tucker 2008
Economic importance	Plant: hedge plant Part not specified: medicinal	Duke 1993; Selvamani et al. 2008

(*Continued*)

TABLE 2.32 (*Continued*)
Capparis sepiaria

		References
Local medicinal uses	Seed: antidote (snakebite) Leaf: swelling of body with oozing pus Root: earache, mumps; magic (gives good luck); veterinary (gallsickness in stock) Stem, root bark: aphthae; gout; dropsy Part not specified: alterative; tonic; skin problems; fever	Belayneh et al. 2012, Chopra et al. 1950, Corrigan et al. 2011, Duke 1993, Hebbar et al. 2004, Mishra et al. 2007, Watt and Breyer-Brandwijk 1962, *Wealth of India* 1948–1976
Toxicity		

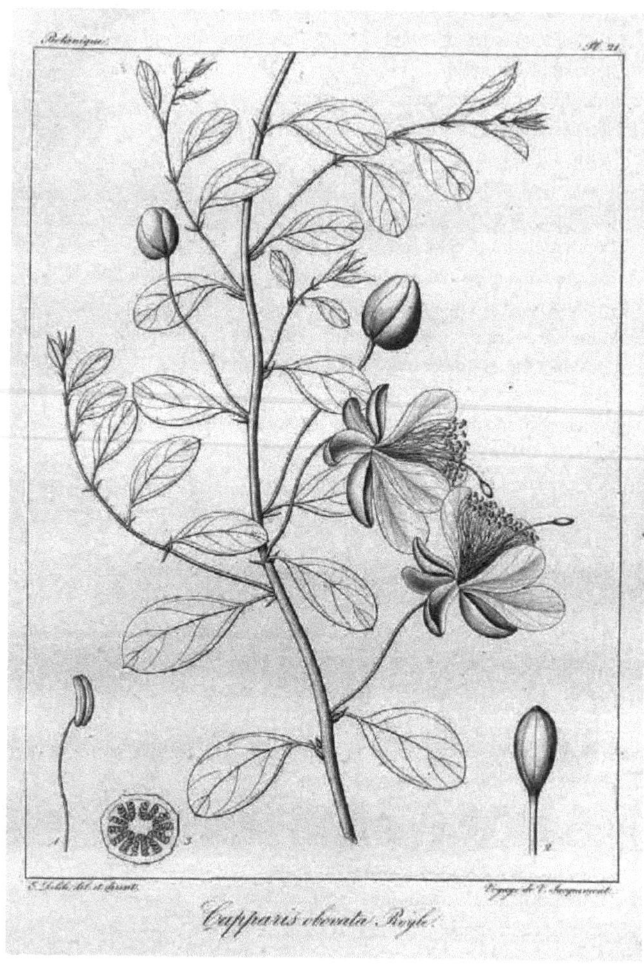

FIGURE 2.40 *Capparis sicula* Duhamel (as *C. obovata* Royle). (Jacquemont, Victor. 1841–1844. *Voyage dans l'Inde pendant les Années 1828 à 1832*. Vol. 4(3), t. 21. Paris: Typ. Firmin-Didot. Via http://www.plantillustrations.org. 2009. [Online collection of illustrations.] http://www.plantgenera.org/illustration.php?id_illustration=107542 accessed 5 February 2013.)

TABLE 2.33
Capparis sicula
C. sicula* ssp. *sicula

		References
Synonyms	*Capparis sicula* Duhamel (basionym): *C. ovata* Desf. subsp. *sicula* (Veillard) Pugnaire *C. spinosa* subsp. *canescens* (Cosson) A. Bolós & O. Bolós	Greuter et al. 1984/1986, Hanelt et al. 2001, Heywood 1996, IOPI 1991, IPNI 2012
Common names	Arabic: kabar English: caper-bush, caper-plant, Sicili caper French: câprier German: Kapernstrauch, sizilianische Kaper Greek: kappari Hebrew: tsalaf sitsili Hindi: kabra Italian: cappari, capparo, cappero, chiappara, chiappara spinusa Portugese: alcaparras, alcaparreira Russian: kapersy Spanish: alcaparra, alcaparrilla (flower buds), alcaparro, alcaparrones (fruits), caparra, caparrón (flower bud), mata panera, melón (fruit), tapanera (plant), tápano, tapara, tápena, tapenera, tapera (fruit), taperera (plant), táperes Turkish: kapari, kebere, keditirnagi; kebere çiçegi (flowers), kebera kökü (roots)	Başer et al. 1986, Danin 2006–, Duhamel du Monceau 1800, Ferrández and Sanz 1993, Hanelt et al. 2001, Martínez-Lirola et al. 1997, Mesa 1996, Obón et al. 1991, Pitrè 1971, Rivera et al. 2003, Seidemann 2005, 80
Habitat and distribution	Disturbed habitats; Mediterranean woodlands and shrublands, semisteppe shrublands, shrub steppes, deserts and extreme deserts From the Atlantic coasts of southern Europe and North Africa to the eastern Mediterranean coast *Southwestern Europe*: Balearic Islands, Spain *Southeastern Europe*: Yugoslavia, Albania, Crete and Karpathos, Greece, Italy, Sicily *Northern Africa*: Morocco, Algeria, Tunisia, Libya *Western Asia*: Cyprus, Israel and Jordan, Syria, Turkey Cultivated as crop plant in Morocco, Italy, and Spain	Danin 2006–, Greuter et al. 1984/1986, IOPI 1991, Rivera et al. 2003
Economic importance	*C. sicula* Veill. in Duham. subsp. *sicula*: Flower buds: condiment Flower: food (snacks) Fruit, unripe: condiment Shoots: condiment Shoots with thorns: food (salads)	Arnold 1985, *Cyprus Food* 2013, Mesa 1996, Rivera et al. 2003
Local medicinal uses	*C. sicula* Veill. in Duham. subsp. *sicula*: Flower: tonic; diuretic; blood purifier in allergic diseases Fruit: antirheumatic, antiodontalgic, headaches; piles; aphrodisiac Leaf: antirheumatic, antiodontalgic; aphrodisiac Twigs: for ulcers Root: aperitif; tonic; astringent; baldness; inflammations, antirheumatic, toothache; antiemetic, antidiarrheal; hemorrhoids; veterinary (horses: wounds) Root bark: healing wounds; toothache; intermittent fevers	Arnold 1985, Başer et al. 1986, González-Tejero 1989, Martínez-Lirola et al. 1997, Obón et al. 1991, Öztürk and Özçelik 1991, Pitrè 1971, Rivera et al. 2003
Toxicity		

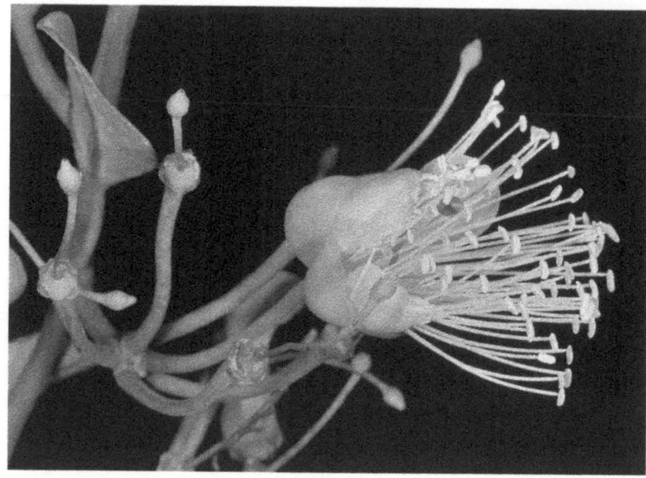

FIGURE 2.41 *Capparis sikkimensis* subsp. *formosana*. (12.5.2011, Taiwan: by Ming-I Weng.)

TABLE 2.34
Capparis sikkimensis
C. sikkimensis* subsp. *formosana

		References
Synonyms	*Capparis sikkimensis* Kurz: *C. cathcartii* Hemsley ex Gamble *Capparis sikkimensis* Kurz subsp. *formosana* (Hemsl.) Jacobs.: *C. formosana* Hemsl. *C. kanehirae* Hayata ex Kanehira	IPNI 2012, Zhang and Tucker 2008
Common names	*C. sikkimensis* Kurz: Chinese: xi jin shan gan *C. sikkimensis* Kurz subsp. *formosana* (Hemsley) Jacobs.: Chinese: tai wan shan gan	Zhang and Tucker 2008
Habitat and distribution	*C. sikkimensis* Kurz: Open forests; 1,200–1,800 m *China*: China (Tibet) *Indian Subcontinent*: Bhutan, Himalaya, NE India *Indochina*: Myanmar *C. sikkimensis* Kurz subsp. *formosana* (Hemsl.) Jacobs: Dense montane forests; above 700 m *China*: China *Eastern Asia*: Taiwan, Japan (Ryukyu Islands) *Indochina*: Vietnam	IPNI 2012, Taiwan Plant Names n.d., Zhang and Tucker 2008
Economic importance		
Local medicinal uses		
Toxicity		

TABLE 2.35
Capparis sinaica

		References
Synonyms	*Capparis sinaica* Veillard: *C. cartilaginea* Decaisne *C. galeata* Fresen. *C. spinosa* subsp. *cartilaginea* (Decaisne) Maire & Weiller	Greuter et al. 1984/1986, IOPI 1991
Common names	Arabic: ajra, kabar, lassaf, 'assaf English: mountain pepper	Bibliotheca Alexandrina n.d., Encyclopedia of Life 2007, Hobbs 1876, 88
Habitat and distribution	Desert rocks and cliffs; stony wadis. Altitude range: less than 1,000 m. Eastern and southwestern Africa, southwestern Asia to India. *Northern Africa*: Egypt *Western Asia*: Israel and Jordan, Sinai	Bibliotheca Alexandrina n.d., Encyclopedia of Life 2007, Greuter et al. 1984/1986, IOPI 1991, MEDUSA n.d.
Economic importance	Flower buds: food (pickled) Leaf: fodder to ruminants Fruit: food	Boulos 1983, MEDUSA n.d.
Local medicinal uses	Flower buds: stimulant Leaf: swellings, bruises; rheumatic joints	Boulos 1983, MEDUSA n.d.
Toxicity		

FIGURE 2.42 *Capparis sola* flowers, buds, stem, and leaf. (2.4.2006, Manu, Madre de Dios, Peru: by C. E. Timothy Paine.)

TABLE 2.36
Capparis sola

		References
Synonyms	*Capparis sola* J. F. Macbr. (basionym): *C. acutifolia* J. F. Macbr. *C. sola* var. *longiracemosa* Dugand *Capparidastrum sola* (J. F. Macbr.) Cornejo & Iltis	Brako and Zarucchi 1993, IOPI 1991, IPNI 2012, Jørgensen 2008
Common names	Ese Eja: sosewi Spanish: nina-caspi	Alexiades 1999
Habitat and distribution	Disturbed areas, seasonally inundated areas, terra firme forests; 0–1,000 m *Northern South America*: Surinam, Venezuela *Western South America*: Bolivia, Colombia, Peru *Brazil*: Brazil	Boggan et al. 1997, Brako and Zarucchi 1993, Cornejo and Iltis 2008a, Encyclopedia of Life 2007, IOPI 1991, IPNI 2012, Jørgensen 2008, JSTOR Plant Science n.d.
Economic importance		
Local medicinal uses	Bark: skin problems ("granos"); aching joints Part not specified: tonic; pain; sore throat, colds; contraceptive, abortifacient	Alexiades 1999, de Feo 1990, Friedman et al. 1993, Marles et al. 1988
Toxicity		

FIGURE 2.43 All parts of *Capparis spinosa* are used in ethnomedicine, although it is mostly known in modern societies for its pickled buds, capers. (20.5.2010, Jerusalem, Israel: by Helena Paavilainen.)

FIGURE 2.44 The root and root bark of most of the *Capparis* spp. are the part most appreciated in medicine. At the same time, it is the part that grows slowest and is most difficult to harvest. *Capparis spinosa* starting its spring growth between stones on a wall. (7.4.2010, Jerusalem, Israel: by Helena Paavilainen.)

TABLE 2.37
***Capparis spinosa* L.**
C. spinosa* L. var. *spinosa
C. spinosa* subsp. *spinosa
C. spinosa* subsp. *rupestris
C. spinosa* var. *aegyptia
C. spinosa* var. *deserti
C. spinosa* var. *mucronifolia
C. spinosa* Desf. var. *canescens

		References
Synonyms	*Capparis spinosa* L.: *C. aegyptia* Lam. *C. cordifolia* Lam. *C. herbacea* Willd. *C. leucophylla* DC. *C. mariana* Jacq. *C. murrayana* J. Grah. *C. nepaulensis* DC. *C. obovata* Royle	Greuter et al. 1984/1986, GRIN 1994, Hanelt et al. 2001, Heywood 1996, IOPI 1991, IPNI 2012, Jafri 1985, MMPND 1995, USDA, NRCS 2012

(*Continued*)

TABLE 2.37 (*Continued*)
***C. spinosa* L.**

		References
	C. ovata Desf. *C. ovata* M. Bieb. *C. parviflora* Boiss. *C. rupestris* Sibth. & Sm. *C. rupestris* Sm. *C. sandwichiana* DC. *C. sicula* Duham. *C. spinosa* L. subsp. *rupestris* (Sm.) Nyman *C. spinosa* var. *mariana* (Jacq.) K. Schum. *C. uncinata* Edgew. *C. spinosa* L. subsp. *rupestris* (Sm.) Nyman: *C. inermis* Turra *C. orientalis* Veill. *C. rupestris* Sm. (basionym) *C. spinosa* subsp. *canescens* (Cosson) A. Bolós & O. Bolós: *C. sicula* Veilland *C. spinosa* var. *canescens* Cosson: *C. ovata* var. *canescens* (Cosson) Heywood *C. spinosa* var. *deserti* Zohary (basionym): *C. deserti* (Zohary) Täckh. & Boulos	
Common names	*C. spinosa* L.: Arabic: assaf, azuf, al kabara, kabar, kabbar, kabur, shafallah, lassaf, lussef (Egypt); taylulut (Berber) Bengali: kabra Chinese: ci shan gan, shan gan Croatian: kapari Czech: kapara Danish: kapers Dogri: kabra, kalbara, kappra, ratupka Dutch: kappertjes English: caper(s), caperbush, caper-bush, caper-plant, common caperbush, fabagelle, flinders-rose, Mediterranean caperbush, spiny caperbush Estonian: torkav kappar Finnish: kapris, kaprispensas French: câpres, câprier, câprier épineux; cornichons de câprier (fruit) German: Kapernstaude, echter Kapernstrauch, dorniger Kaperstrauch, Kaper, Kapernstrauch Hebrew: tsalaf, tsalaf qotsani Hindi: kiari, kobra, kabra, ber, bhotiayas-kabara, kabar Hungarian: kapri, kapribogyó, kapricserje, kaporna Italian: cappero, capperone (fruit), cucunci (fruit) Japanese: keepaa, keipaa, keppaa Kannada: kathari mullina gida, katthari mullina gida, maraat moggu, mullu katthari, mullukattari, mullukatthari, nibate, nippatthike Malay: melada	Aldén et al. 2009, Australian Plant Common Name Database s.d., Bailey and Bailey 1976, Burkill 1985, ENVIS 1999, GRIN 1994, Hanelt et al. 2001, Jafri 1985, JSTOR Plant Science n.d., Kress 1995, MMPND 1995, Norton et al. 2009, PFAF 1996, Rehm 1994, Rivera et al. 2003, Seidemann 2005, 81, USDA, NRCS 2012, Wiersema and León 1999, Zhang and Tucker 2008

TABLE 2.37 (*Continued*)
***C. spinosa* L.**

		References
	Marathi: kabar Norwegian: kapers Persian: kabar, kebir, kurak Polish: kapar ciernisty, kapary Portuguese: alcaparras Punjabi: kabarra Russian: kapersy Sanskrit: himsra, kakadani, kartotti Slovakian: kapary Slovenian: kaprovec Spanish: alcaparro, alcaparrón (berries), caparra, tápana Spiti: rohtokpa-martokpa Swahili: mchezo, mruko Swedish: kapris Tagalog: alcaparras Tamachek (Mali): qabbar Telugu: kokilakshmu, kokitha Turkish: kebere Urdu: kabar, bikh kabar Vietnamese: cáp In Pakistan: kabar, khawarg, panetero *C. spinosa* L. var. *inermis* Turra: English: spineless caper, thornless caper bush, caper spurge French: câprier sans épines *C. spinosa* L. ssp. *rupestris* (Sm.) Nyman: French: câprier German: dornenloser Kapernstrauch Italian: capparo, cappari, chiappara Portuguese: alcoparras, alcaparreira Spanish: tàpara, alcaparra (flower buds), gorrinets (fruits)	
Habitat and distribution	*C. spinosa* L. Plains, desert flats, open and sunny areas; near sea level to 1,100 m. Old walls, cliffs and rocky hillsides in the Mediterranean. On rocks, to 3,600 m in the Himalayas. Europe; Mediterranean to Caucasus, eastern Asia, Himalayas; to northern West Africa. Cultivated in the Mediterranean countries, also in southern Russia and North America. *Southwestern Europe*: southern France (incl. Corsica), Portugal, Spain (incl. Baleares) *Southeastern Europe*: Albania, former Yugoslavia, Greece (incl. Crete), Italy (incl. Sardinia, Sicily), Malta *Eastern Europe*: Crimea	APNI 2013, Burkill 1985, Gamble 1902, Greuter et al. 1984/1986, GRIN 1994, Hanelt et al. 2001, Heywood 1996, IOPI 1991, IPNI 2012, JSTOR Plant Science n.d., PFAF 1996, Phillips and Rix 1991, Zhang and Tucker 2008

(*Continued*)

TABLE 2.37 (*Continued*)
***C. spinosa* L.**

		References
	Northern Africa: Morocco, Algeria, Tunisia, Libya, Egypt	
	Western Asia: Sinai	
	Middle Asia: Kazakhstan, Kyrgyzstan, Tajikistan, Turkmenistan, Uzbekistan	
	Caucasus: Armenia, Azerbaijan, Georgia, Russian Federation (Ciscaucasia, Dagestan)	
	Western Asia: Turkey, Cyprus, Israel, Jordan, Lebanon, Syria, Iraq, Iran, Afghanistan	
	Arabian Peninsula: Bahrain, Qatar, Saudi Arabia, United Arab Emirates, Yemen	
	China: China	
	Indian Subcontinent: Pakistan, India, Nepal	
	Malesia: Indonesia, Philippines	
	Papuasia: Solomon Islands	
	Australia: Australia	
	Southwestern Pacific: Vanuatu	
	Northwestern Pacific: Guam	
	Northern South America: Guyana (?)	
	C. spinosa L. subsp. *spinosa*	
	Southwestern Europe: France (incl. Corsica), Spain (incl. Baleares)	
	Southeastern Europe: Albania, Croatia, Slovenia, Yugoslavia, Greece (incl. Crete), Italy (incl. Sicily)	
	Northern Africa: Morocco, Algeria, Tunisia, Libya, Egypt	
	Middle Asia: Turkmenistan	
	Western Asia: Turkey, Cyprus, Israel, Sinai, Jordan, Lebanon, Syria, Iraq, Iran, Afghanistan	
	Indian Subcontinent: India, Bangladesh, Nepal	
	Northern South America: Guyana (?)	
	C. spinosa L. subsp. *rupestris* (Sm.) Nyman	
	Mediterranean area to Caucasus, Turkestan, Tibet, India, SE Asia.	
	Southwestern Europe: S France, Spain (incl. Baleares)	
	Southeastern Europe: Albania, Croatia, Slovenia, Greece (incl. Crete), Italy (incl. Sardinia, Sicily), Malta	
	Northern Africa: Morocco, Algeria, Tunisia, Libya, Egypt	
	Middle Asia: Turkestan	
	Western Asia: Turkey	
	China: Tibet	
	Indian Subcontinent: India	
	Northern South America: Guyana (?)	

TABLE 2.37 (*Continued*)
***C. spinosa* L.**

		References
Economic importance	Flower buds: food (pickle, sauces, condiments, spices, flavorings); stimulant Fruits: food, condiment (pickled) Leaf: food (condiment); medicinal, veterinary Branch tips: condiment (pickled) Shoots, young: food Root: cosmetic Root bark: medicinal Plant: medicinal; environmental boundary/barrier/support; ornamental; grazing (camels, goats) Part not specified: agrihorticulture; veterinary medicine, medicine	Bown 1995, Burkill 1985, Chiej 1984, Duke et al. 2002, Facciola 1990, Gamble 1902, GRIN 1994, Hanelt et al. 2001, Hill 1812, Huxley et al. 1992, Jafri 1985, JSTOR Plant Science n.d., Komarov, et al. 1934–1964, Kunkel 1984, McGuffin et al. 2000, PFAF 1996, Remington et al. 1918, Sonder 1894, Sturtevant and Hedrick 1919, Townsend and Guest 1966–1980, Tutin et al. 1993
Local medicinal uses	Flower buds: stimulant, refreshing, tonic for arteriosclerosis; scurvy; skin diseases (wet and dry eczema); sciatica; headache; eye infections, eye disease, prevention of cataracts; chills, colds, cough, respiratory problems; kidney stone; diuretic, renal disinfectant, against diminishing of urine; diabetes; stomach pain, gastrointestinal infections, diarrhea, dysentery; laxative Flower: eczema; rheumatism; hypertension; diabetes Fruit: cooling; tranquilizing; stimulant; antiscorbutic; rheumatism, gout, sciatica; colds of the back; colds; hypertension, dropsy; diabetes, diabetic complications; stomach pain, carminative; painful menstruation; aphrodisiac Seed: ulcers; scrofula; ganglions; female sterility, dysmenorrhea Leaf: skin diseases (wet and dry eczema); mucosae; headache; toothache; eye diseases, improves eyesight; rheumatism, gout; respiratory problems, colds; kidney stone; diminishing of urine; diabetes; gastrointestinal infections, diarrhea, dysentery, carminative; aphrodisiac; veterinary (camels: itch, external parasites, skin infections) Stem: antidiarrheic; febrifuge Stem bark: bitter; diuretic; increases appetite Bark: tonic; astringent; diuretic; for chest problems, expectorant; rheumatism, gout; jaundice; splenomegaly; laxative; veterinary (camel: scabies) Root: tonic, tonic for arteriosclerosis; astringent; cosmetic, rose-colored rashes, capillary weaknesses; eye problems; rheumatism; chills; jaundice; splenomegaly; renal disinfectant, diuretic; intestinal problems; dysmenorrhea; sterility; blennorrage; aphrodisiac; veterinary (camel: scabies)	Bellakhdar 1997, Bellakhdar et al. 1991, Ben Haj Jilani et al. 2007, Boulos 1983, Bown 1995, Burkill 1985, Chevallier 1996, Chiej 1984, Chopra et al. 1986, Curasson 1947, Eddouks et al. 2002, El-Hilaly et al. 2003, Hammiche and Maiza 2006, Hill 1812, Jouad et al. 2001, JSTOR Plant Science n.d., Maiza et al. 1995, Nanyingi et al. 2008, PFAF 1996, Remington et al. 1918, Sher and Alyemeni 2010, Sonder 1894, Sturtevant and Hedrick 1919, Tahraoui et al. 2007, Ziyyat et al. 1997

(*Continued*)

TABLE 2.37 (*Continued*)
***C. spinosa* L.**

		References
	Root bark: bitter; astringent; analgesic; tonic; depurative; appetizer; diuretic, anemia; antihypertensive, vasoconstrictive, against capillary weakness and easy bruising; skin problems (rashes, dry skin); swollen joints, gout, rheumatism; upper respiratory tract infection, expectorant; dropsy; laxative, gastrointestinal infections, antidiarrheic, indigestion; jaundice, obstructions of liver; obstructions of spleen; emmenagogue; antihemorrhoidal; anthelmintic; febrifuge Plant: arthritis, rheumatism; vaginal thrush; helminthiases Part not specified: stimulant, tonic, alterative; astringent; diuretic; scurvy; glands; antispasmodic; paralysis; rheumatism, gout; headache; toothache; colds; expectorant; hepatitis; sclerosis (spleen), splenitis; laxative; emmenagogue; veterinary medicine	
Toxicity	Part not specified: causes allergic contact dermatitis	Angelini et al. 1991, FDA Poisonous Plant Database n.d., Mitchell and Rook 1979, PFAF 1996, Sher and Alyemeni 2010

TABLE 2.38
Capparis stylosa

		References
Synonyms	*Capparis stylosa* DC.: *C. divaricata* Lam.	IPNI 2012, The Plant List 2010
Common names	Marathi: pachúnda Tamil: toaratti Telugu: badreni	Gamble 1915, 45
Habitat and distribution	Dry forest, up to 1,500 feet *Indian Subcontinent*: India	JSTOR Plant Science n.d., Gamble 1915, 45
Economic importance		
Local medicinal uses		
Toxicity	Piscicidal	Ambedkar and Muniyan 2009

TABLE 2.39
Capparis tenera

		References
Synonyms	*Capparis tenera* Dalzell: *C. tenera* var. *dalzellii* J. D. Hooker & Thomson	IPNI 2012, Zhang and Tucker 2008
Common names	Chinese: bao ye shan gan In Myanmar: sun let thè	Brandis 1906, Zhang and Tucker 2008
Habitat and distribution	Forests; 1,200–1,300 m Africa, Indian Ocean islands and: *China*: southern Xizang, western Yunnan *Indian Subcontinent*: northeastern India, Sri Lanka *Indochina*: Myanmar, northern Thailand	Encyclopedia of Life 2007, Zhang and Tucker 2008
Economic importance	Plant: hedge plant	Brandis 1906
Local medicinal uses		
Toxicity		

FIGURE 2.45 Fruit, leaves, stems, bark, and roots of *Capparis tomentosa* are all used in ethnomedical practice. (Left, 18.11.2008; right, 28.10.2008: by Linda Loffler via Swaziland's Flora Database, Swaziland National Trust Commission 2013, http://www.sntc.org.sz/flora/photo.asp?phid=2730 and http://www.sntc.org.sz/flora/photo.asp?phid=2733, respectively.)

TABLE 2.40
Capparis tomentosa

		References
Synonyms	*Capparis tomentosa* Lam.: *Boscia tomentosa*, sensu O. B. Mill. *C. alexandrae* Chiov. *C. biloba* Hutch. & Dalz. *C. corymbifera* E. Mey. ex Sond. *C. persicifolia* A. Rich. *C. polymorpha* A. Rich. *C. polymorpha* Guill. & Pers. *C. puberula* DC. *C. volkensii* Gilg	Burkill 1985, Elffers et al. 1964, Fici et al. 1993, GRIN 1994, Hutchinson and Dalziel 1954, IPNI 2012, JSTOR Plant Science n.d., MMPND 1995, Oliver 1868, Roux 2003, Wild 1960

(*Continued*)

TABLE 2.40 (*Continued*)
Capparis tomentosa

		References
Common names	Afrikaans: wollerige kapperbos Amarinya: gemaro Arabic ('Maure') (Senegal, Mauritania): diemar Arabic-Shuwa (Nigeria): mardo Balanta (Senegal): purage Berom (Nigeria): fwí Dagaari (Upper Volta): wagua Dagbani (Ghana): sansangwa Diola (Senegal): fungok, kânog English: African caper, woolly caperbush French: câprier d'Afrique, câprier de brousse Fula-Fulfulde (Ivory Coast): dalévi, gumi Fula-Fulfulde (Nigeria): ngumi daleewi Fula-Pulaar (Mali): dalévi, gumi Fula-Pulaar (Senegal): bugi baley, gubi, gumibalewi Grusi (Upper Volta): galo, galu Hausa (Niger): haujeri Hausa (Nigeria): cii záảкíí̊ кabdóódo, haujari, haujarin raƙumi, ja ni baibai,ˀкabdarai, kabdodo,ˀкábdóódo,ˀкàbdoódò,ˀкadabebe, kaùdoódò,ˀкadabebe,ˀкaudodo Kanuri (Niger, Nigeria): zájì Manding-Bambara (Senegal): donkori, dukari Manding-Mandinka (Gambia): bambaиin, belibelo Manding-Maninka (Guinea, Ivory Coast): diatabeli kilifara Maninka (Senegal): ɗatabeli kilifara Moore (Upper Volta): kalengoré, lemboitéka, silikoré Non (Nyominka) (Senegal): gufor, igufor, ingufor Serer (Senegal): ngubor, ngufor Serer-Non (Senegal): bufa, nguo, nguo Shona (Zimbabwe): gonashindi, chikatavuwa Somali: kor qabad, liimo-daanyer, gomor Sonenke-Sarakole (Senegal): teko Tiigrinya: andal Wolof (Senegal): kareń, kérén Unidentified African Names: kowangee, mbakhuzi, mkongachare, mkozi, mkwangwachale, mkwangwachare mwaka Unidentified names: haujeri, hekkabit	Burkill 1985, Duke et al. 2002, Encyclopedia of Life 2007, FDA Poisonous Plant Database n.d., Fici et al. 1993, GRIN 1994, Hyde et al. 2013a, JSTOR Plant Science n.d., MMPND 1995, USDA, NRCS 2012, Vasisht and Kumar 2004, 46
Habitat and distribution	Common in the drier types of woodland and bushland, dry grassy savanna, in riverine fringes. Often on termite mounds. Altitude range 0–2,500 m. Widespread through tropical Africa, from Senegal to Eritrea and southward to South Africa *Western Tropical Africa*: Ghana, Guinea, Mali, Niger, Nigeria, Senegal	Burkill 1985, Elffers et al. 1964, Fici et al. 1993, GRIN 1994, Hutchinson and Dalziel 1954, IPNI 2012, JSTOR Plant Science n.d., Oliver 1868, Roux 2003, Wild 1960

TABLE 2.40 (*Continued*)
Capparis tomentosa

		References
	West-Central Tropical Africa: Burundi, Cameroon, Central African Republic, Congo, Rwanda *Northeastern Tropical Africa*: Eritrea, Ethiopia, Somalia, Chad *East Tropical Africa*: Kenya, Tanzania, Uganda *Southern Tropical Africa*: Angola, Malawi, Mozambique, Zambia, Zimbabwe *Southern Africa*: Botswana, Namibia, South Africa, Swaziland *Arabian Peninsula*: Saudi Arabia, Yemen *Northern South America*: Guyana (?)	
Economic importance	Fruit: food, food (goats) (but also considered poisonous); medicine Leaves: food, fodder for cattle, camels (but also considered poisonous); medicine, eye medicine Stem: medicine Bark: magic (against "evil eye") Root: medicine, eye medicine, antidotes; magic (against evil eye; against haunting) Root bark: medicine Part not specified: agrihorticulture: fodder; medicine, veterinary medicine; social: religion, superstitions, magic (against spells, against evil eye, for protection; for rainfall, for warding off storms); in sayings, aphorisms	Burkill 1985, Bussmann 2006, Duke et al. 2002, Giday et al. 2007, GRIN 1994, JSTOR Plant Science n.d., McGuffin et al. 2000, Sturtevant and Hedrick 1919, Tabuti et al. 2003, Teklehaymanot 2009, Teklehaymanot and Giday 2007, Teklehaymanot et al. 2007, Wild 1960, Wilson and Woldo Gebre 1979, Zerabruk and Yirga 2012
Local medicinal uses	Fruit: laxative; sterility; hernia Leaf: generally healing; antisudorific; wounds, burns; anticonvulsive; lameness; headache; eye medicine; tonsil problems; respiratory problems, asthma; emetic, stomach disease, antidiarrheic; urinary schistosomiasis; venereal diseases (blennorrage); herpes; fever; antidote (poison) Stem: generally healing; wounds; venereal diseases (blennorrage) Bark: conjunctivitis; pectoral/cardiac pain; leprosy Root: antisudorific; hemostatic; wounds; itch; muscle spasm; rheumatism, articular pain; eye medicine (ophthalmia, conjunctivitis, cataract); headache, migraine; meningitis; mental disorder; toothache; oral candidiasis; anthrax; tuberculosis; respiratory problems, cough; kidneys, diuretics, urinary disease; laxatives, antidiarrheic, stomach pain, colics; before childbirth to prevent neonatal problems, against miscarriage, sterility, chronic endometry, uterine fibroids, menorrhagia; hemorrhoids; venereal diseases (urethral discharges, syphilis, chancres, orchites, gonorrhea, blennorrage); cryptoccocal infections; bilharzia; urinary schistosomiasis; internal parasites; lice; malaria; herpes, herpes simplex, herpes zoster; leprosy; antidotes (venomous stings, bites, poison, snake bite); hernia; veterinary medicine (cattle: diarrhea) Root ashes: veterinary medicine (animals: stomach troubles; cattle: diarrhea; cows: sore teats) Root bark: generally healing; abscesses; cough; prolapse of rectum	Adjanohoun et al. 1989, 1993, Arnold and Gulumian 1984, Bah et al. 2006, Bandeira et al. 2001, Bonnefoux 1885–1937, Bouquet and Debray 1974, Burkill 1985, Bussmann 2006, Buwa and van Staden 2006, Chhabra et al. 1989, Chinemana et al. 1985, Chinsembu and Hedimbi 2010, Duke et al. 2002, Gelfand et al. 1985, Giday et al. 2007, GRIN 1994, Haerdi et al. 1964, Hamill et al. 2003, Harjula 1980, Hutchings and van Staden 1994, JSTOR Plant Science n.d., Kerharo and Adam 1964a,b, 1974, Kerharo and Bouquet 1950, Kisangau et al. 2007, Kokwaro 1976, Lemordant 1971, Lulekal et al. 2008, Malgras 1992,

(*Continued*)

TABLE 2.40 (*Continued*)
Capparis tomentosa

		References
	Part not specified: veterinary medicine (cattle: stomach troubles; cows: sore teats); magic (against spells, for protection; for rainfall; for warding off storms); madness, psychosis; leprosy; wound; swelling; rheumatism; myalgia; otitis; toothache; sore throat, colds, cough, bronchitis, pneumonia, asthma; bronchial troubles; respiratory disease; jaundice; laxative; emmenagogue; sterility; venereal disease; internal parasites; malaria; antidote (poisoning); hernia	McGuffin et al. 2000, Muthee et al. 2011, Nwude and Ebong 1980, Pujol 1990, Samuelsson et al. 1991, Tabuti et al. 2003, Tadesse 1994, Tapsoba and Deschamps 2006, Teklehaymanot 2009, Teklehaymanot et al. 2007, Vasisht and Kumar 2004, 98, Wondimu et al. 2007
Toxicity	Sources disagree strongly on the issue of toxicity; some consider different parts of *C. tomentosa* as edible and even fodder for animals. The following presents some of the partly contradictory claims: Flower: toxic Fruit: not edible; possibly toxic; poisonous; poisonous to humans, camels, horned stock (except goats) Leaf: poisonous to goats, sheep, Zebu calves Root: possibly toxic; decoction highly poisonous in large quantities; very poisonous Parts not specified: injurious to skin; poisonous; poisonous to mammals; fatal to camels and horned stock (except goats)	Ahmed and Adam 1980, Ahmed et al. 1981, 1993, Burkill 1985, FDA Poisonous Plant Database n.d., Fici et al. 1993, GRIN 1994, JSTOR Plant Science n.d., Kokwaro 1976, Malgras 1992, Mitchell and Rook 1979, Mukhwana 1994, Salih et al. 1980, Shommein et al. 1980, Wild 1960, Wood 1997

FIGURE 2.46 *Capparis zeylanica* shows interesting potential through its ethnomedical use. All the parts of the plants, except the flower, are used. (11.3.2008, India; by A. Lalithamba.)

TABLE 2.41
Capparis zeylanica

		References
Synonyms	*Capparis zeylanica* L.: *C. brevispina* DC. *C. hastigera* Hance *C. hastigera* var. *obcordata* Merrill & F. P. Metcalf *C. horrida* L. f. *C. swinhoei* Hance.	GRIN 1994, IPNI 2012, Jafri 1985, MMPND 1995, Zhang and Tucker 2008
Common names	Bengali: asria, hingshra, kakadoni, kalikera, kijr, rohini Chinese: niu yan jing English: Ceylon caper, Indian caper Hindi: ardanda, bauri, gitoran, govindaphal, hins, his, jhiris, khalis Kannada: aathundi kaayi, anthundi kaayi, anthundikaayi, athendri, govinda phala Marathi: ardandi, govindi, taramati, vaagati, vaganti, wag, wagati Oriya: osaro Tamil: adondai, atanday, atandam, tondai Telugu: adonda, aradonda, arhonda, ari donda, aridonda Sanskrit: govindi, granthila, grdhranakhi, kantakalata, karambha, katukandari, kinkani, vyakhra nakhi	ENVIS 1999, Gamble 1915, 46, GRIN 1994, MMPND 1995, *Wealth of India* 1985–1992, Zhang and Tucker 2008
Habitat and distribution	Forest margins, thickets, limestone slopes or sandy soil, scattered grasslands. Altitude range below 700 m. From Pakistan to Malesia and Indochina *China*: China *Indian Subcontinent*: Pakistan, India, Sri Lanka, Bangladesh, Nepal *Indochina*: Cambodia, Laos, Myanmar, Thailand, Vietnam *Malesia*: Indonesia, Philippines	GRIN 1994, IPNI 2012, Jafri 1985, Zhang and Tucker 2008
Economic importance	Part not specified: medicinal	GRIN 1994, Parrotta 2001, *Wealth of India* 1985–1992
Local medicinal uses	Leaf: spasmolyte; blisters, boils; pneumonia; colic; cholera Fruit: antidote (snakebite) Stem: spasmolyte Bark: blisters, boils; pneumonia; colic; cholera Root: blisters, boils; pneumonia; colic; cholera Plant: rheumatism; swelling of testicles; antidote to snakebites Part not specified: anodyne; antiperspirant; diuretic; sedative; swelling; hemiplegia; neuralgia; cholagogue; stomachic; piles; venereal diseases	Ashton et al. 1997, Batugal et al. 2004, Chopra et al. 1950, Duke 1993, GRIN 1994, Jain and Tarafder 1970, Mishra et al. 2007, Parrotta 2001, Patil et al. 2011, Tyagi 2005, 115, *Wealth of India* 1948–1976, 1985–1992
Toxicity		

REFERENCES

Adjanohoun, E., V. Adjakidje, M.R.A. Ahyi, et al. 1989. *Contribution aux Études ethnobotaniques et floristiques en république populaire du Bénin.* Paris: Agence de Coopération Culturelle et Technique (ACCT).

Adjanohoun, E., et al. 1993. *Contribution to Ethnobotanical and Floristic Studies in Uganda.* Lagos: OUA/CSTR.

Ahmed, O.M.M. and S.E.I. Adam. 1980. The toxicity of *Capparis tomentosa* in goats. *J Comp Pathol* 90(2): 187–95.

Ahmed, O.M.M., S.E.I. Adam, and G.T. Edds. 1981. The toxicity of *Capparis tomentosa* in sheep and calves. *Vet Hum Toxicol* 23(6): 403–9.

Ahmed, S.A., A.E. Amin, S.E. Adam, and H.J. Hapke. 1993. By toxic effects of the dried leaves and stem of *Capparis tomentosa* on Nubian goats. *DTW. Dtsch Tierarztl Wochenschr* 100(5): 192–4.

Ake-Assi, Y.A. 1992. Contribution au recensement des espèces végétales utilisées traditionnellement sur le plan zootechnique et vétérinaire en Afrique de l'Ouest. PhD thesis. Lyon, France: Université Claude Bernard.

Aldén, B., S. Ryman, and M. Hjertson. 2009. *Våra kulturväxters namn: ursprung och användning*. Stockholm: Formas. USDA, ARS, National Genetic Resources Program. Germplasm Resources Information Network (GRIN) [Online database]. National Germplasm Resources Laboratory, Beltsville, MD. http://www.ars-grin.gov/cgi-bin/npgs/html/stdlit.pl?Vara%20kulturvaxt%20namn (accessed 8 February 2013).

Alexiades, M.N. 1999. Ethnobotany of the Ese Eja: plants, health, and change in an Amazonian society. PhD thesis. New York: City University of New York.

Ali, S.A., T.H. Al-Amin, A.H. Mohamed, and A.A. Gameel. 2009. Hepatoprotective activity of aqueous and methanolic extracts of *Capparis decidua* stems against carbon tetrachloride induced liver damage in rats. *J Pharmacol Toxicol* 4(4): 167–72.

Al-Khalil, S. 1995. A survey of plants used in Jordanian traditional medicine. *Int J Pharmacognosy* 33(4): 317–23.

Alpini, P. and R. de Fenoyl. 1980. *Plantes d'Égypte: 1581–1584*. [Le Caire]: Institut français d'archéologie orientale du Caire.

Al-Said, M.S. 1993. Traditional medicinal plants of Saudi Arabia. *Am J Chin Med* 21(3–4): 291–8.

Ambedkar, G. and M. Muniyan. 2009. Piscicidal activity of methanolic extract of *Capparis stylosa* on the freshwater fish *Channa punctatus* (Bloch). *Internet J Toxicol* 6(1).

Andrade, G., E. Esteban, L. Velasco, M.J. Lorite, and E.J. Bedmar. 1997. Isolation and identification of N_2-fixing microorganisms from the rhizosphere of *Capparis spinosa* (L.). *Plant Soil* 197(1): 19–23.

Angelini, G., G.A. Vena, R. Filotico, C. Foti, and M. Grandolfo. 1991. Allergic contact dermatitis from *Capparis spinosa* L. applied as wet compresses. *Contact Dermatitis* 24(5): 382–3.

APNI. 2013. Australian Plant Name Index (APNI) [Online database.] Australian Centre for Plant Biodiversity Research, Canberra. http://www.anbg.gov.au/cgi-bin/apni (accessed 11 November 2012 via International Organization for Plant Information [IOPI], http://bgbm3.bgbm.fu-berlin.de/IOPI/GPC/results.asp).

Arenas, P. 1987. Medicine and magic among the Maka Indians of the Paraguayan Chaco. *J Ethnopharmacol* 21(3): 279–95.

Arenas, P. and G.F. Scarpa. 2007. Edible wild plants of the Chorote Indians, Gran Chaco, Argentina. *Bot J Linn Soc* 153(1): 73–85.

Arenas, P. and M.E. Suárez. 2007. Woods employed by Gran Chaco Indians to make fire drills. *Candollea* 62: 27–40.

Arnold, H.J. and M. Gulumian. 1984. Pharmacopoeia of traditional medicine in Venda. *J Ethnopharmacol* 12(1): 35–74.

Arnold, N. 1985. Contribution a la connaissance ethnobotanique et medicinale de la flore de Chypre. PhD thesis. Paris: Université René Descartes.

Arslan, R. and N. Bektas. 2010. Antinociceptive effect of methanol extract of *Capparis ovata* in mice. *Pharm Biol* 48: 1185–90.

Ashton, M.S., S. Gunatilleke, N. De Zoysa, N. Gunatilleke, M.D. Dassanayake, and S. Wijesundara. 1997. *A Field Guide to the Common Trees and Shrubs of Sri Lanka*. Colombo, Sri Lanka: WHT for the Wildlife Heritage Trust of Sri Lanka.

Australian Plant Common Name Database. s.d. [Online database.] Integrated Botanical Information System (IBIS). Australian National Botanic Gardens. http://www.anbg.gov.au/common.names/USDA, ARS, National Genetic Resources Program. Germplasm Resources Information Network (GRIN) [Online database]. National Germplasm Resources Laboratory, Beltsville, MD. http://www.ars-grin.gov/cgi-bin/npgs/html/stdlit.pl?Aust%20Pl%20Common%20Names (accessed 8 February 2013).

Bah, S., D. Diallo, S. Dembélé, and B.S. Paulsen. 2006. Ethnopharmacological survey of plants used for the treatment of schistosomiasis in Niono district, Mali. *J Ethnopharmacol* 105: 387–99.

Bailey, L.H. and E.Z. Bailey. 1976. *Hortus Third: A Concise Dictionary of Plants Cultivated in the United States and Canada*. New York: Macmillan. USDA, ARS, National Genetic Resources Program. Germplasm Resources Information Network (GRIN) [Online database]. National Germplasm Resources Laboratory, Beltsville, MD. http://www.ars-grin.gov/cgi-bin/npgs/html/stdlit.pl?Hortus%203 (accessed February 2013).

Balar, C. and A. Nakum. 2010. Herbal ingredients for the treatment of arthritis. Indian Patent CODEN: INXXBQ.

Bandeira, S.O., F. Gaspar, and F.P. Pagula. 2001. Ethnobotany and healthcare in Mozambique. *Pharm Biol* 39(1, Supplement 1): 70–3.

Baquar, S.R. 1989. *Medicinal and Poisonous Plants of Pakistan*. Karachi, Pakistan: Printas.

Barbera, G. and R. Di Lorenzo. 1982. La coltura specializzata del cappero nell'isola di Pantelleria. *L'Informatore Agrario* 32: 22112–217.

Barbera, G. and R. Di Lorenzo. 1984. The caper culture in Italy. *Acta Hortic* 144: 167–71.

Barboza, G.E., J.J. Cantero, C. Núñez, A. Pacciaroni and L. Ariza Espinar. 2009. Medicinal plants: a general review and a phytochemical and ethnopharmacological screening of the native Argentine flora. *Kurtziana* 34(1–2): 7–365.

Başer, K.H.C., G. Honda, and W. Miki. 1986. *Herb Drugs and Herbalists in Turkey*. Tokyo: Institute for the Study of Languages and Cultures of Asia and Africa.

Batugal, P.A., J. Kanniah, S.Y. Lee, and J.T. Oliver (International Plant Genetic Resources Institute. Regional Office for Asia, the Pacific and Oceania, Serdang [Malaysia]), eds., and International Plant Genetic Resources Institute (IPGRI), Regional Office for Asia, the Pacific and Oceania (IPGRI-APO), Serdang (Malaysia). 2004. *Medicinal Plants Research in Asia. V. 1. The Framework and Project Workplans*. http://www.bioversityinternational.org/publications/pdf/944.pdf.

Bebawi, F.F. and L. Neugebohrn. 1991. *A Review of Plants of Northern Sudan: With Special Reference to Their Uses*. Eschborn, Germany: Deutsche Gesellschaft für technische Zusammenarbeit (GTZ).

Beentje, H.J., J. Adamson, D. Bhanderi, et al. 1994. *Kenya Trees, Shrubs, and Lianas*. Nairobi, Kenya: National Museums of Kenya.

Belayneh, A. A., S. Demissew, and N. F. Bussa. 2012. Medicinal plants' potential and use by pastoral and agro-pastoral communities in Erer Valley of Babile Wereda, eastern Ethiopia. *J Ethnobiol Ethnomed* 8: 42. http://www.ethnobiomed.com/content/8/1/42.

Bellakhdar, J. 1997. *La pharmacopée marocaine traditionnelle: médecine arabe ancienne et savoirs populaires*. Paris: Ibis Press.

Bellakhdar, J., R. Claisse, J. Fleurentin, and C. Younos. 1991. Repertory of standard herbal drugs in the Moroccan pharmacopoea. *J Ethnopharmacol* 35(2): 123–43.

Benchelah, A.C., H. Bouziane, M. Maka, and C. Ouahès. 2000. *Fleurs du Sahara: Voyage Ethno-botanique avec les Touaregs du Tassili*. Paris: Ibi Press.

Ben Haj Jilani, I., Z. Ghrabi-Gammar, and M. Zouaghi. 2007. Valorisation de la biodiversité en plantes médicinales et étude ethnobotanique de la flore du Sud-Ouest du Kef (Tunisie). *Ethnopharmacologia* 39: 36–43.

Bernhard-Smith, A. 1923. *Poisonous Plants of All Countries*. 2nd ed. London: Baillière, Tindall & Cox.

Bibliotheca Alexandrina n.d. LifeDesks. BibAlex. http://lifedesk.bibalex.org/ba/ (accessed 12 December 2012).

Blakelock, R.A. and C.C. Townsend. 1980. Capparidaceae. In C.C. Townsend and E. Guest, eds., *Flora of Iraq*. Volume 4, Part 1. Baghdad: Ministry of Agriculture and Agrarian Reform, pp. 139–45.

Boggan, J., V. Funk, C. Kelloff, M. Hoff, G. Cremers, and C. Feuillet. 1997. *Checklist of the Plants of the Guianas (Guyana, Surinam, French Guiana)*. 2nd ed. Washington, DC: Department of Botany, National Museum of Natural History, Smithsonian Institution.

Bonnefoux, B.M. 1885–1937. Plantes médicinales du Hufla. Manuscrits no. 1 et no. 2 (Congrégation do Espirito Santo), 1885–1937. Via Bossard, E. 1996. *La médecine traditionnelle au centre et a l'ouest de l'Angola*. Lisbon: Ministério da Ciência e da Tecnologia, Instituto de Investigação Científica Tropical.

Boulos, L. 1983. *Medicinal Plants of North Africa*. Algonac, MI: Reference Publications.

Bouquet, A. and M. Debray. 1974. *Plantes médicinales de la Côte d'Ivoire*. Paris: ORSTOM.

Boury, N.J. 1962. Végétaux utilisés dans la médecine africaine dans la région de Richard-Toll (Sénégal). In Adam, J., Les plantes utiles en Afrique occidentale. *Notes Africaines* 93: 14–16.

Bown, Deni. 1995. *Encyclopedia of Herbs and Their Uses*. London: Dorling Kindersley.

Brack Egg, A. 1999. *Diccionario enciclopédico de plantas útiles del Perú*. Cuzco, Peru: Centro de Estudios Regionales Andinos "Bartolomé de las Casas," CBC.

Brako, L. and J.L. Zarucchi. 1993. *Catalogue of the Flowering Plants and Gymnosperms of Peru* [Catálogo de las angiospermas y gimnospermas del Perú]. St. Louis: Missouri Botanical Garden.

Brandis, D. 1874. *Illustrations of the Forest Flora of North-West and Central India*. London: s.n. Via http://www.biolib.de.

Brandis, D. 1906. *Indian Trees: An Account of Trees, Shrubs, Woody Climbers, Bamboos, and Palms Indigenous or Commonly Cultivated in the British Indian Empire*. London: Constable.

Broun, A.F. and R.E. Massey. 1929. *Flora of the Sudan: With a Conspectus of Groups of Plants and Artificial Key to Families*. London: s.n.

Brummitt, R.K., F. Pando, S. Hollis, N.A. Brummitt, et al. 2001. *World Geographical Scheme for Recording Plant Distributions*. Pittsburgh, PA: Published for the International Working Group on Taxonomic Databases for Plant Sciences (TDWG) by the Hunt Institute for Botanical Documentation, Carnegie Mellon University.

Bryant, A.T. 1966. *Zulu Medicine and Medicine-Men*. Cape Town, South Africa: Struik.

Burkill, H.M. 1985. *The Useful Plants of West Tropical Africa*. Vol. 1. Kew, UK: Royal Botanic Gardens.

Burkill, I.H., W. Birtwistle, F.W. Foxworthy, J.B. Scrivenor, and J.G. Watson. 1966. *A Dictionary of the Economic Products of the Malay Peninsula*. Kuala Lumpur, Malaysia: Published on behalf of the governments of Malaysia and Singapore by the Ministry of Agriculture and cooperatives.

Bussmann, R.W. 2006. Ethnobotany of the Samburu of Mt. Nyirun South Turkana, Kenya. *J Ethnobiol Ethnomed* 2: 35.

Bussmann, R.W. and D. Sharon. 2007. *Plantas de los Cuatro Vientos: Las Plantas Mágicas y Medicinales del Perú* [Plants of the Four Winds: The Magic and Medicinal Plants of Peru]. Trujillo, Peru: Ed. Graficart.

Buwa, L.V. and J. van Staden. 2006. Antibacterial and antifungal activity of traditional medicinal plants used against venereal diseases in South Africa. *J Ethnopharmacol* 103: 139–42.

Chakravarti, S. and A. Venkatasubban. 1932. Chemical investigation of Indian medicinal plants. I. Preliminary chemical examination of the root bark of *Capparis horrida*. *J Annamalai Univ* 1: 176–80.

Chaudhari, S.R., M.J. Chavan, and R.S. Gaud. 2004. Phytochemical and pharmacological studies on the roots of *Capparis sepiaria*. *Indian J Pharm Sci* 66(4): 454–7.

Chevalier, A. 1938. *Flore vivante de l'Afrique occidentale française (inclus Togo, Cameroun Nord, Oubangui-Chari-Tchad, Sahara français). T. 1, Gymnospermes et premières familles d'Angiospermes (Casnarimées aux Buxacées)*. Paris: Muséum National d'Histoire Naturelle, Laboratoire d'Agronomie Coloniale.

Chevallier, A. 1996. *The Encyclopedia of Medicinal Plants*. London: Dorling Kindersley.

Chhabra, S.C., R.L.A. Mahunnah, and E.N. Mshiu. 1989. Plants used in traditional medicine in eastern Tanzania. II. Angiosperms (Capparidaceae to Ebenaceae). *J Ethnopharmacol* 25(3): 339–59.

Chiej, R. 1984. *The Macdonald Encyclopedia of Medicinal Plants* [English translation by Sylvia Mulcahy.] London: Macdonald.

Chinemana, F., R.B. Drummond, S. Mavi, and I. De Zoysa. 1985. Indigenous plant remedies in Zimbabwe. *J Ethnopharmacol* 14: 159–72.

Chinsembu, K.C. and M. Hedimbi. 2010. An ethnobotanical survey of plants used to manage HIV/AIDS opportunistic infections in Katima Mulilo, Caprivi region, Namibia. *J Ethnobiol Ethnomed* 6: 25.

Chopra, R.N., I.C. Chopra, K.L. Handa, and L.D. Kapur. 1950. *Chopra's Indigenous Drugs of India*. Calcutta: Dhur.

Chopra, R.N., S.L. Nayar, and I.C. Chopra. 1986. *Glossary of Indian Medicinal Plants (Including the Supplement)*. New Delhi: Council of Scientific and Industrial Research.

Collet, H. 1902. *Flora Simlensis, a Handbook of the Blossom Plants of Simla and the Neighbourhood*. London: Thacker, Spink.

Conforti, F., M.C. Marcotullio, F. Menichini, G.A. Statti, L. Vannutelli, G. Burini, F. Menichini, and M. Curini. 2011a. The influence of collection zone on glucosinolates, polyphenols and flavonoids contents and biological profiles of *Capparis sicula* ssp. *sicula*. *Food Sci Technol Int* 17(2): 87–97.

Conforti, F., S. Modesto, F. Menichini, G.A. Statti, D. Uzunov, U. Solimene, P. Duez, and F. Menichini. 2011b. Correlation between environmental factors, chemical composition, and antioxidative properties of caper species growing wild in Calabria (South Italy). *Chem Biodivers* 8(3): 518–31.

Cornejo, X. and H.H. Iltis. 2008a. The reinstatement of Capparidastrum (Capparaceae). *Harvard Papers in Botany* 13(2): 229–36.

Cornejo, X. and H.H. Iltis. 2008b. Two new genera of Capparaceae: *Sarcotoxicum* and *Mesocapparis* stat. nov., and the reinstatement of *Neocalyptrocalyx*. *Harvard Papers in Botany* 13(1): 103–16.

Cornforth, J.W. and A.J. Henry. 1952. Isolation of L-stachydrine from the fruit of *Capparis tomentosa*. *J Chem Soc* 1952: 601–3.

Corrigan, B.M., B.-E. Van Wyk, C.J. Geldenhuys, and J.M. Jardine. 2011. Ethnobotanical plant uses in the KwaNibela Peninsula, St. Lucia, South Africa. *S Afr J Bot* 77: 346–59.

Costa, M.R.G.F., M.S. de S. Carneiro, E.S. Pereira, J.A. Magalhães, N. de L. Costa, L.B. de Morais Neto, W. de J.E. Mochel Filho, A.P.A. Bezerra, and F.B. Moreira. 2011. Use of the native forage hay of Brazilian's northeast in the feeding of sheep and goats. *PUBVET* 5(7) [no pagination]. http://www.cabdirect.org/abstracts/20113177246.html [Original article: Costa, M.R.G.F., et al. 2011. Utilização do feno de forrageiras lenhosas nativas do Nordeste brasileiro na alimentação de ovinos e caprinos. *PUBVET*, Londrina 5(7): Ed. 154, Art. 1035. http://www.pubvet.com.br/imagens/artigos/2032011–073442-costa1035.pdf].

Craib, W.G. 1912. *Contributions to the Flora of Siam*. [Aberdeen]: Printed for the University of Aberdeen.

Cunningham, E. and C. Janson. 2007. Integrating information about location and value of resources by white-faced saki monkeys (*Pithecia pithecia*). *Anim Cogn* 10(3): 293–304.

Curasson, M.G. 1947. *Le chameau et ses maladies*. Paris: Éditions Vigot Frères.

Cyprus Food. 2013. In A Window on Cyprus. http://www.windowoncyprus.com/cyprus_food.htm. (accessed 3 March 2013).

Czerepanov, S.K. 1996. Vascular plants of Russia and adjacent countries. International Organization for Plant Information (IOPI) [Online database]. http://bgbm3.bgbm.fu-berlin.de/IOPI/GPC/results.asp (accessed 11 October 2011).

Dale, I.R. and P.J. Greenway. 1961. *Kenya Trees and Shrubs*. Nairobi: Buchanan's Kenya Estates; London: Hatchards.

Dangi, K.S. and S.N. Mishra. 2011a. Antioxidative and β cell regeneration effect of *Capparis aphylla* stem extract in streptozotocin induced diabetic rat. *Biol Med (Aligarh)* 3(3): 82–91.

Dangi, K.S. and S.N. Mishra. 2011b. *Capparis aphylla* extract compound as antidiabetic and antipathogenic agent. Indian Patent IN 2009DE00742 A.

Danin, A., ed. 2006 (continuously updated). Flora of Israel online. Hebrew University of Jerusalem, Jerusalem, Israel. http://flora.huji.ac.il/browse.asp (accessed 20 June 2011).

de Feo, V. 1990. Medicinal and magical plants in the northern Peruvian Andes. *Fitoterapia* 63: 417–40.

Delessert, J.B. and A.P. de Candolle. 1837. *Icones selectae plantarum*. Vol. 3, t. 10. Via www.plantillustrations.org 2009. [Online Collection of Illustrations]. URL: www.plantillustrations.org (accessed February 5, 2013).

Deng, Y., 2010. Method for manufacturing extract of *Capparis acutifolia* for treating Freund's complete adjuvant (fca) arthritis. Chinese Patent CN 101780124.

Dioscorides, P. 1902. *Des Pedanios Dioscurides aus Anazarbos Arzneimittellehre in fünf Büchern* (trans. and comm. J. Berendes). Stuttgart, Germany: Enke.

Dragendorff, G. 1898. *Die Heilpflanzen der verschiedenen Völker und Zeiten, ihre Anwendung, wesentlichen Bestandtheile und Geschichte; ein Handbuch für Ärzte, Apotheker, Botaniker und Droguisten*. Stuttgart, Germany: Enke.

Duhamel du Monceau, H.-L. 1800. *Traité des arbres et arbustes que l'ou cultive en France en plein terre*. Vol. 1, 2nd ed. Paris: s.n.

Duke, J.A. 1993. Dr. Duke's Phytochemical and Ethnobotanical Databases [Online database]. USDA–ARS–NGRL, Beltsville Agricultural Research Center, Beltsville, MD. http://www.ars-grin.gov/duke/ethnobot.html (accessed 12 May 2011).

Duke, J.A., M.J. Bogenschutz-Godwin, J. duCellier, and P.-A.K. Duke 2002. *Handbook of Medicinal Herbs*. Boca Raton, FL: CRC Press.

Eddouks, M., M. Maghrani, A. Lemhadri, M.-L. Ouahidi, and H. Jouad. 2002. Ethnopharmacological survey of medicinal plants used for the treatment of diabetes mellitus, hypertension and cardiac diseases in the south-east region of Morocco (Tafilalet). *J Ethnopharmacol* 82: 97–103.

efloraofindia. 2007. [Online discussion group and database.] https://sites.google.com/site/efloraofindia/ (accessed 3 January 2013).

Elffers, J., R.A. Graham, G.P. DeWolf, C.E. Hubbard, and E. Milne-Redhead. 1964. *Flora of Tropical East Africa: Capparidaceae*. [London]: Crown agents for oversea governments and administrations.

El-Gadi, A. and S. Bshana. 1986. *Usages of Some Plants in Libyan Folk-Medicine*. Tripoli: s.n.

El Ghazali, G.E.B. 1986. *Medicinal Plants of the Sudan. Part 1. Medicinal Plants of Erkowit*. Khartoum, Sudan: Medical and Aromatic Plants Institute, National Council for Research.

El Ghazali, G.E.B., M.S. El Tohami, and A.A.B. El Egami. 1994. *Medicinal Plants of the Sudan. Part 3. Medicinal Plants of the White Nile Provinces*. Khartoum, Sudan: Medical and Aromatic Plants Institute, National Council for Research.

El Ghazali, G.E.B., M.S. El Tohami, A.B.B. El Egami, W.S. Abdalla, and M.G. Mohammed. 1997. *Medicinal Plants of the Sudan. Part IV. Medicinal Plants of Northern Kordofan*. Khartoum, Sudan: Omdurman Islamic University Press.

El-Hilaly, J., M. Hmammouchi, and B. Lyoussi. 2003. Ethnobotanical studies and economic evaluation of medicinal plants in Taounate province (northern Morocco). *J Ethnopharmacol* 86: 149–58.

Encyclopedia of Life. 2007. [Online database.] http://www.eol.org (accessed 15 January 2012).

ENVIS Centre on Medicinal Plants. 1999. [Online database.] http://envis.frlht.org.in (accessed 15 July 2012).

Facciola, S. 1990. *Cornucopia: a source book of edible plants*. Vista, CA: Kampong.

Fahn, A., D. Heller, and M. Avishai. 1998. *The Cultivated Plants of Israel*. Tel Aviv, Israel: Hakibbutz Hameuchad.

FDA Poisonous Plant Database. n.d. U.S. Food and Drug Administration, U.S. Department of Health and Human Services. http://www.accessdata.fda.gov/scripts/plantox/ (accessed 11.12.2012).

Ferrández, J.V. and J.M. Sanz. 1993. *Las plantas en la medicina popular de la comarca de Monźon (Huesca)*. Huesca, Spain: Instituto de Estudios Altoaragoneses, Diputacíon de Huesca.

Fici, S., M. Thulin, and L.E. Kers. 1993. Capparaceae. In M. Thulin, ed., *Flora of Somalia. Vol. 1, Pteridophyta; Gymnospermae; Angiospermae (Annonaceae-Fabaceae)*. Kew, UK: Royal Botanic Gardens, pp. 37–60.

Filipoy, A. 1994. Medicinal plants of the Pilaga of Central Chaco. *J Ethnopharmacol* 44(3): 181–93.

Folkers, K. 1938. Preliminary studies of the botanical components of Tecuna and Java curare. *J Am Pharm Assoc (1912–1977)* 27: 689–93.

Franzan, M. 2001. Capperi! *Informatore* (April). http://www.coopfirenze.it/informatori/notizie/capperi-3676 (accessed 5 March 2013).

Freiburghaus, F., E.N. Ogwal, M.H.H. Knunya, R. Kaminsky, and R. Brun. 1996. In vitro antitrypanosomal activity of African plants used in traditional medicine in Uganda to treat sleeping sickness. *Trop Med Int Health* 1(6): 765–71.

Friedman, J., D. Bolotin, M. Rios, P. Mendosa, Y. Cohen, and M.J. Balick. 1993. A novel method for identification and domestication of indigenous useful plants in Amazonian Ecuador. In J. Janick and J.E. Simon, eds., *Proceedings of the Second National Symposium, New Crops: Exploration, Research, and Commercialization, Indianapolis, Indiana, October 6–9, 1991*. New York: Wiley, pp. 167–74.

Gaind, K.N., K.S. Gandhi, T.R. Juneja, A. Kjaer, and B.J. Nielsen. 1975. 4,5,6,7-Tetrahydroxydecyl isothiocyanate derived from a glucosinolate in *Capparis grandis*. *Phytochemistry* 14(5–6): 1415–8.

Gamble, J.S. 1902. *A Manual of Indian Timbers; An Account of the Growth, Distribution, and Uses of the Trees and Shrubs of India and Ceylon with Descriptions of Their Wood-Structure*. London: Low, Marston.

Gamble, J.S. 1915. *Flora of the Presidency of Madras*. London: West, Newman and Adlard.

Ganguly, J.K. and R.N. Kaul. 1969. *Wind Erosion Control*. New Delhi, India: Indian Council of Agricultural Research.

Garcia, D.E., M.G. Medina, L.J. Cova, J. Humbria, A. Torres, and P. Moratinos. 2008. Goats preference for fodder species with wide distribution in Trujillo State, Venezuela. *Arch Zootec* 57(220): 403–13.

Gast, M. 2000. *Moissons du désert: utilisation des ressources naturelles en période de famine au Sahara central*. Paris: Ibis Press.

GBIF. Global Biodiversity Information Facility. 2001. Home page. http://www.gbif.org (accessed 6 December 2012).

Gelfand, M., S. Mavi, R.B. Drummond, and B. Ndemera. 1985. *The Traditional Medical Practitioner in Zimbabwe: His Principles of Practice and Pharmacopoeia*. Gweru, Zimbabwe: Mambo Press.

Ghazanfar, S. 1994. *Handbook of Arabian Medicinal Plants*. Boca Raton, FL: CRC Press.

Ghule, B.V., G. Murugananthan, and P.G. Yeole. 2007. Analgesic and antipyretic effects of *Capparis zeylanica* leaves. *Fitoterapia* 78(5): 365–9.

Giday, M., T. Teklehaymanot, A. Animut, and Y. Mekonnen. 2007. Medicinal plants of the Shinasha, Agew-awi and Amhara peoples in northwest Ethiopia. *J Ethnopharmacol* 110(3): 516–25.

González-Tejero, M.R. 1989. Investigaciones etnobotánicas en la provincia de Granada. PhD thesis. Granada, Spain: Universidad de Granada.

Goodman, S.M and J.I. Hobbs. 1988. The ethnobotany of the Egyptian Eastern desert: a comparison of common plant usage between two culturally distinct Bedouin groups. *J Ethnopharmacol* 23 (1): 73–89.

Greuter, W., H.M. Burdet, and G. Long. 1984/1986. *Med-Checklist*. Vols. 1 and 3. Geneva, Switzerland: Éd. des Conservatoire et Jardin botanique de la Ville Genève; Berlin: Botanischer Garten & Botanisches Museum Berlin-Dahlem.

GRIN. Germplasm Resources Information Network. 1994. [Online database.] USDA, ARS, National Genetic Resources Program. National Germplasm Resources Laboratory, Beltsville, MD. http://www.ars-grin.gov/npgs/(accessed 22 July 2011).

Guerau, C., N. Torres, and J. Escandell. 1981. *Nova aportacío al coneixement de les plantes d'Eivissa i Formentera*. Eivissa, Spain: Institut d'Estudis Eivissencs.

Haerdi, F., J. Kerharo, and J.G. Adam. 1964. *Afrikanische Heilpflanzen: Die Eingeborenen-Heilpflanzen des Ulanga-Distriktes Tanganjikas (Ostafrika)*. Basel, Switzerland: Verlag für Recht und Gesellschaft.

Hamed, A.R., K.A. Abdel-Shafeek, N.S. Abdel-Azim, et al. 2007. Chemical Investigation of some *Capparis* species growing in Egypt and their antioxidant activity. *Evid Based Complement Alternat Med* 4(1): 25–8.

Hamill, F.A., S. Apio, N.K. Mubiru, R. Bukenya-Ziraba, M. Mosango, O.W. Maganyi, and D.D. Soejarto. 2003. Traditional herbal drugs of Southern Uganda, II: Literature analysis and antimicrobial assays. *J Ethnopharmacol* 84(1): 57–78.

Hammiche, V. and K. Maiza. 2006. Traditional medicine in Central Sahara: pharmacopoeia of Tassili N'ajjer. *J Ethnopharmacol* 105(3): 358–67.

Hammouda, F.M., M.M.S. El-Nasr, and A.M. Rizk. 1975. Constituents of Egyptian *Capparis* species. *Pharmazie* 30(11): 747–8.

Hanelt, P., R. Büttner, and R. Mansfeld, eds. 2001. *Mansfeld's Encyclopedia of Agricultural and Horticultural Crops (except Ornamentals).* Berlin: Springer.

Harjula, R. 1980. *Mirau and His Practice: A Study of the Ethnomedicinal Repertoire of a Tanzanian Herbalist.* London: Tri-Med.

Hartwell, J.L. 1967–1971. Plants used against cancer: a survey. *Lloydia* 32(1): 78–107; 32(2): 153–205; 32(3): 247–96; 33(1): 97–194; 33(3): 288–392; 34(1): 103–60; 34(2): 204–55; 34(3): 310–61; 34(4): 386–425.

Hashimoto, Y. 1967. Alkaloidal extract from *Capparis baducca.* Japanese Patent JP 42010922 B4.

Hebbar, S.S., V.H. Harsha, V. Shripathi, and G.R. Hegde. 2004. Ethnomedicine of Dharwad district in Karnataka, India: plants used in oral health care. *J Ethnopharmacol* 94(2–3): 261–6.

Heywood, V. 1996. *Flora Europaea* (ESFEDS Database). International Organization for Plant Information (IOPI) [Online database]. http://bgbm3.bgbm.fu-berlin.de/IOPI/GPC/results.asp (accessed 13 October 2011).

Hill, J. 1812. *The Family Herbal: Or an Account of All Those English Plants, Which Are Remarkable for Their Virtues, and of the Drugs Which Are Produced by Vegetables of Other Countries, with Their Descriptions and Their Uses, as Proved by Experience.* Bungay, UK: Brightly and Kinnersley. Via Kress, H. 1995. Henriette's Herbal home page. http://www.henriettesherbal.com (accessed 15 December 2011).

Hobbs, C.E. 1876. *C.E. Hobbs' Botanical Hand-book of Common Local, English, Botanical and Pharmacopœial Names Arranged in Alphabetical Order, of Most of the Crude Vegetable Drugs, etc., in Common Use: Their Properties, Productions and Uses in an Abbreviated Form: Especially Designed as a Reference Book for Druggists and Apothecaries.* Boston: Roberts.

Hooper, D. and H. Field. 1937. *Useful Plants and Drugs of Iran and Iraq.* Chicago: Field Museum of Natural History.

Howard, R.A. 1974–1989. *Flora of the Lesser Antilles: Leeward and Windward Islands.* Jamaica Plain, MA: Arnold Arboretum, Harvard University. USDA, ARS, National Genetic Resources Program. Germplasm Resources Information Network (GRIN) [Online database]. National Germplasm Resources Laboratory, Beltsville, MD. http://www.ars-grin.gov/cgi-bin/npgs/html/stdlit.pl?F%20LAnt (accessed 7 February 2013).

Hutchings, A. and J. van Staden. 1994. Plants used for stress-related ailments in traditional Zulu, Xhosa and Sotho medicine. Part 1: Plants used for headaches. *J Ethnopharmacol* 43(2): 89–124.

Hutchinson, J. and J.M. Dalziel. 1954. *Flora of West Tropical Africa; The British West African Territories, Liberia, the French and Portuguese Territories South of Latitude l8 N. to Lake Chad, and Fernando Po.* Vol. 1, Part 1. London: Crown Agents for Oversea Governments and Administrations.

Huxley, A.J., M. Griffiths, and M. Levy. 1992. *The New Royal Horticultural Society Dictionary of Gardening.* London: Macmillan.

Hyde, M.A., B.T. Wursten, and P. Ballings. 2013a. Flora of Zimbabwe. [Online database.] http://www.zimbabweflora.co.zw (accessed 5 February 2013).

Hyde, M.A., B.T. Wursten, P. Ballings, and S. Dondeyne. 2013b. Flora of Mozambique. [Online database.] http://www.mozambiqueflora.com (accessed 21 February 2013).

Inocencio, C., D. Rivera, F. Alcaraz, F.A. Tomás-Barberán. 2000. Flavonoid content of commercial capers (*Capparis spinosa*, *C. sicula* and *C. orientalis*) produced in Mediterranean countries. *Eur Food Res Technol* 212: 70–4.

IOPI. The International Organization for Plant Information (IOPI). 1991. Home page. http://plantnet.rbgsyd.nsw.gov.au/iopi/iopihome.htm (accessed 10 September 2012).

IPNI. The International Plant Names Index. 2012. Home page. http://www.ipni.org (accessed 5 September 2012).

ITIS. Integrated Taxonomic Information System. n.d. [Online database.] http://www.itis.gov/(accessed 12 May 2012).

Jabbar, S.A. and G.A. Hassan. 1993. Blood pressure lowering effect of the extract of aerial parts of *Capparis aphylla* is mediated through endothelium-dependent and independent mechanisms. *Clin Exp Hypertens* 33(7): 470–7.

Jacobson, R.L. and Y. Schlein. 1999. Lectins and toxins in the plant diet of *Phlebotomus papatasi* (Diptera: Psychodidae) can kill *Leishmania major* promastigotes in the sandfly and in culture. *Ann Trop Med Parasitol* 93(4): 351–6.

Jacquemont, V. 1841–1844. *Voyage dans l'Inde pendant les années 1828 à 1832*. Vol. 4(3), t. 21. Paris: Typ. Firmin-Didot. Via http://www.plantillustrations.org. 2009. [Online collection of illustrations]. http://www.plantillustrations.org (accessed 5 February 2013).

Jafri, S.M.H. 1977. *Capparis*. In S. Ali, S. Jafri, and A. El-Gadi, eds., *Flora of Libya. Vol. 12, Capparaceae*. Tripoli, Libya: Al-Faateh University, Department of Botany, pp. 2–8.

Jafri, S.M.H. 1985. Moraceae. In Flora of Pakistan, eFloras. [Online database.] St. Louis: Missouri Botanical Garden; Cambridge, MA: Harvard University Herbaria. http://www.efloras.org (accessed 24 July 2009).

Jain, S.K. and C.R. Tarafder. 1970. Medicinal plant-lore of the Santals. *Econ Bot* 24(3): 241–78.

Jankowski, W.J. and T. Chojnacki. 1991. Long-chain polyisoprenoid alcohols in leaves of *Capparis* species. *Acta Biochim Pol* 38(2): 265–76.

Ji, Y., F. Dong, S. Gao, and M. Yu. 2011. Study on *Capparis spinosa* L polysaccharide (CSPS) induced HepG2 apoptosis by controlling Ca^{2+} path. *Adv Mater Res* 282: 203–8.

Ji, Y., F. Dong, S. Gao, and X. Zou. 2008a. Apoptosis induced by *Capparis spinosa* polysaccharide in human HepG2. *Zhongcaoyao* 39(9): 1364–7.

Ji, Y.B., L. Yu, W. Wang, and X. Zou. 2008b. Effect on mitochondrial membrane potential and Ca^{2+} concentration in HepG-2 by CSEO. *Chin J Nat Med* 6(6): 474–8.

Johns, T., G.M. Faubert, J.O. Kokwaro, R.L.A. Mahunnah, and E.K. Kimanani. 1995. Anti-giardial activity of gastrointestinal remedies of the Luo of East Africa. *J Ethnopharmacol* 46(1): 17–23.

Johns, T., J.O. Kokwaro, and E.K. Kimanani. 1990. Herbal remedies of the Luo of Siaya District, Kenya: establishing quantitative criteria for consensus. *Econ Bot* 44(3): 369–81.

Jørgensen, P. 2008. Bolivia checklist. In eFloras [Online database]. St. Louis: Missouri Botanical Garden; Cambridge, MA: Harvard University Herbaria. http://www.efloras.org/flora_page.aspx?flora_id = 40 (accessed 18 May 2010).

Jouad, H., M. Haloui, H. Rhiouani, J. El Hilaly, and M. Eddouks. 2001. Ethnobotanical survey of medicinal plants used for the treatment of diabetes, cardiac and renal diseases in the North centre region of Morocco (Fez-Boulemane). *J Ethnopharmacol* 77(2–3): 175–82.

JSTOR Plant Science. n.d. [Online database.] http://plant.jstor.org (accessed 10 September 2010).

Juneja, T.R., K.N. Gaind, and A.S. Panesar. 1970. *Capparis decidua*: study of isothiocyanate glucoside. *Res Bull Panjab Univ Sci* 21(3–4): 519–21.

Kakrani, H.K.N. and A.K. Saluja. 1994. Traditional treatment through herbs in Kutch district, Gujarat State, India. Part II. Analgesic, anti-inflammatory, antirheumatic, antiarthritic plants. *Fitoterapia* 65(5): 427–30.

Kanaujia, A., R. Duggar, S.T. Pannakal, et al. 2010. Insulinomimetic activity of two new gallotannins from the fruits of *Capparis moonii*. *Bioorg Med Chem* 18: 3940–5.

Kerharo, J. and J.G. Adam. 1964a. Les plantes médicinales, toxiques et magiques des Niominka et des Socé des Iles du Saloum (Sénégal). *Acta Trop* Suppl. 8: 279–334.

Kerharo, J. and J.G. Adam. 1964b. Plantes médicinales et toxiques des Peuls et des Toucouleurs du Sénégal. *J Agric Trop Bot Appl* 11: 384–444, 543–99.

Kerharo, J. and J.-G. Adam. 1974. *La pharmacopée sénégalaise traditionnelle: plantes médicinales et toxiques*. Paris: Éditions Vigot Frères.

Kerharo, J. and A. Bouquet. 1950. *Plantes médicinales et toxiques de la Côte-d'Ivoire-Haute-Volta; mission d'étude de la pharmacopée indigène en A. O. F.* Paris: Éditions Vigot Frères.

Kirtikar, K.R., B.D. Basu, and E. Blatter. 1987. *Indian Medicinal Plants*. Vol. 1. Dehradun, India: International Books Distributors.

Kisangau, D.P., K.M. Hosea, H.V.M. Lyaruu, C.C. Joseph, Z.H. Mbwambo, P.J. Masimba, C.B. Gwandu, L.N. Bruno, K.P. Devkota, and N. Sewald. 2009. Screening of traditionally used Tanzanian medicinal plants for antifungal activity. *Pharm Biol* 47 (8): 708–16.

Kisangau, D.P., H.V. Lyaruu, K.M. Hosea, and C.C. Joseph. 2007. Use of traditional medicines in the management of HIV/AIDS opportunistic infections in Tanzania: a case in the Bukoba rural district. *J Ethnobiol Ethnomed* 3: 29.

Kitada, K., K. Shibuya, M. Ishikawa, et al. 2009. Enhancement of oral moisture using tablets containing extract of *Capparis masaikai* Levl. *J Ethnopharmacol* 122: 363–6.

Kittur, M.H., C.S. Mahajanshetti, and G. Lakshminarayana. 1993. Characteristics and composition of *Trichosanthes bracteata*, *Urena sinuata* and *Capparis divaricata* seeds and oils. *J Oil Technol Assoc India* 25(2): 39–41.

Kjaer, A. and A. Schuster. 1971. Glucosinolates in *Capparis flexuosa* of Jamaican origin. *Phytochemistry* 10(12): 3155–60.

Kjaer, A. and H. Thomsen. 1962. Isothiocyanates. XLVI. Glucocappasalin, a new naturally occurring 1-thioglucoside. *Acta Chem Scand* 16: 2065–6.

Kjaer, A., H. Thomsen, and S.E. Hansen. 1960. Isothiocyanates. XXXVIII. Glucocapangulin, a novel isothiocyanate-producing glucoside. *Acta Chem Scand* 14(5): 1226–7.

Kjaer, A. and W. Wagnieres. 1965. Isothiocyanates. LIII. 3-Methyl-3-butenylglucosinolate, a new isothiocyanate-producing thioglucoside. *Acta Chem Scand* 19(8): 1989–91.

Kokwaro, J.O. 1976. *Medicinal Plants of East Africa*. Kampala, Uganda: East African Literature Bureau.

Kokwaro, J.O. 1987 [1988]. Some common African herbal remedies for skin diseases: with special reference to Kenya medicinal and poisonous plants of the tropics. In W. Greuter and B. Zimmer, eds., *Proceedings of the XIV International Botanical Congress, Berlin, 24 July–1 August 1987*. Königstein, Germany: Koeltz Scientific Books.

Komarov, V.L., et al. 1934–1964. *Flora SSSR*. Moskow: Izdatel′stvo Akademii Nauk SSSR. USDA, ARS, National Genetic Resources Program. Germplasm Resources Information Network (GRIN) [Online database]. National Germplasm Resources Laboratory, Beltsville, MD. http://www.ars-grin.gov/cgi-bin/npgs/html/stdlit.pl?F%20USSR (accessed 4 March 2013).

Kress, H. 1995. Henriette's Herbal home page. http://www.henriettesherbal.com (accessed 15 December 2011).

Kress, W.J. and J.H. Lace. 2003. *A Checklist of the Trees, Shrubs, Herbs, and Climbers of Myanmar (Revised from the Original Works by J.H. Lace ... [et al.] on the List of Trees, Shrubs, Herbs and Principal Climbers, etc., Recorded from Burma]*. Washington, DC: Department of Systematic Biology–Botany, National Museum of Natural History.

Kunkel, G. 1984. *Plants for Human Consumption: An Annotated Checklist of the Edible Phanerogams and Ferns*. Koenigstein, Germany: Koeltz Scientific Books.

Lehmann, J. 2001. Medicinal veterinary plants in Sub-Saharan Africa. http://pc4.sisc.ucl.ac.be/prelude.html. Via Rivera, D., C. Inocencio, C. Obón, and F. Alcaraz. 2003. Review of food and medicinal uses of *Capparis* L. subgenus *Capparis* (Capparidaceae). *Econ Bot* 57(4): 515–34.

Le Maout, E. and J. Decaisne. 1873. *A General System of Botany, Descriptive and Analytical* (English translation by Mrs. Hooker, with additional material by J.D. Hooker. Illustrations by L. Steinheil and A. Riocreux). London: Longmans, Green. Via Watson, L. and Dallwitz, M.J. 1992 onward. The families of flowering plants: descriptions, illustrations, identification, and information retrieval. Version: 18 May 2012. http://delta-intkey.com.

Lemordant, D. 1971. Contribution à l'ethnobotanique éthiopienne. *J Agric Trop Bot Appl* 18: 1–35, 142–79.

Levizou, E., P. Drilias, and A. Kyparissis. 2004. Exceptional photosynthetic performance of *Capparis spinosa* L. under adverse conditions of Mediterranean summer. *Photosynthetica* 42(2): 229–35.

Li, Y.Q., S.L. Yang, H.R. Li, and L.Z. Xu. 2008. Two new alkaloids from *Capparis himalayensis*. *Chem Pharm Bull* 56(2): 189–91.

Lindsay, R.S. and F.N. Hepper. 1978. *Medicinal Plants of Marakwet, Kenya*. Kew, UK: Royal Botanic Gardens.

Liogier, A.H. 1974. *Diccionario botanico de nombres vulgares de la española*. Santo Domingo, Dominican Republic: Impresora UNPHU.

Liu, K.C., C.J. Chou, and W.C. Pan. 1977. Studies on the constituents of the stems of *Capparis formosana* Hemsl. *Taiwan Yaoxue Zazhi* 28(1–2): 2–5.

Liu, X., S. Maeda, Z. Hu, T. Aiuchi, K. Nakaya, and Y. Kurihara. 1993. Purification, complete amino acid sequence and structural characterization of the heat-stable sweet protein, mabinlin II. *Eur J Biochem* 211(1–2): 281–7.

Luecha, P., K. Umehara, T. Miyase, and H. Noguchi. 2009. Antiestrogenic constituents of the Thai medicinal plants *Capparis flavicans* and *Vitex glabrata*. *J Nat Prod* 72(11): 1954–9.

Lulekal, E., E. Kelbessa, T. Bekele, and H. Yineger. 2008. An ethnobotanical study of medicinal plants in Mana Angetu District, southeastern Ethiopia. *J Ethnobiol Ethnomed* 4: 10.

Madagascar Catalogue. 2012. Catalogue of the Vascular Plants of Madagascar. St. Louis: Missouri Botanical Garden; Antananarivo, Madagascar: Missouri Botanical Garden, Madagascar Research and Conservation Program. http://www.efloras.org/madagascar (accessed 24 July 2012).

Mahasneh, A.M., J.A. Abbas, and A.A. El-Oqlah. 1996. Antimicrobial activity of extracts of herbal plants used in the traditional medicine of Bahrain. *Phytother Res* 10(3): 251–3.

Maikhuri, R.K. and A.K. Gangwar. 1993. Ethnobiological notes on the Khasi and Garo tribes of Meghalaya, Northeast India. *Econ Bot* 47(4): 345–57.

Maire, R. 1933. *Mission scientifique du Hoggar ... Etudes sur la flore et la végétation du Sahara central. Vol. 2*. Algiers, Algeria: Impr. La Typo-Litho.

Maiza, K.R., R.A. Brac De La Perriere, and V. Hammiche. 1995. Pharmacopée traditionnelle saharienne. *Bull Méd Pharm Afr* 9(1): 71–8.

Malgras, D. 1992. *Arbres et arbustes guérisseurs des savanes maliennes*. Paris: ACCT.

Mandaville, J.P. 1990. *Flora of Eastern Saudi Arabia*. London: Kegan Paul International jointly with the National Commission for Wildlife Conservation and Development, Riyadh, Saudi Arabia.

Manniche, L. 1989. *An Ancient Egyptian Herbal*. Austin: University of Texas Press.

Marles, R.J., D.A. Neill, and N.R. Farnsworth. 1988. A contribution to the ethnopharmacology of the lowland Quichua people of Amazonian Ecuador. *Revista de la Academia Colombiana de Ciencias Exactas, Fisicas y Naturales* 16(63): 111–20.

Martínez, M. 1933. *Las plantas medicinales de México*. Mexico City: Ediciones Botas.

Martínez-Lirola, M.J., M.R. González-Tejero, and J. Molero. 1997. *Investigaciones etnobotánicas en el Parque Natural de Cabo de Gata-Níjar (Almería)*. Almería, Spain: Sociedad Almeriense de Historia Natural.

Martínez-Lirola, M.J., M.R. González-Tejero, and J. Molero-Mesa. 1996. Ethnobotanical resources in the province of Almeria, Spain: Campos de Nijar. *Econ Bot* 50(1): 40–56.

McGuffin, M., J.T. Kartesz, A.Y. Leung, and A.O. Tucker. 2000. *Herbs of Commerce*. 2nd ed. [Silver Spring]: American Herbal Products Association. USDA, ARS, National Genetic Resources Program. Germplasm Resources Information Network (GRIN) [Online database]. National Germplasm Resources Laboratory, Beltsville, MD. http://www.ars-grin.gov/cgi-bin/npgs/html/stdlit.pl?Herbs%20Commerce%20ed2 (accessed 4 March 2013).

MEDUSA. The Medusa Database. n.d. Home page http://medusa.maich.gr and references contained therein (accessed 12 November 2012).

Mesa, S. 1996. Estudio etnobotánico y agroecológico de la comarca de la Sierra de Mágina (Jaén). PhD thesis. Madrid: Universidad Complutense.

Metcalfe, C.R. and L. Chalk. 1957. *Anatomy of the Dicotyledons: Leaves, Stem, and Wood in Relation to Taxonomy with Notes on Economic Uses*. Vol. 1. Oxford, UK: Oxford University Press.

Miller, A.G., M. Morris, and S. Stuart-Smith. 1988. *Plants of Dhofar, the Southern Region of Oman: Traditional, Economic and Medicinal Uses*. [Muscat]: Office of the Adviser for Conservation of the Environment.

Mishra, S.N., P.C. Tomar, and N. Lakra. 2007. Medicinal and food value of *Capparis*—a harsh terrain plant. *Indian J Tradit Knowledge* 6(1): 230–8.

Mitchell, J.C. and A.J. Rook. 1979. *Botanical Dermatology: Plants and Plant Products Injurious to the Skin*. Vancouver, BC: Greengrass.

MMPND. Multilingual Multiscript Plant Name Database (MMPND). 1995. [Online database.] Ed. M.H. Porcher. School of Agriculture and Food Systems, Faculty of Land and Food Resources, University of Melbourne, Australia. http://www.plantnames.unimelb.edu.au/ (accessed 22 July 2011).

Mohan, R.T.S. and A.M. Suganthi. 1998. Antibacterial activity of the root extracts of *Capparis stylosa*. *Orient J Chem* 14(1): 137–8.

Moller, M.S.G. 1961. Custom, pregnancy and child rearing in Tanganyika. *J Trop Pediatr African Child Health* 7 (3): 66–78.

Monteil, V. and C. Sauvage. 1953. *Contribution à l'étude de la flore du Sahara occidental*. Paris: Éditions Larose.

Mostacero, J., F. Mejïa, and O. Gamarra. 2002. *Taxonomia de las fanerogamas utiles del Peru*. Vol. 1. Trujillo, Peru: Editora Normas Legales.

Mukhwana, E.J. 1994. Epidemiological aspects of plant poisoning in camels (*Camelus dromedarius*) in Kenya. In S.M. Colegate and P.R. Dorling, eds., *Plant-Associated Toxins: Agricultural, phytochemical and ecological aspects*. Wallingford, UK: CAB International, pp. 31–4.

Muthee, J.K., D.W. Gakuya, J.M. Mbaria, P.G. Kareru, C.M. Mulei, and F.K. Njonge. 2011. Ethnobotanical study of anthelmintic and other medicinal plants traditionally used in Loitoktok district of Kenya. *J Ethnopharmacol* 135(1): 15–21.

Nadkarni, K.M. 1976. *Indian Materia Medica*. Vol. 1. Bombay: Popular Prakashan.

Nanyingi, M.O., J.M. Mbaria, A.L. Lanyasunya, C.G. Wagate, K.B. Koros, H.F. Kaburia, R.W. Munenge, and W.O. Ogara. 2008. Ethnopharmacological survey of Samburu district, Kenya. *J Ethnobiol Ethnomed* 4: 14.

Nègre, R. 1961. *Petite flore des régions arides du Maroc occidental*. Paris: Centre national de la recherche scientifique.

New World Fruits Database. 2012. [Online Database.] Biodiversity International. http://www.bioversityinternational.org/databases/new_world_fruits_database/search.html.

Norton, J., S. Abdul Majid, D. Allan, M. Al Safran, B. Böer, and R. Richer. 2009. *An Illustrated Checklist of the Flora of Qatar*. Gosport, UK: Browndown. USDA, ARS, National Genetic Resources Program. Germplasm Resources Information Network (GRIN) [Online database]. National Germplasm Resources

Laboratory, Beltsville, MD. http://www.ars-grin.gov/cgi-bin/npgs/html/stdlit.pl?Ill%20L%20Qatar (accessed 8 February 2013).

Nwude, N., and O.O. Ebong. 1980. Some plants used in the treatment of leprosy in Africa. *Lepr Rev* 51: 11–8.

Obón, C., D. Rivera, J. Alvarez, and M.D. Vicente Meseguer. 1991. *Las plantas medicinales de nuestra reg̃ion.* Murcia, Spain: Agencia Regional para el Medio Ambiente y la Naturaleza.

Oliveira, A.F.M., S.T. Meirelles, and A. Salatino. 2003. Epicuticular waxes from caatinga and cerrado species and their efficiency against water loss. *An Acad Bras Cienc* 75(4): 431–9.

Oliver, D. 1868. Capparidaceae. In D. Oliver et al., eds., *Flora of Tropical Africa*. Vol. 1. London: Reeve, pp. 73–101.

Olmez, Z., A. Gokturk, and S. Gulcu. 2006. Effects of cold stratification on germination rate and percentage of caper (*Capparis ovata* Desf.) seeds. *J Environ Biol* 27(4): 667–70.

Osborn, D.J. 1968. Notes on medicinal and other uses of plants in Egypt. *Econ Bot* 22(2): 165–77.

Özcan, M. and A. Akgul. 1998. Influence of species, harvest date, and size on composition of capers (*Capparis* spp.) flower buds. *Nahrung* 42(2): 102–5.

Ozkur, O., F. Ozdemir, M. Bor, and I. Turkan. 2009. Physiochemical and antioxidant responses of the perennial xerophytes *Capparis ovata* Desf. to drought. *Environ Exp Bot* 66(3): 487–92.

Özturk, M., and M. Özçelik. 1991. *Dogu anadolunun faydali bitkileri. Useful Plants of East Anatolia.* Ankara, Turkey: Siskav.

Parrotta, J.A. 2001. *Healing Plants of Peninsular India*. Wallingford, UK: CAB International.

Parsa, A. 1951. *Capparis*. In A. Parsa, ed., *Flore de l'Iran (La Perse). Vol. 1, Introduction and Ranunculaceae-Ampelidaceae*. Teheran, Iran: Mazaher, pp. 918–23.

Patil, S.B., N.S. Naikwade, C.S. Magdum, and V.B. Awale. 2011. Some medicinal plants used by people of Sangli District, Maharashtra. *Asian J Pharm Res* 1(2): 42–3.

Peckholt, T. 1898. Heil- und Nutzpflanzen aus der Familie Capparidaceae. *Ber Dtsch Pharm Ges* 8: 40–6.

Peled, Y. 1997. *Ha-tzalaf ki-veit gidul.* Jerusalem, Israel: Ha-merkaz ha-israeli le-hora'at ha-mada'im, Ha-universita ha-'ivrit b-Yerushalayim.

Peltier, J.P. 2006. Plant Biodiversity of South-Western Morocco. [Online website.] http://www.teline.fr (accessed 13 March 2013).

Pennington, T.D., C. Reynel, A. Daza, and R. Wise. 2004. *Illustrated Guide to the Trees of Peru*. Sherborne, UK: Hunt.

Petelot, A. 1952–1954. *Les plantes médicinales du Cambodge, du Laos et du Viêtnam*. Saigon, Vietnam: Centre de recherches scientifiques et techniques.

PFAF. Plants for a Future Database. 1996. [Online database.] Plants for a Future, Lostwithiel, Cornwall, England. http://www.pfaf.org/ (accessed 12 July 2010).

Phillips, R. and M. Rix. 1991. *Perennials*. Vols.1–2. London: Pan Books.

Pitrè, G. 1971. *Sicilian Folk Medicine*. [Lawrence]: Coronado Press.

Pittier, H. 1926. *Manual de las plantas usuales de Venezuela*. Caracas, Venezuela: Litografia del comercio.

Plantillustrations.org. 2009. Home page [Online collection of illustrations]. http://www.plantillustrations.org (accessed 5 February 2013).

The Plant List. 2010. Home page. Version 1. http://www.theplantlist.org/ (accessed 12 November 2012).

Prasad, S., D. Mudappa, and T.R.S. Raman. 2013. The Anamalais Woody Plant Database [Online database]. http://kalyanvarma.net/plantdb/info?genus=Capparis&species=grandis (accessed 11 January 2013).

Prelude. Prelude Medicinal Plants Database. 1993. [Online database.] Royal Museum for Central Africa (RMCA), Tervuren, Belgium. http://www.africamuseum.be/collections/external/prelude (accessed 29 July 2011).

Pugnaire, F.I. and E. Esteban. 1991. Nutritional adaptations of caper shrub (*Capparis ovata* Desf.) to environmental stress. *J Plant Nutr* 14(2): 151–61.

Pujol, J. 1990. *The Herbalist Handbook: African Flora, Medicinal Plants*. Durban, South Africa: NaturAfrica.

Qaiser, M. and S.A. Qadir. 1972. Contribution to the autecology of *Capparis decidua*. II. Effect of edaphic and biotic factors on growth and abundance. *Pak J Bot* 4(2): 137–56.

Quisumbing, E.A. 1951. *Medicinal Plants of the Philippines*. Manila: Bureau of Printing.

Raffeneau-Delile, A. 1813. *Flore d'Egypte: explanation des planches*. Plates, t. 26. Paris: Imprimerie Impériale. http://www.plantillustrations.org. 2009. [Online collection of illustrations.] http://www.plantillustrations.org (accessed 5 February 2013).

Ramachandran, V.S. and N.C. Nair. 1981. Ethnobotanical observations on Irulars of Tamil Nadu (India). *J Econ Tax Bot* 2: 183–90.

Ramirez, V.R., L.J. Mostacero, A.E. Garcia, C.F. Mejia, P.F. Pelaez, C.D. Medina, and C.H. Miranda. 1988. *Vegetales empleados en medicina tradicional norperuana*. Trujillo, Peru: Banco Agrario del Peru and Universidad Nacional de Trujillo.

Ravikumar, K., P.S. Udayan, and S.P. Subramani. 2004. Notes on distribution of *Capparis rotundifolia* Rottler (Capparaceae)—in southern India. *Indian Forester* 130(3): 313–7.

Rehm, S. 1994. *Multilingual Dictionary of Agronomic Plants*. Dordrecht, Netherlands: Kluwer Academic. USDA, ARS, National Genetic Resources Program. Germplasm Resources Information Network (GRIN) [Online database]. National Germplasm Resources Laboratory, Beltsville, MD. http://www.ars-grin.gov/cgi-bin/npgs/html/stdlit.pl?Dict%20Rehm (accessed 8 February 2013).

Remington, J.P., H.C. Wood, et al. 1918. *The Dispensatory of the United States of America*. Philadelphia: Lippincott. Via Kress, H. 1995. Henriette's Herbal home page. http://www.henriettesherbal.com (accessed 15 December 2011).

Renfrew, J. 1987. Fruits from ancient Iraq: the paleoethnobotanical evidence. *Bull Sumerian Agric* 3: 157–61.

Rhizopoulou, S. and G.K. Psaras. 2003. Development and structure of drought-tolerant leaves of the Mediterranean shrub *Capparis spinosa* L. *Ann Bot* 92(3): 377–83.

Rivera, D., C. Inocencio, C. Obón, and F. Alcaraz. 2003. Review of food and medicinal uses of *Capparis* L. subgenus *Capparis* (Capparidaceae). *Econ Bot* 57(4): 515–34.

Rodríguez, E., M. Mora, and W. Aguilar. 1996. Inventario Florístico de El Algarrobal de Moro (Provincia de Chepén, Departamento de La Libertad, Perú) y su importancia económica. *Rebiol* 16(1–2): 57–65.

Rodríguez Rodríguez, E.F., R.W. Bussmann, S.J. Arroyo Alfaro, S.E. López Medina, and J. Briceño Rosario. 2007. *Capparis scabrida* (Capparaceae) una especie del Perú y Ecuador que necesita planes de conservación urgente. *Arnaldoa* 14(2): 269–82.

Roig y Mesa, J.T. 1945. *Plantas medicinales, aromáticas o venenosas de Cuba*. Havana, Cuba: [Editorial Guerrero Casamayor y cía].

Rolland, E. 1967. *Flore populaire; ou, Histoire naturelle des plantes dans leurs rapports avec la linguistique et le folklore*. Vol. 1. Paris: Maisonneuve et Larose.

Roux, J.P. 2003. *Flora of South Africa*. JSTOR Plant Science [Online database]. http://plants.jstor.org/ (accessed 24 July 2012).

Salih, Y.M., O.F. Idris, A.G.A. Wahbi, and A.A. Yousif. 1980. Toxicity of *Capparis tomentosa* to sheep and goats. *Sudan J Vet Res* 2: 13–21.

Samuelsson, G., M.H. Farah, P. Claeson, M. Hagos, M. Thulin, O. Hedberg, A.M. Warfa, A.O. Hassan, A.H. Elmi, A.D. Abdurahman, A.S. Elmi, Y.A. Abdi, and M.H. Alin. 1991. Inventory of plants used in traditional medicine in Somalia. I. Plants of the families Acanthaceae-Chenopodiaceae. *J Ethnopharmacol* 35 (1): 25–63.

Sánchez-Monge y Parellada, E. 1981. *Diccionario de plantas agrícolas*. Madrid: Ministerio de Agricultura, Servicio de Publicaciones Agrarias.

Satyanarayana, T., A.A. Mathews, and E.M. Chinna. 2010. Prevention of carbon tetrachloride induced hepatotoxicity in rats by alcohol extract of *Capparis zeylanica* stem. *J Pharm Chem* 4(2): 37–9.

Satyanarayana, T., A.A. Mathews, and P. Vijetha. 2008. Phytochemical and pharmacological review of some Indian *Capparis* species. *Pharmacogn Rev* 2(4): Suppl. 36–45.

Sawadogo, M., A.M. Tessier, and P. Delaveau. 1981. Chemical study of *Capparis corymbosa* Lam. roots. *Plantes Med Phytother* 15(4): 234–9.

Scarpa, G.F. 2009. Etnobotánica médica de los indígenas Chorote y su comparación con la de los Criollos del Chaco semiárido (Argentina). *Darwiniana* 47(1): 92–107.

Schmeda-Hirschmann, G. 1994. Plant resources used by the Ayoreo of the Paraguayan Chaco. *Econ Bot* 48(3): 252–8.

Schraudolf, H. 1988. Indole glucosinolates of *Capparis spinosa*. *Phytochemistry* 28(1): 259–60.

Schweinfurth, G. 1912. *Arabische Pflanzennamen aus Aegypten, Algerien und Jemen*. Berlin: Dietrich Reimer.

Seidemann, J. 2005. *World Spice Plants*. Berlin: Springer.

Selvamani, P., S. Latha, K. Elayaraja, P.S. Babu, J.K. Gupta, et al. 2008. Antidiabetic activity of the ethanol extract of *Capparis sepiaria* L leaves. *Indian J Pharm Sci* 70(3): 378–80.

Shah, H.H. 1962. Study on tuberculostatic activity of Rudanti (*Capparis moonii*). *Indian J Med Sci* 16: 343–6.

Sher, H. and M.N. Alyemeni. 2010. Ethnobotanical and pharmaceutical evaluation of *Capparis spinosa* L, validity of local folk and Unani system of medicine. *J Med Plants Res* 4(17): 1751–6. http://www.academicjournals.org/JMPR/PDF/pdf2010/4Sept/Alyemeni%20and%20Sher.pdf

Shommein, A.M., O.F. Idris, and Y.M. Salih. 1980. Pathological studies in domestic ruminants experimentally intoxicated with crude extract of *Capparis tomentosa* leaves. *Sudan J Vet Res* 2: 57–60.

Sibthrop, J. and J.E. Smith. 1825. *Flora Graeca* (drawings). Vol. 5, t. 87. [s.n.]: [s.n.]. Via http://www.plantillustrations.org. 2009. [Online collection of illustrations.] http://www.plantillustrations.org (accessed 5 February 2013).

Simon, C. and M. Lamla. 1991. Merging pharmacopoeia: understanding the historical origins of incorporative pharmacopoeial processes among Xhosa healers in Southern Africa. *J Ethnopharmacol* 33(3): 237–42.

Sita, G. 1978. Traitement traditionnel de quelques maladies en pays Bissa (République de Haute-Volta). *Bull Agric Rwanda* 11(1): 24–34.

Siu-cheong, C. and L. Ning-hon, eds. 1978–1986. *Chinese Medicinal Herbs of Hong Kong*. [Hong Kong: Commercial Press.]

Soltan, M.M. and A.K. Zaki. 2009. Antiviral screening of forty-two Egyptian medicinal plants. *J Ethnopharmacol* 126(1): 102–7.

Sonder, W. 1894. Capparideae. Juss. In W.H. Harvey, O.W. Sonder, and W.T. Thiselton-Dyer, eds., *Flora Capensis: Being a Systematic Description of the Plants of the Cape Colony, Caffraria, and Port Natal (and Neighbouring Territories)*. Vol. 1. Kent, UK: Reeve.

Standley, P.C. 1920–1926. *Trees and Shrubs of Mexico*. Washington, DC: Smithsonian Institution.

Standley, P.C. and J.A. Steyermark. 1952. *Flora of Guatemala*. Chicago: Field Museum of Natural History.

Sturtevant, E.L. and U.P. Hedrick. 1919. *Sturtevant's Edible Plants of the World*. Albany, NY: Lyon, state printers. Via Kress, H. 1995. Henriette's Herbal home page. http://www.henriettesherbal.com (accessed 15 December 2011).

Su, D.M., Y.H. Wang, S.S. Yu, D.Q. Yu, Y.C. Hu, W.Z. Tang, G.T. Liu, and W.J. Wang. 2007. Glucosides from the roots of *Capparis tenera*. *Chem Biodivers* 4(12): 2852–62.

Sundara, R. 1993. Capparaceae. In B.D. Sharma and N.P. Balakrishnan, eds., *Flora of India. 2, Papaveraceae–Caryophyllaceae*. Calcutta: Botanical Survey of India, pp. 248–335.

Swaziland National Trust Commission. Swaziland's Flora Database. 2013. http://www.sntc.org.sz/flora (accessed 14 March 2013).

Tabuti, J.R.S., K.A. Lye, and S.S. Dhillion. 2003. Traditional herbal drugs of Bulamogi, Uganda: plants, use and administration. *J Ethnopharmacol* 88(1): 19–44.

Tadesse, D. 1994. Traditional use of some medicinal plants in Ethiopia. In J.H. Seyani and A.C. Chikuni, eds., *Proceedings of the XIIIth Plenary Meeting of AETFAT, Zomba, Malawi, 2–11 April 1991 = Comptes rendus de la Treizième Réunion plenière de l'AETFAT, Zomba, Malawi, 2–11 avril 1991: Plants for the People*. Zomba, Malawi: National Herbarium and Botanic Gardens of Malawi, pp. 273–93.

Tahraoui, A., J. El-Hilaly, Z.H. Israili, and B. Lyoussi. 2007. Ethnopharmacological survey of plants used in the traditional treatment of hypertension and diabetes in south-eastern Morocco (Errachidia province). *J Ethnopharmacol*. 110(1): 105–17.

Taiwan Plant Names, eFloras. n.d. [Online database.] St. Louis: Missouri Botanical Garden; Cambridge, MA: Harvard University Herbaria. http://www.efloras.org (accessed 24 July 2010).

Tapsoba, H. and J.P. Deschamps. 2006. Use of medicinal plants for the treatment of oral diseases in Burkina Faso. *J Ethnopharmacol* 104(1–2): 68–78.

Teklehaymanot, T. 2009. Ethnobotanical study of knowledge and medicinal plants use by the people in Dek Island in Ethiopia. *J Ethnopharmacol* 124(1): 69–78.

Teklehaymanot, T. and M. Giday. 2007. Ethnobotanical study of medicinal plants used by people in Zegie Peninsula, Northwestern Ethiopia. *J Ethnobiol Ethnomed* 3: 12.

Teklehaymanot, T., M. Giday, G. Medhin, and Y. Mekonnen. 2007. Knowledge and use of medicinal plants by people around Debre Libanos monastery in Ethiopia. *J Ethnopharmacol* 111(2): 271–83.

Terra, G.J.A. 1966. *Tropical Vegetables. Vegetable Growing in the Tropics and Subtropics Especially of Indigenous Vegetables*. Amsterdam, Netherlands: Koninklijk Instituut voor de Tropen.

Thoen, D. and A. Thiam. 1990. Utilisations des plantes ligneuses et sub-ligneuses par les populations de la région sahélienne du lac de Guiers (Sénégal). *Bull Méd Trad Pharm* 4(2): 169–78.

Thomé, O.W. 1885–1905. *Flora von Deutschland, Österreich und der Schweiz in Wort und Bild für Schule und Haus*. Gera-Untermhaus, Germany: Köhler. Via http://www.biolib.de.

Timberlake, J.R. 1987. Ethnobotany of the Pokot of Northern Kenya. Unpublished report. East African Herbarium Center for Economic Botany, Royal Botanic Gardens, Kew, UK.

Townsend, C.C. and E. Guest, eds. 1966–1980. *Flora of Iraq*. Baghdad, Iraq: Ministry of Agriculture and Agrarian Reform.

Tropicos®. Tropicos, botanical information system at the Missouri Botanical Garden. 2013. [Online database.] http://www.tropicos.org (accessed 3 January 2013).

Trotter, A. 1915. *Flora economica della Libia*. Rome: Tipografia dell'Unione editrice.

Tutin, T.G., J.R. Akeroyd, M.E. Newton, R.R. Mill, et al. 1993. *Flora Europaea. Vol. 1, Psilotaceae to Platanaceae*. 2nd ed. Cambridge: Cambridge University Press.

Tyagi, D.K. 2005. *Pharma Forestry: Field Guide to Medicinal Plants*. New Delhi, India: Atlantic.

Ungarelli, G. 1985. *Le piantea romatichee medicinali nei nomi nell'uso e nella tradizione popolare Bolognese*. Bologna, Italy: Arnaldo Forni.

USDA, NRCS. 2012. The PLANTS Database. National Plant Data Team, Greensboro, NC. http://plants.usda.gov (accessed 2 February 2012).

Uphof, J.C.Th. 1968. *Dictionary of Economic Plants*. Lehre, Germany: Cramer.

Vardhana, R. 2008. *Direct Uses of Medicinal Plants and Their Identification*. New Delhi, India: Sarup.

Vasileva, B. 1969. *Plantes medicinales de Guinee*. Conakry, Guinea: [s.n.].

Vasisht, K. and V. Kumar. 2004. *Compendium of Medicinal and Aromatic Plants. Vol. 1, Africa*. United Nations Industrial Development Organization and the International Centre for Science and High Technology. http://www.scribd.com/doc/82620011/Compendium-of-Medicinal-and-Aromatic-Plants-Volume-1 (accessed 12 November 2012).

Villasenor, I.M. 2007. Bioactivities of iridoids. *Antiinflammactory & Antiallergy Agents Med Chem* 6(4): 307–14.

von Reis Altschul, S. 1973. *Drugs and Foods from Little-Known Plants; Notes in Harvard University Herbaria*. Cambridge, MA: Harvard University Press.

Wang, X., W. Chen, J. Xing, et al. 2009. Method for manufacturing antiinflammatory and analgesic cataplasma of *Capparis heyneana*. Chinese Patent CN 2009-10113221.

Wasuwat, S. 1967. *A List of Thai Medicinal Plants*. Bangkok, Thailand: Applied Scientific Research Corporation of Thailand (ASRCT).

Watson, L. and Dallwitz, M.J. 1992 onward. The families of flowering plants: descriptions, illustrations, identification, and information retrieval. Version: 19 December 2012. http://delta-intkey.com

Watt, J.M. and M.G. Breyer-Brandwijk. 1962. *The Medicinal and Poisonous Plants of Southern and Eastern Africa*. 2nd ed. London: Livingstone.

Wealth of India. 1948–1976. *Wealth of India: Dictionary of Indian Raw Materials*. New Delhi, India: Publications and Information Directorate, Council of Scientific and Industrial Research.

Wealth of India. 1985–1992. *The Wealth of India: A Dictionary of Indian Raw Materials and Industrial Products*. New Delhi, India: Publications and Information Directorate, Council of Scientific and Industrial Research. USDA, ARS, National Genetic Resources Program. Germplasm Resources Information Network (GRIN) [Online database]. National Germplasm Resources Laboratory, Beltsville, MD. http://www.ars-grin.gov/cgi-bin/npgs/html/stdlit.pl?Wealth%20India%20RM%20ed2 (accessed 8 February 2013).

Wiersema, J.H. and B. León. 1999. *World Economic Plants: A Standard Reference*. Boca Raton, FL: CRC Press. USDA, ARS, National Genetic Resources Program. Germplasm Resources Information Network (GRIN) [Online database]. National Germplasm Resources Laboratory, Beltsville, MD. http://www.ars-grin.gov/cgi-bin/npgs/html/stdlit.pl?World%20Econ%20Pl (accessed 8 February 2013).

Wight, R. 1846. *Icones plantarum indiae orientalis*. Vol. 1, t. 173. Madras, India: published by J.B. Pharoah for the author, 1840–1853. Via http://www.plantillustrations.org. 2009. [Online collection of illustrations]. http://www.plantillustrations.org (accessed 5 February 2013).

Wild, H., 1960. Capparidaceae. In A.W. Exell and H. Wild, eds., *Flora zambesiaca. Volume one, Mozambique, Federation of Rhodesia and Nyasaland, Bechuanaland protectorate*. London: Crown agents for oversea governments and administrations, pp. 195–245.

Wilson, R.T. and M. Woldo Gebre. 1979. Medicine and magic in Central Tigre: a contribution to the ethnobotany of the Ethiopian plateau. *Econ Bot* 33(1): 29–34.

Wondimu, T., Z. Asfaw, and E. Kelbessa. 2007. Ethnobotanical study of medicinal plants around "Dheeraa" town, Arsi Zone, Ethiopia. *J Ethnopharmacol* 112(1): 152–61.

Wood, J.R.I. 1997. *A Handbook of the Yemen Flora*. Kew, UK: Royal Botanic Gardens.

Wu, J.H., F.R. Chang, K.I. Hayashi, H. Shiraki, C.C. Liaw, Y. Nakanishi, K.F. Bastow, D. Yu, I.S. Chen, and K.H. Lee. 2003. Antitumor agents. Part 218: Cappamensin A, a new in vitro anticancer principle, from *Capparis sikkimensis*. *Bioorg Med Chem Lett* 13(13): 2223–5.

www.BioLib.de. n.d. Group of Kurt Stüber at the Max Planck Institute for Plant Breeding Research (http://caliban.mpiz-koeln.mpg.de/~stueber), Max-Planck-Gesellschaft zur Förderung der Wissenschaften, Munich (accessed 22 January 2013).

Zerabruk, S. and G. Yirga. 2012. Traditional knowledge of medicinal plants in Gindeberet district, Western Ethiopia. *S Afr J Bot* 78: 165–9.

Zhang, M. and G.C. Tucker. 2008. Capparaceae. In Flora of China Vol. 7, eFloras [Online database]. St. Louis: Missouri Botanical Garden; Cambridge, MA: Harvard University Herbaria. http://www.efloras.org (accessed 12 May 2011).

Ziyyat, A., A. Legssyer, H. Mekhfi, A. Dassouli, M. Sehrouchni, and W. Benjelloum. 1997. Phytotherapy of hypertension and diabetes in oriental Morocco. *J Ethnopharmacol* 58: 45–54.

Zohary, M. 1966–1986. *Flora Palaestina*. Jerusalem: Israel Academy of Sciences and Humanities.

Zorn, J. and N.J.F. von Jacquin. 1786–1787. *Dreyhundert auserlesene amerikanische Gewächse*. Vol. 2, t. 166. Nuremberg, Germany: [s.n.]. Via http://www.plantillustrations.org. 2009. [Online collection of illustrations.] http://www.plantillustrations.org (accessed 5 February 2013).

Section I

Chemistry

The chemistry of plants is frequently divided into "primary metabolites" and "secondary metabolites," and because the concept of the latter impinges so much on medicinal chemistry, it is worth discussing these concepts at the outset. By *metabolites* in general, we refer to the products of biochemical reactions whose formation is catalyzed by proteins, specifically enzymes. Something is also known about metabolites that assist the enzymes in their catalyses; these are known as coenzymes and are often vitamins. Metabolites are thus not proteins, and proteins are not metabolites. Nonetheless, the two are tightly intertwined since the production of metabolites gives the enzymes their "reason to be" because the enzymes *catalyze* the reactions that produce metabolites.

Primary metabolites are directly involved in the organism's movement of energy throughout itself, the breakdown of sugars for energy, or in plants, production of sugars from energy via photosynthesis. Secondary metabolites are not directly involved in these processes but are manufactured at considerable economic costs for achieving specific biological objectives, often related to organisms in the opposite kingdom. Such compounds are used by the plant to communicate with other plants and attract pollinators or repel enemies from within its own kingdom but mainly from other kingdoms, be they microbes or animals. Although compounds from both categories, primary metabolites and secondary metabolites, are included in the ensuing presentation, the interest is naturally on secondary metabolites since these are the most specific, sophisticated from the standpoint of evolution, and hold the most potential for being physiologically and medically active. Essentially, the study of plant metabolites for medicinal usage is about secondary metabolites.

Nonetheless, the area separating primary from secondary metabolites may be blurred. This is especially true concerning vitamins, which may function as coenzymes in vital processes of the plant. For examples, tocols, the class of compounds comprising vitamin E, often exert physiological effects in animals comparable to secondary metabolites, but their production occurs in the plant as might be expected from primary metabolites.

The chapters that follow each take up a different class of compounds that occur in *Capparis*. There is no special logic to the order of these chapters other than general interest and aesthetic sense.

3 Alkaloids

Alkaloids are among the most physiologically active classes of compounds found in plants, often affecting neurological and other vital physiological functions in animals. By definition, these are always organic compounds consisting of carbon atom skeletons that also include nitrogen.

In *Capparis*, the most reported, and first reported, alkaloid is stachydrine, which is the betaine of proline and is present in all parts of *C. spinosa*. Stachydrine is also present in a medicinal herb, *Leonurus heterophyllus* Sweet, used in traditional Chinese medicine to promote blood circulation and to mitigate blood stasis. Recent investigations showed pure stachydrine to significantly reduce anoxic damage in HUVECs (human umbilical vein endothelial cells) through a mechanism involving reoxygenation tissue factor (TF) (Yin et al. 2010). *In vivo*, cerebral artery occlusion injury in rats responded favorably to intraperitoneal (i.p.) injections of the alkaloid-rich fraction of this herb as well (Liang et al. 2011), suggesting a possible role for *Capparis* alkaloid-rich fraction in the medical management of stroke.

Such a *Capparis* alkaloid-rich fraction would include not only stachydrine but also at least 20 other alkaloids, for which the data are summarized in Table 3.1. Although alkaloids occur in all parts of *Capparis* individuals, as in most plants the highest concentration of alkaloids is in their roots (Figure 3.1). These other alkaloids, many unique to the *Capparis* genus, are in several cases spermidine alkaloids, characterized, like the related spermine alkaloids, by the presence of macrocyclic lactam rings (Badawi et al. 1973) (i.e., cyclic amides). First noted as crystals in semen by Anton Van Leeuwenhoek in the seventeenth century, spermine is an example of a polyamine that is important in all eukaryotic cells for promoting growth and regulating cellular metabolism. Other important polyamines present in plants and animals as well as microorganisms are putrescine, cadaverine, and spermidine.

FIGURE 3.1 Bark and roots of *Capparis spinosa* contain high concentrations of alkaloids. (Haifa, Israel, January 2010: by Ephraim Lansky.)

Polyamines in plants, including spermine and spermidine, may be diverted to alkaloid production, as of nicotine in tobacco, and for this reason, such alkaloids receive the name spermine and spermidine alkaloids, after the polyamines from which they are formed (Ghosh 2000). The diversion of polyamines to alkaloids is the heart of the energetic price the plant pays for this trade-off.

Capparis alkaloids contained in complex *Capparis* extracts have also been employed as arrow poisons and, besides stachydrine, may exert pronounced bradycardic effects. The caper plant is somewhat tolerant of affronts to its buds and fruits, but attacks on its roots are met with chemicals of more potent physiological activity and in higher concentrations. These alkaloids, however, present the drug developer with special opportunities for compounds with antibacterial, antiparasitic, cytotoxic (Samoylenko et al. 2009), antipyretic, antipoison (Murata et al. 2010), antihypertensive (Hikino et al. 1983), and nonviral templates for gene therapy owing to the multiple cationic nature of the polyamines, their ease of penetration to DNA, and tendency to produce nanometer-size complexes suitable for transfecting cells (Blagbrough et al. 2003).

A list of the alkaloids found in the *Capparis* genus, their structures when available, and some notes regarding their uses, methods of purification, and identification as well as the parts of the plants in which they were found is given in Table 3.1.

TABLE 3.1
Alkaloids of *Capparis* spp.

Compound	Notes
Alkaloids in general	Quantified in *C. spinosa* as 0.02% in leaves, 0.074% in fruits (Rakhimova et al. 1978); in *C. aegyptia*, *C. deserti*, and *C. leucophylla* (Hammouda et al. 1975); in root bark, stem, leaf, and flower of *C. sepiaria* (Juneja et al. 1970); from *C.* sp. suggested as possible component in Tecuna curare arrow poison (Folkers and Unna 1939); from *C. sola* possessed curare-like action (Folkers 1938); unknown alkaloid from 100 g *C. horrida* root bark (used as cholera cure, sedative, stomachic, anti-idriotic) extd. by cold percolation w/1.6% HCl, then addn. of KI_3 to obtain the periodide (2 g), m. above 300°, rx w/SO_2 → iodide w/NH_4OH → alkaloid, m. above 260°, → a picrate, decomposing at 225° (Chakravarti and Venkatasubban 1932); in fruits of *C. spinosa* (Zhang et al. 2007); in leaves of *C. spinosa* (Zhou et al. 2010); in stem of *C. decidua* (Rathee et al. 2009)
15-N-Acetyl capparisine 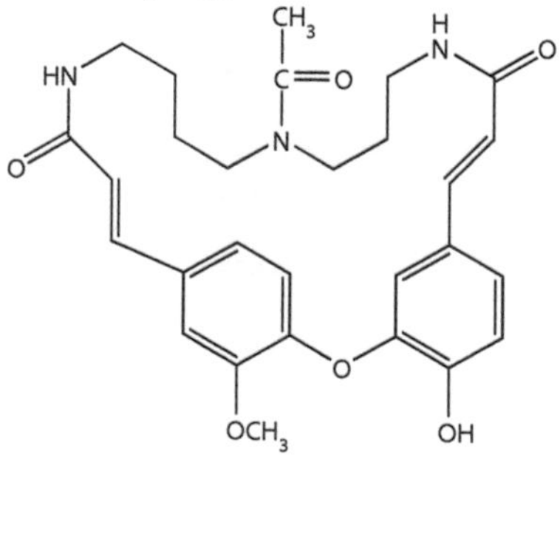	*C. decidua* root bark, spermidine alkaloid, structure via spectral, chem. data, 2D NMR (Ahmad et al. 1992)

TABLE 3.1 (*Continued*)
Alkaloids of *Capparis* spp.

Compound	Notes
14-N-Acetylisocodonocarpine 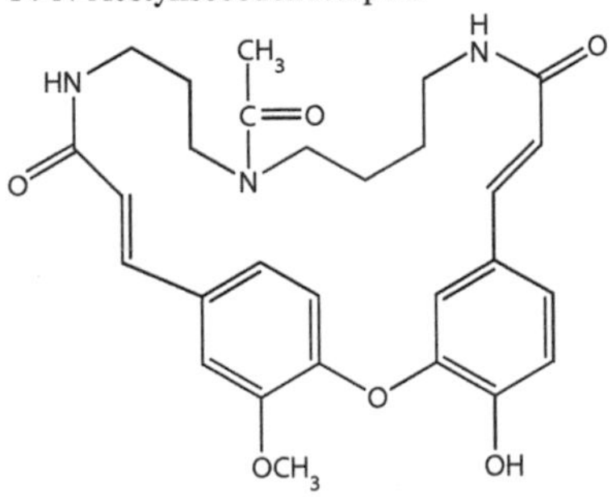	*C. decidua* root bark, spermidine alkaloid, structure via spectral, chem. data, 2D NMR (Ahmad et al. 1992)
Cadabicine 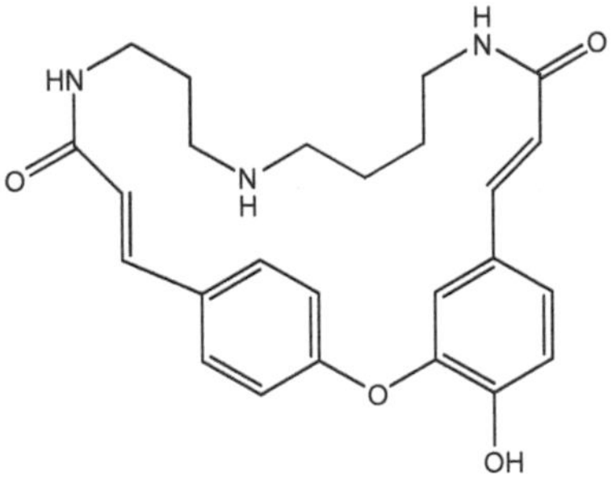	From *C. spinosa* (Khanfar et al. 2003)
Cadabicine 26-O-β-D-glucoside hydrochloride 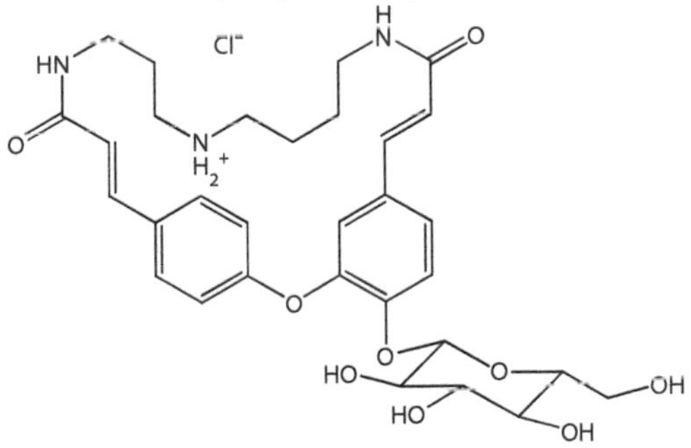	*C. spinosa* root spermidine alkaloid, structure by spectroscopic anal., incl. 1D and 2D NMR, ^{1}H-^{1}H COSY, HSQC, HMBC (Fu et al. 2008)

(*Continued*)

TABLE 3.1 (*Continued*)
Alkaloids of *Capparis* spp.

Compound	Notes
Capparasinine 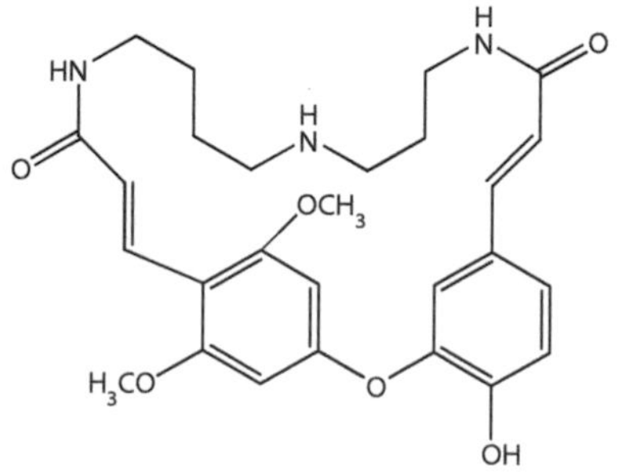	From *C. decidua* root bark, spermidine alkaloid, positional isomer of capparidisine (Ahmad et al. 1987)
Capparidisine 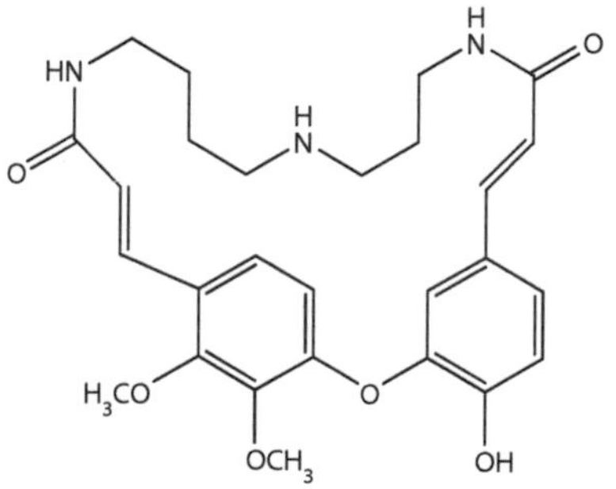	*C. decidua* bark, root, spermidine alkaloid dose-dependent depressant effect on heart rate and coronary flow in isolated rabbit's heart, max. fall in coronary flow at 1 μg/mL, heart rate ↑ at 2 ng dose w/dose dependent ↓ of 128 and 32 ng in force of contraction and heart rate, resp. (Rashid et al. 1989), first isolated from *C. decidua* root bark, structure confirmed by spectral analysis (Ahmad et al. 1985)
Cappariline	Isol. from fresh roots of *C. aphylla* m. 188–91° (Manzoor-i-Khuda and Jeelani 1968)
Capparin A 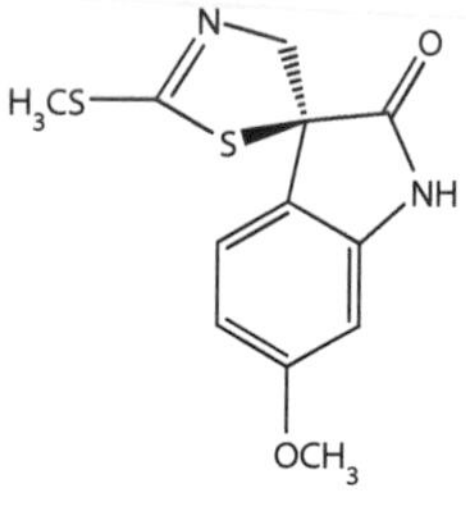	Isol. from whole plant of *C. himalayensis*, structure elucidated by spectral methods and confirmed by X-ray crystallography (Li et al. 2008)
Capparin B 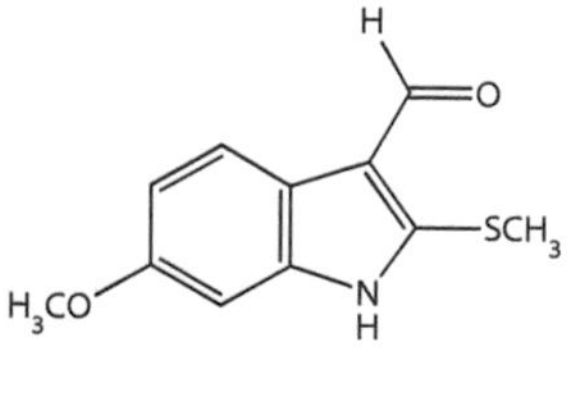	Isol. from whole plant of *C. himalayensis*, structure elucidated by spectral methods (Li et al. 2008)

TABLE 3.1 (*Continued*)
Alkaloids of *Capparis* spp.

Compound	Notes
Capparine	Isol. from fresh roots of *C. aphylla*, m. 236–8° (Manzoor-i-Khuda and Jeelani 1968)
Capparinine	Isol. from fresh roots of *C. aphylla* m. 229–31° (Manzoor-i-Khuda and Jeelani 1968)
Capparisine 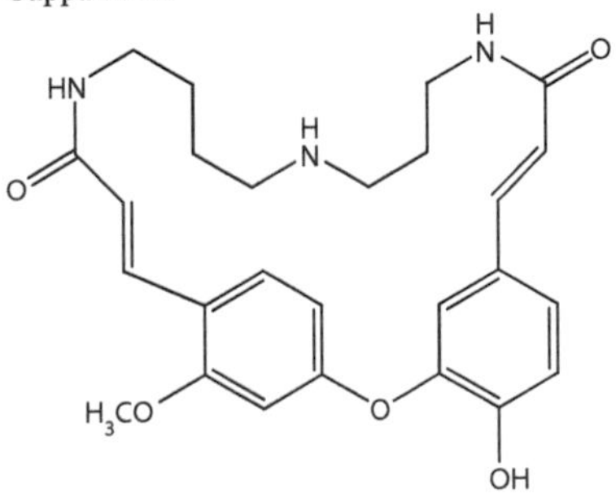	Spermidine alkaloid isol. from EtOH ext. of *C. decidua* root bark (Ahmad et al. 1986)
Capparisine A 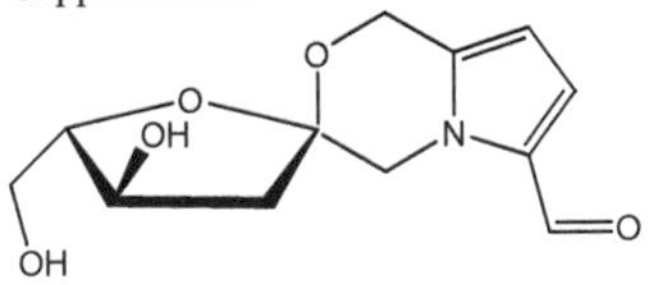	*C. spinosa* fruits solvent sepn., column chromatog., preparative HPLC, structure by spectroscopic anal., stereochemistry proved by X-ray crystallog., no inhibition of human hepatocytes line HL-7702 apoptosis induced by Act D (200 ng/mL), TNF-α (20 ng/mL) w/ high-content screening assay (Yang et al. 2010)
Capparisine B 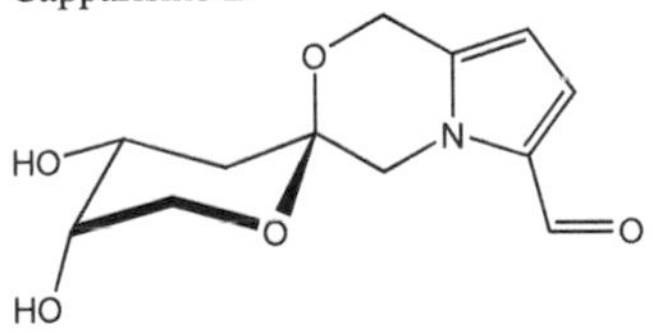	*C. spinosa* fruits solvent sepn., column chromatog., preparative HPLC, structure by spectroscopic anal., stereochemistry proved by X-ray crystallog., no inhibition of human hepatocyte line HL-7702 apoptosis induced by Act D (200 ng/mL), TNF-α (20 ng/mL) w/ high-content screening assay (Yang et al. 2010)
Capparisine C 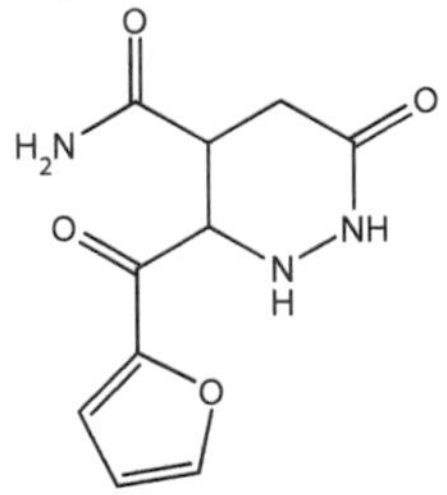	*C. spinosa* fruits solvent sepn., column chromatog., preparative HPLC, structure by spectroscopic anal., no inhibition of human hepatocyte line HL-7702 apoptosis induced by Act D (200 ng/mL), TNF-α (20 ng/mL) w/ high-content screening assay (Yang et al. 2010)

(*Continued*)

TABLE 3.1 (*Continued*)
Alkaloids of *Capparis* spp.

Compound	Notes
Capparispine 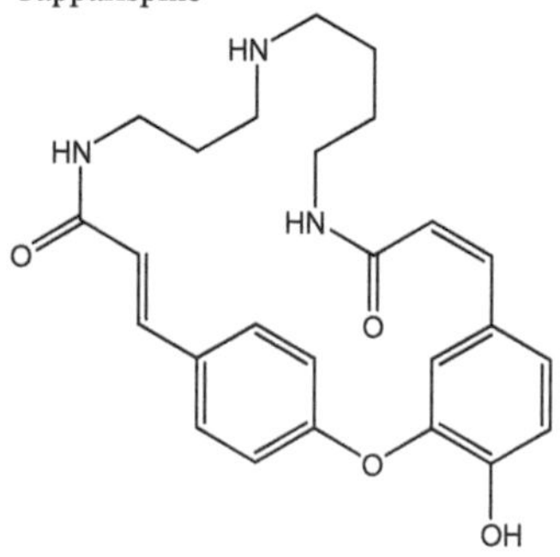	*C. spinosa* root spermidine alkaloid, structure by spectroscopic anal., incl. 1D and 2D NMR, ^{1}H-^{1}H COSY, HSQC, HMBC (Fu et al. 2008)
Capparispine 26-O-β-D-glucoside 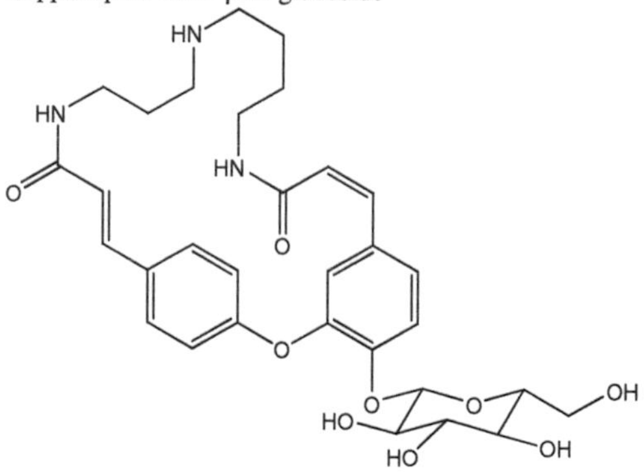	*C. spinosa* root spermidine alkaloid, structure by spectroscopic anal., incl. 1D and 2D NMR, ^{1}H-^{1}H COSY, HSQC, HMBC (Fu et al. 2008)
2-(5-Hydroxymethyl-2-formylpyrrol-1-yl) propionic acid lactone 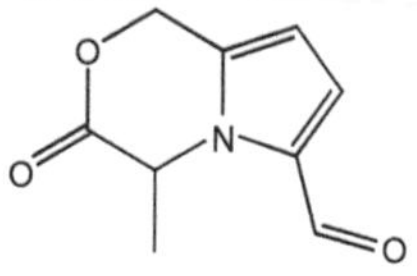	*C. spinosa* fruits solvent sepn., column chromatog., preparative HPLC, structure by spectroscopic anal., no inhibition of human hepatocyte line HL-7702 apoptosis induced by Act D (200 ng/mL), TNF-α (20 ng/mL) w/high-content screening assay (Yang et al. 2010)
Isocodonocarpine 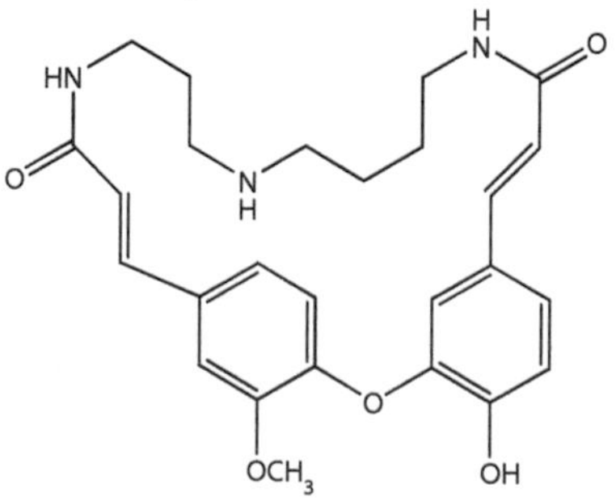	Isolated from *C. decidua* root bark, spermidine alkaloid, structure by spectral studies including 2D NMR (Ahmad et al. 1989)

TABLE 3.1 (*Continued*)
Alkaloids of *Capparis* spp.

Compound	Notes
N-(3′-Maleimidyl)-5-hydroxymethyl-2-pyrrole formaldehyde 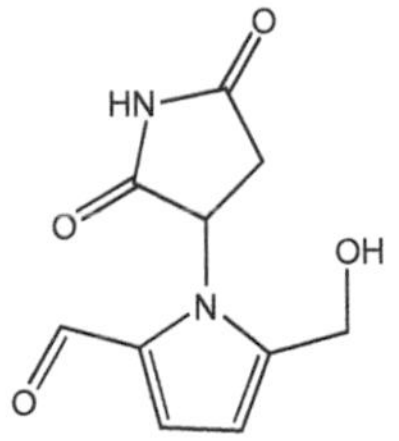	*C. spinosa* fruits solvent sepn., column chromatog., preparative HPLC, structure by spectroscopic anal., no inhibition of human hepatocyte line HL-7702 apoptosis induced by Act D (200 ng/mL), TNF-α (20 ng/mL) w/high-content screening assay (Yang et al. 2010)
Stachydrine 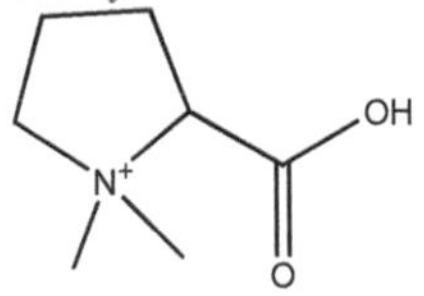	In *C. spinosa* as hydrochloride in multiple parts detd. by dual-wavelength TLC scanning, solvent n-butanol-hydrochloric acid-Et acetate (8:3:1); wavelength λs = 525 nm, λR = 700 nm resp., av. recovery 96.83%, relative std. deviation 2.00% in linearity of 2.03–16.24 μg, conc: roots > buds > fruits > stems > leaves (Jiang et al. 2010); in *C. spinosa* var. *mucronifolia* extd. by 80% methanol, purified by column chromatog., ident. by ^{1}H NMR and IR (Afsharypuor and Jazy 1999); 87.43% of total alkaloids in *C. spinosa*, alkaloids comprising 0.91% of root bark and 0.86% of seeds (Sadykov and Khodzhimatov 1981); in *C. aegyptia*, *C. deserti*, and *C. leucophylla* (Hammouda et al. 1975); from *C. spinosa* roots 5 mg/kg i.v. into dogs → ↓ no. thrombocytes and ↑ their adhesiveness and aggregation, ↓ rate and time of hemorrhage if injected 1–30 mg/kg s.c. into rats or as the HCl salt 100 mg/kg 2x/day into rabbits for 6 days (Mansurov and Mansurov 1974); separated from aq. extract of *C. spinosa* cortex and leaf, yielding 1.2% and 1.5% (of dry wt.) light-yellow noncryst. total base, treatment w/alc. HCL soln. → optically inactive white needle crystals (m.p. 228–9°, $C_7H_{14}O_2NCl$), w/paper chromatog. in BuOH (water-satd.)-HCl (5:1) a isolated w/Rf 0.5, addn. of oxalic acid to a concd. soln. of the hydrochlorides of the satd. alc. soln., the oxalate formed (m.p. 104–5°), picrate (m.p. 194–5°) by same method, shaking aq. soln. of hydrochloride w/AgOH → free base (m.p. 103–5°) to sep. out, after drying at 100° in vacuo, m.p. 225–6° (Mukhamedova et al. 1969); from *C. tomentosa* fruit L-stachydrine isol. as periodide from neutral or acid soln., as HCl salt, m. 222°, and picrate, m. 199–200° (Cornforth and Henry 1952); from *C. moonii* fruits, 0.5% from EtOH ext., m. 232–3°, hydrate m. 116–17° HCl salt m. 220°, complexes w/$AuCl_3$, 215°; picric acid, 195°, picrolonic acid, 190°, ammonium reineckate, 160°, heating to 230–60° → dl-Me hygrinate (Kanthamani et al. 1960); in *C. spinosa* fruit (Fu et al. 2007); as Me stachydrine from EtOAc ext. of *C. spinosa* fruit (Yu et al. 2011)

REFERENCES

Afsharypuor, S. and A.A. Jazy. 1999. Stachydrine and volatile isothiocyanates from the unripe fruit of *Capparis spinosa* L. *DARU: Daru* 7(2): 11–13.

Ahmad, V.U., S. Arif, A.U.R. Amber, and K. Fizza. 1987. Capparisinine, a new alkaloid from *Capparis decidua. Liebigs Ann Chem.* 2: 161–162.

Ahmad, V.U., S. Arif, A.U.R. Amber, M.A. Nasir, and K.U. Ghani. 1986. A new alkaloid from root bark of *Capparis decidua. Z Naturforsch B* 41B(8): 1033–1035.

Ahmad, V.U., S. Arif, A.U.R. Amber, K. Usmanghani, and C.A. Miana. 1985. A new spermidine alkaloid from *Capparis decidua. Heterocycles* 23(12): 3015–20.

Ahmad, V.U., N. Ismail, and A.U.R. Amber. 1989. Isocodonocarpine from *Capparis decidua. Phytochemistry* 28(9): 2493–5.

Ahmad, V.U., N. Ismail, S. Arif, and A.U.R. Amber. 1992. Two new N-acetylated spermidine alkaloids from *Capparis decidua. J Nat Prod* 55(10): 1509–12.

Badawi, M.M., K. Bernauer, P. van den Broek, D. Gröger, A. Guggisberg, S. Johne, I. Kompis, F. Schneider, H.J. Veith, M. Hesse, and H. Schmid. 1973. Macrocyclic spermidine and spermine alkaloids. *Pure Appl Chem* 33(1): 81–107.

Blagbrough, I.S., A.J. Geall, and A.P. Neal. 2003. Polyamines and novel polyamine conjugates interact with DNA in ways that can be exploited in non-viral gene therapy. *Biochem Soc Trans* 31(2): 397–406.

Chakravarti, S. and A. Venkatasubban. 1932. Chemical investigation of Indian medicinal plants. I. Preliminary chemical examination of the root bark of *Capparis horrida. J Annamalai Univ* 1: 176–80.

Cornforth, J.W. and A.J. Henry. 1952. Isolation of L-stachydrine from the fruit of *Capparis tomentosa. J Chem Soc* 1952: 601–3.

Folkers, K. 1938. Preliminary studies of the botanical components of Tecuna and Java curare. *J Am Pharm Assoc (1912–1977)* 27: 689–93.

Folkers, K. and K. Unna. 1939. Chazuta curare, its botanical components, and other plants of curare interest. *Arch Int Pharmacodyn Ther* 61: 370–9.

Fu, X.P., H.A. Aisa, M. Abdurahim, A. Yili, S.F. Aripova, and B. Tashkhodzhaev. 2007. Chemical composition of *Capparis spinosa* fruit. *Chem Nat Compounds* 43(2): 181–3.

Fu, X.P., T. Wu, M. Abdurahim, Z. Su, X.L. Hou, H.A. Aisa, and H. Wu. 2008. New spermidine alkaloids from *Capparis spinosa* roots. *Phytochem Lett* 1(1): 59–62.

Ghosh, B. 2000. Polyamines and plant alkaloids. *Indian J Exp Biol* 38(11): 1086–1091.

Hammouda, F.M., M.M.S. El-Nasr, and A.M. Rizk. 1975. Constituents of Egyptian *Capparis* species. *Pharmazie* 30(11): 747–8.

Hikino, H., K. Ogata, C. Konno, and S. Sato. 1983. Hypotensive actions of ephedradines, macrocyclic spermine alkaloids of ephedra roots. *Planta Med* 48(8): 290–3.

Jiang, X.J., Q.Y. Meng, M.X. Yu, and H.J. Bai. 2010. Determination of stachydrine hydrochloride in different parts of *Capparis spinosa* L. by dual wavelength TLC scanning. *Guangpu Shiyanshi* 27(5): 1959–63.

Juneja, T.R., K.N. Gaind, and C.L. Dhawan. 1970. *Capparis sepiaria. Res Bull Panjab Univ Sci* 21(1–2): 23–6.

Kanthamani, S., C.R. Narayanan, and K. Venkataraman. 1960. Isolation of l-stachydrine and rutin from the fruits of *Capparis moonii. J Sci Ind Res (B)* 19B: 409–10.

Khanfar, M.A., S.S. Sabri, M.A. Zarga, and K.P. Zeller. 2003. The chemical constituents of *Capparis spinosa* of Jordanian origin. *Nat Prod Res* 17(1): 9–14.

Li, Y.Q., S.L. Yang, H.R. Li, and L.Z. Xu. 2008. Two new alkaloids from *Capparis himalayensis. Chem Pharm Bull* 56(2): 189–91.

Liang, H., P. Liu, Y. Wang, S. Song, and A. Ji. 2011. Protective effects of alkaloid extract from *Leonurus heterophyllus* on cerebral ischemia reperfusion injury by middle cerebral ischemic injury (MCAO) in rats. *Phytomedicine* 18(10): 811–18.

Mansurov, M.M. and Z.M. Mansurov. 1974. Effect of the alkaloid stachydrine on adhesiveness and aggregation of thrombocytes and bleeding time. *Med Zh Uzb* 2: 51–55.

Manzoor-i-Khuda, M., and N.A. Jeelani. 1968. Chemical constituents of *Capparis aphylla*. II. Isolation of capparine, cappariline, and capparinine. *Pak J Sci Ind Res* 11(3): 250–2.

Mukhamedova, K.S., S.T. Akramov, and S.Y. Yunusov. 1969. [Stachydrine from *Capparis spinosa*.] *Him Prir Soedin* 5(1): 67.

Murata, T., T. Miyase, and F. Yoshizaki. 2010. Cyclic spermidine alkaloids and flavone glycosides from *Meehania fargesii. Chem Pharm Bull (Tokyo)* 58(5): 696–702.

Rakhimova, A.K., R.A. Abdullaev, and D.Y. Guseinov. 1978. Chemical-biological characteristics of *Capparis spinosa* from Azerbaidzhan. *Azerbaidzhanskii Meditsinskii Zhurnal* 55(2): 70–5.

Rashid, S., F. Lodhi, M. Ahmad, and K. Usmanghani. 1989. Preliminary cardiovascular activity evaluation of capparidisine, a spermidine alkaloid from *Capparis deciduas*. *Pak J Pharmacol* 6(1–2): 61–6.

Rathee, S., O.P. Mogla, S. Sardana, M. Vats, and P. Rathee. 2009. Pharmacognostical and phytochemical evaluation of stem of *Capparis decidua* (Forsk.) Edgew. *Pharmacognosy J* 1(2): 75–81.

Sadykov, Y.D. and M. Khodzhimatov. 1981. Alkaloids of *Capparis spinosa* L. *Dokl Akad Nauk Tadzhikskoi SSR* 24(10): 617–20.

Samoylenko, V., M.R. Jacob, S.I. Khan, J. Zhao, B.L. Tekwani, J.O. Midiwo, L.A. Walker, and I. Muhammad. 2009. Antimicrobial, antiparasitic and cytotoxic spermine alkaloids from *Albizias chimperiana*. *Nat Prod Commun* 4(6): 791–6.

Yang, T., C.H. Wang, G.X. Chou, T. Wu, X.M. Cheng, and Z.T. Wang. 2010. New alkaloids from *Capparis spinosa*: structure and X-ray crystallographic analysis. *Chemistry* 123(3): 705–10.

Yin, J., Z.W. Zhang, W.J. Yu, J.Y. Liao, X.G. Luo, and Y.J. Shen. 2010. Stachydrine, a major constituent of the Chinese herb *Leonurus heterophyllus* Sweet, ameliorates human umbilical vein endothelial cells injury induced by anoxia-reoxygenation. *Am J Chin Med* 38(1): 157–71.

Yu, L., C.P. He, X.M. Zhang, N. Yu, and L.Q. Xie. 2011. Extraction and antioxidant activity of chemical compositions in *Capparis spinosa* L. *Harbin Shangye Daxue Xuebao, Ziran Kexueban* 27(4): 524–7.

Zhang, Y., W. Chen, X. Peng, F. Wang, B. Han, and F. Jiang. 2007. [Preliminary experiment on the chemical constituents of *Capparis spinosa* L. fruits.] *Shihezi Daxue Xuebao, Ziran Kexueban* 25(4): 481–3.

Zhou, F., A. Meiliwan, and A. Hajiakber. 2010. Preliminary study on chemical components of leaves of *Capparis spinosa* L. *Guoyi Guoyao* 21(3): 515–6.

FIGURE 2.7 The hard *Capparis aegyptia* leaves, as part of their adaptation system, have the bluish tinge typical of desert plants. (16.5.2011, near Kibbutz Ketura, Arava, Israel: by Helena Paavilainen.)

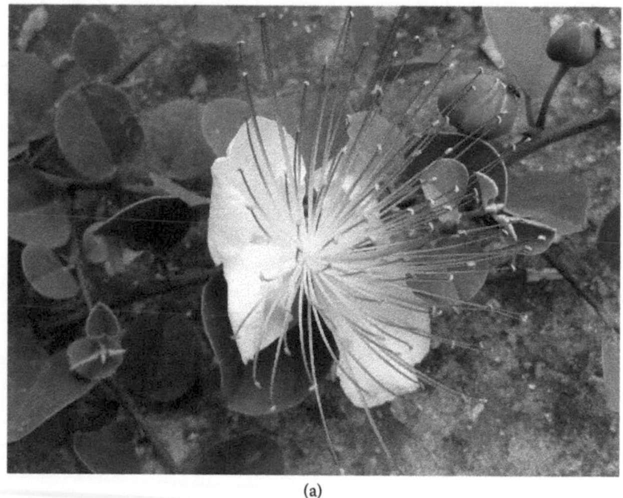

(a)

FIGURE 2.16 (a) Examples of the variability of flowers within genus *Capparis: Capparis spinosa* L. (23.5.2010, Hadassah Ein Kerem, Jerusalem, Israel: by Helena Paavilainen.)

(d)

FIGURE 2.16 (d) Examples of the variability of flowers within genus *Capparis: Capparis decidua.* (15.3.2010, near Shirur, Pune, Maharashtra, India: by Abhijeet Shiral.)

FIGURE 2.17 Considerable variation may also exist between the flowers of the same plant: *Capparis cartilaginea* with flowers of different age showing differing colors that range between white and red. (16.5.2011, Kibbutz Ketura, Southern Arava, Israel: by Helena Paavilainen.)

(a)

FIGURE 2.18 (a) Variations in fruit within genus *Capparis: Capparis cartilaginea*, ripe red fruit. (8.8.2011, Kibbutz Ketura, Southern Arava, Israel: by Helena Paavilainen.)

(b)

FIGURE 2.18 (b) Variations in fruit within genus *Capparis: Capparis spinosa*, unripe green fruit. (25.7.2010, Hadassah Ein Kerem, Jerusalem, Israel: by Helena Paavilainen.)

FIGURE 2.28 The most important parts of *Capparis cartilaginea* for traditional use are the fruit (for food) and the leaves and stems (medicinally). The ripe fruit are red. (8.8.2011, Kibbutz Ketura, Southern Arava, Israel: by Helena Paavilainen.)

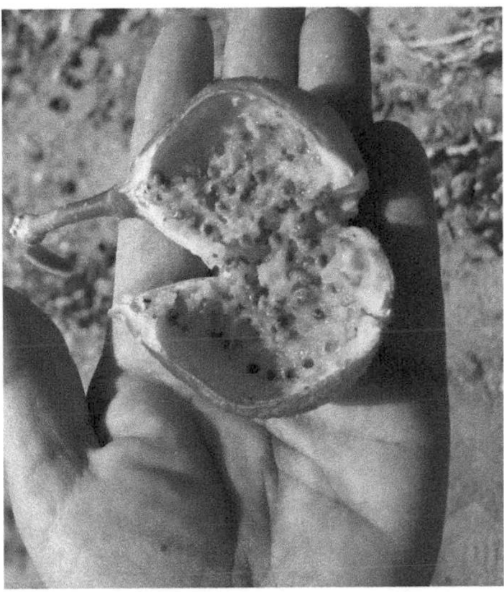

FIGURE 2.29 The yellow inside of ripe split fruits of *Capparis cartilaginea* presented by Dr. Elaine Soloway. Note the seeds. (8.8.2011, Kibbutz Ketura, Southern Arava, Israel: by Helena Paavilainen.)

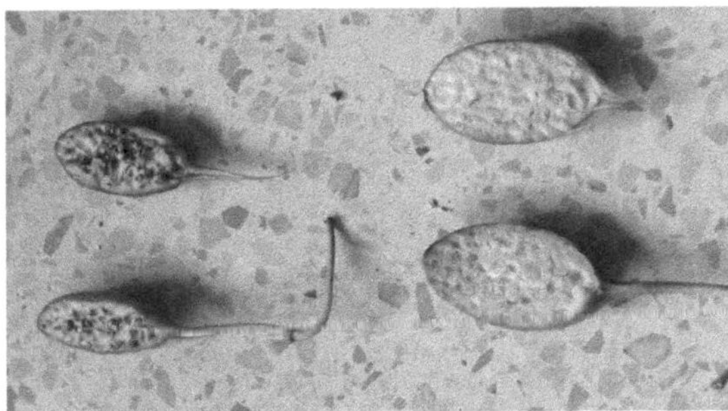

FIGURE 4.2 Unripe fruit of *Capparis spinosa*. Notice the difference both in seed color and in the consistency of the pulp between different stages of ripening. (27.7.2010, Jerusalem, Israel: by Helena Paavilainen.)

FIGURE 16.1 Aerial parts of *Capparis decidua*, including fruit, have a hypolipidemic effect. (Mauritania: by Sébastien Sant, in J.P. Peltier. 2006. Plant Biodiversity of South-Western Morocco, http://www.teline.fr, accessed 13 March 2013.)

4 Fatty Acids

Capers (i.e., *Capparis* plants) hold significant fat, mainly in the seeds, with around 10–15% (Figure 4.1). Most of this consists of fatty acids, and probably most of these are stored as triacylglycerols. The overall bulk of this pure fat is in the form of the common fatty acids, especially linoleic, oleic, palmitic, stearic, and also lauric, linolenic, myristic, and sterculic acids. There is an impressive occurrence in *C. zeylanica* seed oil of 30% ricinoleic acid, also the main component of castor seed oil, which possesses substantial anti-inflammatory activity in the rat carrageenan-induced paw edema *in vivo* assay (Vieira et al. 2000), although the range of its action *in vivo* is analogous to that of capsaicin, with biphasic, proinflammatory, and anti-inflammatory response (Vieira et al. 2001). Ricinoleic acid is also suitable as all or part of a carrier of drugs, including the anticancer taxols (Shikanov et al. 2004, 2008) and the antibiotic gentamicin (Brin et al. 2009, Krasko et al. 2007), for injection into deep body spaces.

Other fatty acids in *Capparis* seed are less known; one fatty acid found in *C. zeylanica* roots may be unique to the genus (Haque and Haque 2011). Liu et al. (2008) investigated *C. spinosa* seeds (Figure 4.2) with gas chromatography/mass spectrometry (GC/MS) and reported methyl esters of several conjugated, specifically octadecenoic, 18-carbon fatty acids. (Conjugated fatty acids contain two double carbon-carbon bonds in between which is a single carbon-carbon bond.) In general, conjugated fatty acids, especially 18-carbon fatty acids, are well known for exerting numerous chemopreventive effects *in vivo* and clinically, including reduction of body fat, attenuation of cardiovascular diseases and cancer, modulation of immunity and inflammation, and improvement of bone mass (Dilzer and Park 2012).

The richness of *Capparis* seed oil provides a natural base for solving numerous lipophilic components, such as sterols, tocols, and caretenoids in pharmaceutical preparations. Lipophilic components for such preparations could be sourced from *Capparis* spp. seeds, buds, flowers, leaves, and stems, or they could be lipophilic compounds derived from other plant or animal genera. Details of the experimental work extant on fatty acids in *Capparis* are summarized in Table 4.1.

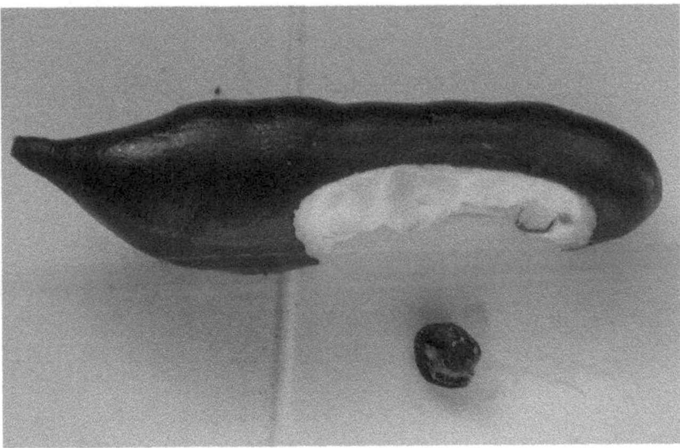

FIGURE 4.1 Red *Capparis sola* fruit (nibbled) and seed. (Manu, Madre de Dios, Peru: by C. E. Timothy Paine.)

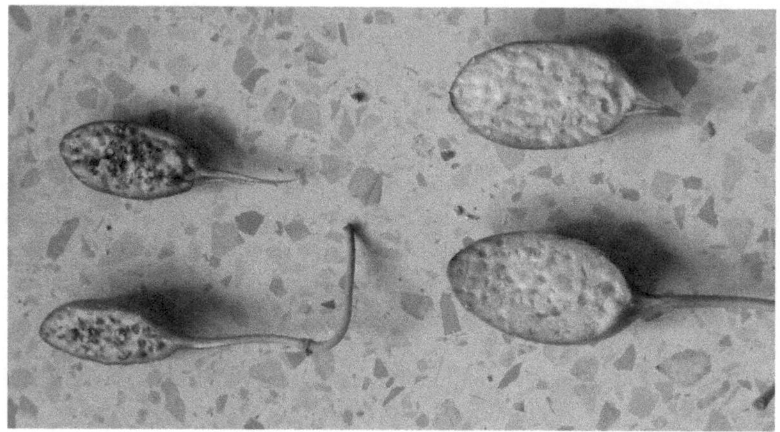

FIGURE 4.2 (See color insert.) Unripe fruit of *Capparis spinosa*. Notice the difference both in seed color and in the consistency of the pulp between different stages of ripening. (27.7.2010, Jerusalem, Israel: by Helena Paavilainen.)

TABLE 4.1
Fatty Acids of *Capparis* spp.

Compound	Notes
Fatty acids in general 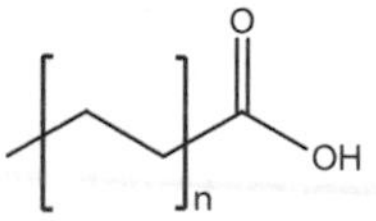	From *C. spinosa*, 7 Tunisian regions, oil 23.25 to 33.64% dry wt., FA main component, oleic 45.82%, linoleic 25.37%, palmitic 15.93%, stearic 4.06% (Tlili et al. 2009); in *C. spinosa* L. fruit by supercrit. CO_2 extn., by GC-MS, 11 FA detctd., incl. oleic, linoleic, and palmitic acids, 94.81% of total (Ren et al. 2009); 33% of total storage in Tunisian *C. spinosa* seeds (Tlili et al. 2011); in aerial parts of *C. decidua* (Abdel-Mogib et al. 2000)
Arachidic acid 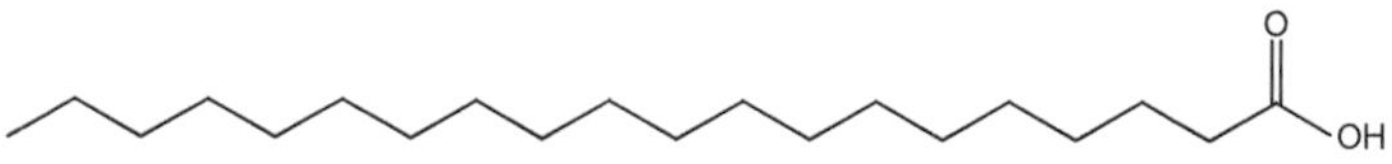	2.0% of seed fat in *C. aphylla* (Sen Gupta and Chakrabarty 1964)
Lauric acid 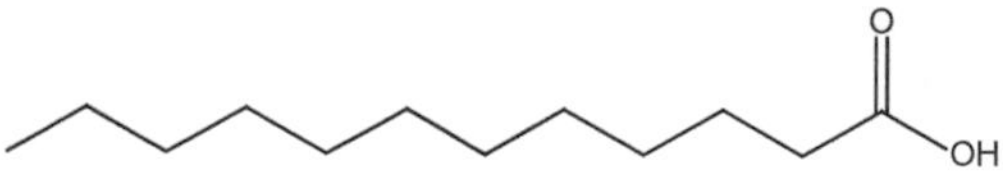	3.5% *C. zeylanica* oil, which is 30% dry wt. of plant matter, by IR, NMR, MS, and GC, LC, chem. degrdns. (Daulatabad et al. 1991)

TABLE 4.1 (*Continued*)
Fatty Acids of *Capparis* spp.

Compound	Notes
Linoleic acid 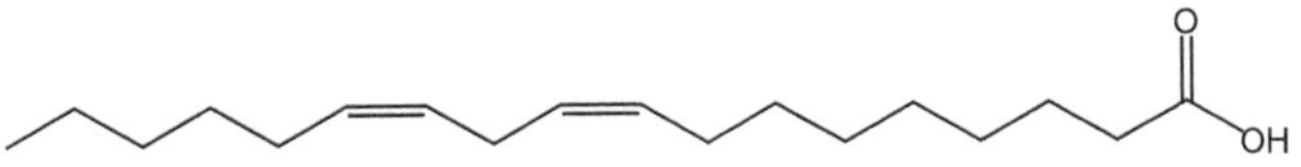	57.21%, of oil in *C. spinosa*, by GC-MS oil 21.2%, FA main component (Nurmamat and Korbanjhon 2011); *C. spinosa* L. fruit by supercrit. CO_2 extn., by GC-MS, 11 FA detctd., incl. oleic, linoleic, and palmitic acids, 94.81% of total (Ren et al. 2009); from *C. spinosa*, 7 Tunisian regions, oil 23.25 to 33.64% dry wt., FA main component, oleic 45.82%, linoleic 25.37%, palmitic 15.93%, stearic 4.06% (Tlili et al. 2009); dominant FA, 26.9–55.3% in *C. ovata* seed oils and for 24.6–50.5% in *C. spinosa* seed oils (Matthäus and Özcan 2005); one of three major FA in *C. spinosa* (Akgul and Özcan 1999); 30.5% *C. zeylanica* oil, which is 30% dry wt. of plant matter, by IR, NMR, MS, and GC, LC, chem. degrdns. (Daulatabad et al. 1991); 11.4% of seed fat of *C. aphylla* (Sen Gupta and Chakrabarty 1964); 44.9–51.1% of *C. spinosa* seed oil (Zabramnyi et al. 1940)
Linolenic acid 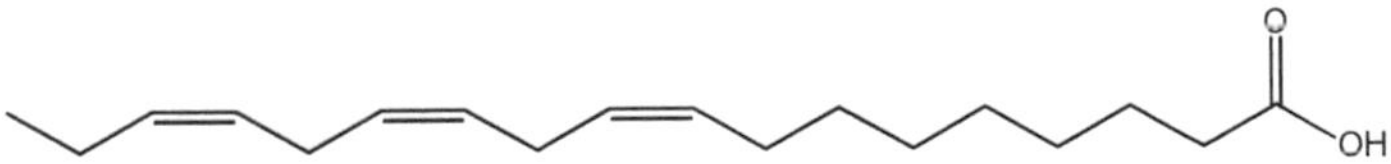	4th most prevalent FA in *C. spinosa* by GC-MS (Nurmamat and Korbanjhon 2011); from *C. spinosa*, 7 Tunisian regions, oil 23.25 to 33.64% dry wt., FA main component, oleic 45.82%, linoleic 25.37%, palmitic 15.93%, stearic 4.06% (Tlili et al. 2009)
Malvalic acid 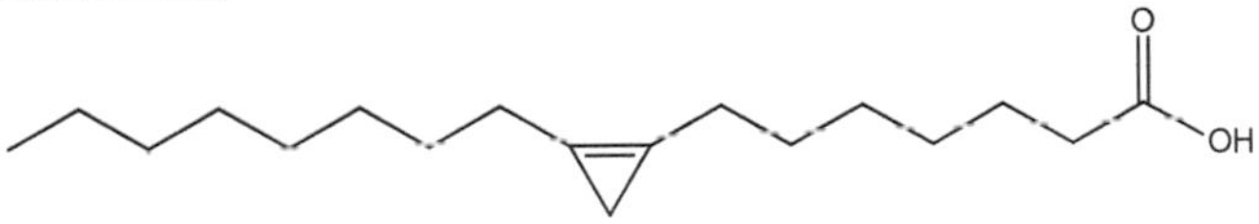	6.2% *C. zeylanica* oil, which is 30% dry wt. of plant matter, by IR, NMR, MS, and GC, LC, chem. degrdns. (Daulatabad et al. 1991)
Myristic acid 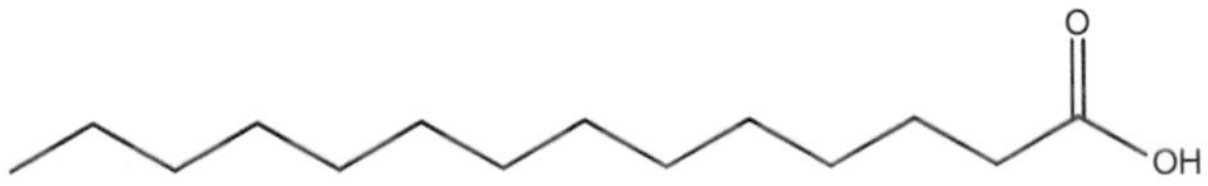	2.5% in *C. zeylanica* oil, which is 30% dry wt. of plant matter, by IR, NMR, MS, and GC, LC, chem. degrdns. (Daulatabad et al. 1991); 0.6% of seed fat of *C. aphylla* (Sen Gupta and Chakrabarty 1964)
(Z,Z)-9,12-Octadecadienoic acid methyl ester 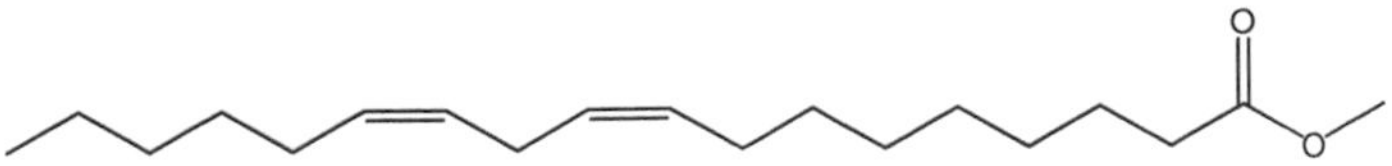	Accelerated solvent extn. and Soxhlet extn., *C. spinosa* seed oil w/hexane and Et ether, GC-MS after esterification (Liu et al. 2008)

(*Continued*)

TABLE 4.1 (*Continued*)
Fatty Acids of *Capparis* spp.

Compound	Notes
(E)-9-Octadecenoic acid methyl ester	Accelerated solvent extn. and Soxhlet extn., *C. spinosa* seed oil w/hexane and Et ether, GC-MS after esterification (Liu et al. 2008)
(Z)-9-Octadecenoic acid methyl ester	Accelerated solvent extn. and Soxhlet extn., *C. spinosa* seed oil w/hexane and Et ether, GC-MS after esterification (Liu et al. 2008)
E-Octadec-7-en-5-ynoic acid	From chloroform ext. of the roots of *C. zeylanica* L., in Long Evan's rats, at 300 µg/rat/day for consecutive 14 days, no detectable abnormalities found in histopathol. of heart, liver, kidney, and lung (Haque and Haque 2011)
Oleic acid 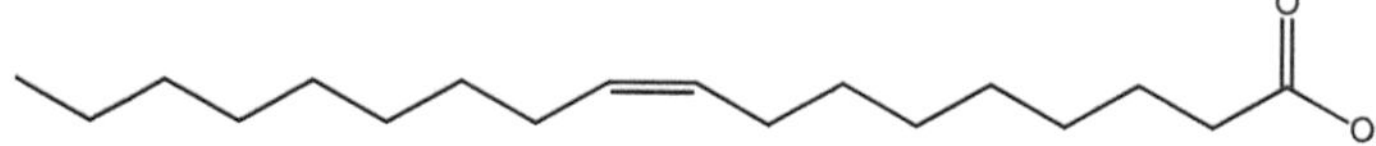	2nd most prevalent FA in *C. spinosa* by GC-MS (Nurmamat and Korbanjhon 2011); *C. spinosa* L. fruit by supercrit. CO_2 extn., by GC-MS, 11 FA detctd., incl. oleic, linoleic, and palmitic acids, 94.81% of total (Ren et al. 2009); from *C. spinosa*, 7 Tunisian regions, oil 23.25 to 33.64% dry wt., FA main component, oleic 45.82%, linoleic 25.37%, palmitic 15.93%, stearic 4.06% (Tlili et al. 2009); in *C. ovata* seed and *C. spinosa* seed oils with its isomer, vaccenic acid, between 10 and 30% (Matthäus and Özcan 2005); one of three major FA in *C. spinosa* (Akgul and Özcan 1999); 9.8% *C. zeylanica* oil, which is 30% dry wt. of plant matter, by IR, NMR, MS, and GC, LC, chem. degrdns. (Daulatabad et al. 1991); 57.2% of seed fat in *C. aphylla* (Sen Gupta and Chakrabarty 1964); 42.4–45.9% of *C. spinosa* oil (Zabramnyi et al. 1940)

TABLE 4.1 (*Continued*)
Fatty Acids of *Capparis* spp.

Compound	Notes
Palmitic acid 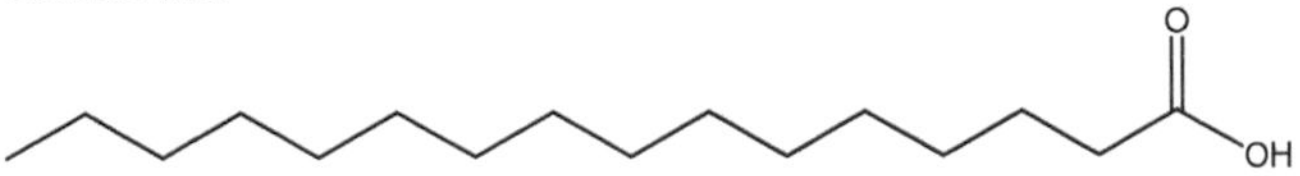	3rd most prevalent FA in *C. spinosa* by GC-MS (Nurmamat and Korbanjhon 2011); *C. spinosa* L. fruit by supercrit. CO_2 extn., by GC-MS, 11 FA detctd., incl. oleic, linoleic, and palmitic acids, 94.81% of total (Ren et al. 2009); from *C. spinosa*, 7 Tunisian regions, oil 23.25 to 33.64% dry wt., FA main component, oleic 45.82%, linoleic 25.37%, palmitic 15.93%, stearic 4.06% (Tlili et al. 2009); one of three major FA in *C. spinosa* (Akgul and Özcan 1999); 11% *C. zeylanica* oil, which is 30% dry wt. of plant matter, by IR, NMR, MS, and GC, LC, chem. degrdns. (Daulatabad et al. 1991); 21.1% of seed fat in *C. aphylla* (Sen Gupta and Chakrabarty 1964)
Palmitoleic acid 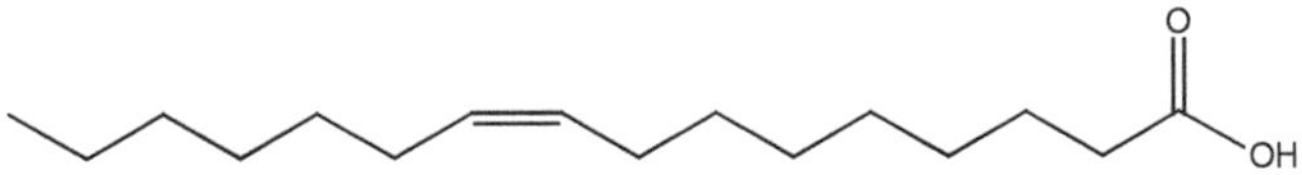	9.8% in *C. divaricata* seed oil (Kittur et al. 1993)
Ricinoleic acid 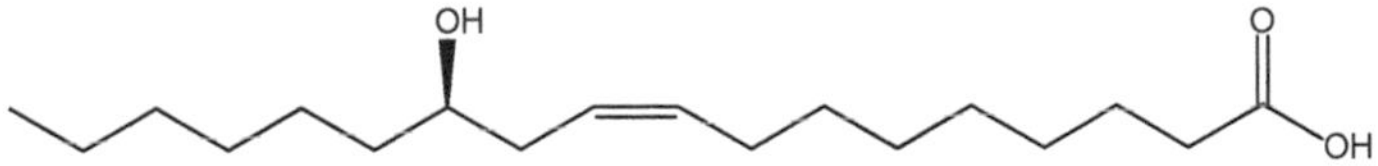	30% *C. zeylanica* oil, which is 30% dry wt. of plant matter, by IR, NMR, MS, and GC, LC, chem. degrdns. (Daulatabad et al. 1991)
Stearic acid 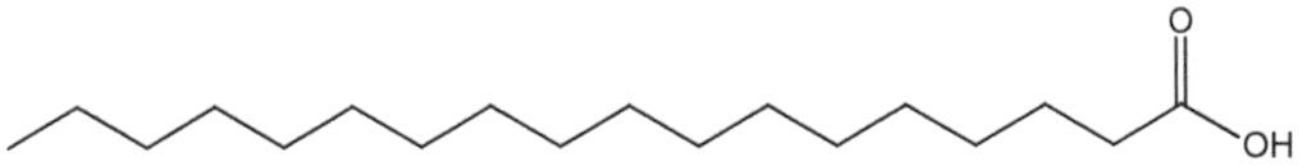	5th most prevalent FA in *C. spinosa* by GC-MS (Nurmamat and Korbanjhon 2011); from *C. spinosa*, 7 Tunisian regions, oil 23.25 to 33.64% dry wt., FA main component, oleic 45.82%, linoleic 25.37%, palmitic 15.93%, stearic 4.06% (Tlili et al. 2009); 1.8% *C. zeylanica* oil, which is 30% dry wt. of plant matter, by IR, NMR, MS, and GC, LC, chem. degrdns. (Daulatabad et al. 1991); 7.7% of seed fat in *C. aphylla* (Sen Gupta and Chakrabarty 1964)
Sterculic acid 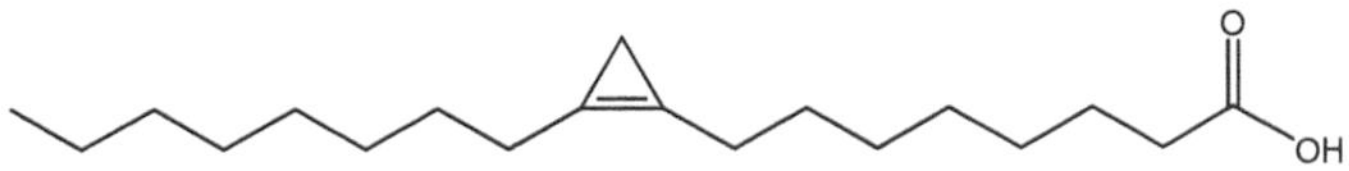	4.7% *C. zeylanica* oil, which is 30% dry wt. of plant matter, by IR, NMR, MS, and GC, LC, chem. degrdns. (Daulatabad et al. 1991)

REFERENCES

Abdel-Mogib, M., S.T. Ezmirly, and S.A. Basaif. 2000. Phytochemistry of *Dipterygium glaucum* and *Capparis decidua*. *J Saudi Chem Soc* 4(1): 103–108.

Akgul, A. and M. Özcan. 1999. Some compositional characteristics of capers (*Capparis* spp.) seed and oil. *Grasas y Aceites (Sevilla)* 50(1): 49–52.

Brin, Y.S., A. Nyska, A.J. Domb, J. Golenser, B. Mizrahi, and M. Nyska. 2009. Biocompatibility of a polymeric implant for the treatment of osteomyelitis. *J Biomater Sci Polym Ed* 20(7–8): 1081–1090.

Daulatabad, C.M.J.D., V.A. Desai, and K.M. Hosamani. 1991. New source of oil with novel fatty acids for industrial utilization. *Ind Eng Chem Res* 30(12): 2596–2598.

Dilzer, A. and Y. Park. 2012. Implication of conjugated linoleic acid (CLA) in human health. *Crit Rev Food Sci Nutr* 52(6): 488–513.

Haque, M. and M.E. Haque. 2011. Sub-acute toxicity study of a novel compound E-octadec-7-en-5-ynoic acid from *Capparis zeylanica* Linn roots. *Agric Biol J North Am* 2(4): 708–712.

Kittur, M.H., C.S. Mahajanshetti, and G. Lakshminarayana. 1993. Characteristics and composition of *Trichosanthes bracteata*, *Urena sinuata* and *Capparis divaricata* seeds and oils. *J Oil Technol Assoc India* 25(2): 39–41.

Krasko, M.Y., J. Golenser, A. Nyska, M. Nyska, Y.S. Brin, and A.J. Domb. 2007. Gentamicin extended release from an injectable polymeric implant. *J Control Release* 117(1): 90–96.

Liu, W.J., H.J. Bai, X.A. Pang, and H.Z. Sun. 2008. GC/MS analysis of fatty acid in *Capparis spinosa* seed oil. *Fenxi Shiyanshi* 27(3): 42–44.

Matthäus, B. and M. Özcan. 2005. Glucosinolates and fatty acid, sterol, and tocopherol composition of seed oils from *Capparis spinosa* var. *spinosa* and *Capparis ovata* Desf. var. *canescens* (Coss.) Heywood. *J Agric Food Chem* 53(18): 7136–7141.

Nurmamat, E. and B. Korbanjhon. 2011. Study on fat and protein components of Uygur folk medicine *Capparis spinosa* L. *Anhui Nongye Kexue* 39(18): 10834–10836.

Ren, Y., J. Xu, J. Zhao, F. Xu, W. Yang, and Y. Liu. 2009. Chemical components of volatile oil and fatty acid by supercritical carbon dioxide fluid extracts from fruit of *Capparis spinosa* L. *Xinjiang Yike Daxue Xuebao* 32(12): 1659–1660, 1663.

Sen Gupta, A., and M.M. Chakrabarty. 1964. Composition of the seed fats of the Capparidaceae family. *J Sci Food Agric* 15(2): 69–73.

Shikanov, A., S. Shikanov, B. Vaisman, J. Golenser, and A.J. Domb. 2008. Paclitaxel tumor biodistribution and efficacy after intratumoral injection of a biodegradable extended release implant. *Int J Pharm* 358(1–2): 114–120.

Shikanov, A., B. Vaisman, M.Y. Krasko, A. Nyska, and A.J. Domb. 2004. Poly(sebacic acid-co-ricinoleic acid) biodegradable carrier for paclitaxel: in vitro release and in vivo toxicity. *J Biomed Mater Res A* 69(1): 47–54.

Tlili, N., T. El Guizani, N. Nasri, A. Khaldi, and S. Triki. 2011. Protein, lipid, aliphatic and triterpenic alcohol content of caper seeds "Capparis spinosa." *J Am Oil Chem Soc* 88(2): 265–270.

Tlili, N., S. Munne-Bosch, N. Nasri, E. Saadaoui, A. Khaldi, and S. Triki. 2009. Fatty acids, tocopherols and carotenoids from seeds of Tunisian caper "Capparis spinosa." *J Food Lipids* 16(4): 452–464.

Vieira, C., S. Evangelista, R. Cirillo, A. Lippi, C.A. Maggi, and S. Manzini. 2000. Effect of ricinoleic acid in acute and subchronic experimental models of inflammation. *Mediators Inflamm* 9(5): 223–228.

Vieira, C., S. Fetzer, S.K. Sauer, S. Evangelista, B. Averbeck, M. Kress, P.W. Reeh, R. Cirillo, A. Lippi, C.A. Maggi, and S. Manzini. 2001. Pro- and anti-inflammatory actions of ricinoleic acid: similarities and differences with capsaicin. *Naunyn Schmiedebergs Arch Pharmacol* 364(2): 87–95.

Zabramnyi, D., A. Ochakovskii, and N. Petrova. 1940. Analysis of *Capparis spinosa* seed oil. *Masloboino-Zhirovaya Promyshlennost* 16(5/6): 57–59.

5 Flavonoids

"Texas tarragon," *Tagetes lucida*, is known as a hallucinogenic smoke when mixed with strong wild tobacco (*Nicotiana rustica*). The flowers yield a deep, bright yellow dye, due in large part to the flavonoid, an O-methyl flavonol, isorhamnetin, which *T. lucida* shares in common with the fruits of *C. spinosa* (Figures 5.1 and 5.2; see Table 5.1). It is possibly this same compound that is responsible for the gentle yellow staining of the fingers and hands after hours of handling caper fruits or buds.

The image is illustrative of flavonoids, small organic molecules whose name derives from the Latin *flavus*, meaning "yellow." Flavonoids are nothing less than the principal dyes in plants and can make colors from yellow to blue, the latter especially in the subset of flavonoids known as anthocyanins

FIGURE 5.1 All the aerial parts of *Capparis spinosa* contain flavonoids, especially the leaves and the buds. (14.5.2010, Jerusalem, Israel: by Helena Paavilainen.)

FIGURE 5.2 The fruit of *Capparis spinosa* also contains considerable amounts of antioxidant flavonoids. (4.8.2011, Old City, Jerusalem, Israel: by Helena Paavilainen.)

(*cyan* = "blue"). These numerous different dyes in plants, the flavonoids, including the isoflavonoids, are responsible for many medicinal effects in animals and mammals, including humans.

More important than their direct antioxidant effects is the role of flavonoids as "biological response modifiers" (Lee et al. 2007, Magrone et al. 2008), namely, their ability to modulate immune responses and inflammation through upregulation or downregulation of specific cytokines, such as the interleukins (ILs) (i.e., the series IL-1, IL-2, … , IL-12, etc.), ultimately resulting in increased release of nitric oxide from cells that cause cascades of health-promoting and healing effects.

In considering the different flavonoids in Table 5.1, keep in mind that some of these molecules, like the first compound, rutin, are bound to sugar moieties and are thus called glycosides. Other compounds, like quercetin, which is rutin without its sugar, are called aglycones. Flavonoids in general and specific compounds such as apigenin in particular exert neuroprotective and anxiolytic effects *in vivo* and clinically (Jäger and Saaby 2011). Also not to be overlooked is the fact that a number of flavonoids (e.g., quercetin, kaempferol, and luteolin) bind to estrogen receptors and are capable of exerting hormone-like effects associated with estrogens (van Elswijk et al. 2004).

Almost across the board, the effects of flavonoids *in vivo* and clinically, including those related to their estrogenic effects, have been almost universally positive, in addition to the widely held belief that flavonoid-rich fruit and vegetable consumption is beneficial. Numerous anticancer activities, for example, have been noted (Spatafora and Tringali 2012), but some potential toxicity of flavonoid consumption has also been described, namely, interference with thyroid hormone synthesis, genotoxicity and carcinogenicity, interference with nonheme iron absorption, and as a cause of a decrease in effective doses of conventional drugs (Mennen et al. 2005, Wang et al. 2007). Biflavonoids (i.e., those containing two flavone backbones rather than one) also occur. In Table 5.1, "salt-stabilized capers" refers to the method of natural lactic fermentation, which is controlled with salt.

TABLE 5.1
Flavonoids of *Capparis* spp.

Compound	Notes
Flavonoids in general [flavone backbone, (2-phenyl-1,4-benzopyrone)] 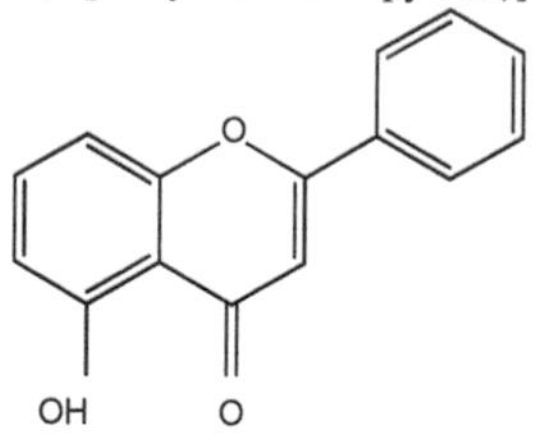	In *C. corymbosa* Lam. roots (Sawadogo et al. 1981); in *C. spinosa* buds, visible spectrophotometry w/detection wavelength set at 500 nm, rutin as the std. sample and $NaNO_3$, $Al(NO_3)_3$ and NaOH composing coloration system, methanol reflux method showed highest extn. rate for total flavonoids, w/ good linear relationship 0 to 32 mg/L ($r = 0.9997$), av. content of total flavonoids 1.13% (Jiang et al. 2010); flavones found in leaves of *C. spinosa* (Zhou et al. 2010); of 75% ethanol ultrasonic extn., 75% ethanol refluxing extn., and 75% ethanol percolation extn. in *C. spinosa*, highest total flavonoids, highest activity against both acetic acid-induced pain and xylene-induced inflammation as follows: refluxing group > percolation group > ultrasonic group (Zhang et al. 2010); macroporous resin for sep. of *C. spinosa* fruit flavonoids, room temp., EtOH 70%; ultrasonic time 60 min, solid-liq. ratio 1:20, static absorption-elution test, H103 resin superior to X-5 and AB-8, total flavonoid 63.98% (Bao et al. 2009); suitable technique for total flavonoids: HPLC detn. of quercetin in *C. spinosa* L. via Choromato bar: Phenmenex Lnna 5N C18 100 choromato bar (250 × 4.60 mm), mobile phase: methanol-H_3PO_4 (50:50), flow rate: 0.07 mL/min, column temp.: 30°, detection wavelength: 360 nm, quercitin ref. of std. curve f(x) = 6.1539e-007x + 0.0019 ($r = 0.999$), av. recovery 95.0%, RSD = 0.3% (Meiliwan et al. 2009); as

TABLE 5.1 (*Continued*)
Flavonoids of *Capparis* spp.

Compound	Notes
	flavones in fruits of *C. spinosa* (Zhang et al. 2007); in stems of *C. spinosa*, potent antioxidant activity by DPPH (Chen et al. 2010); in dried *C. spinosa* leaves, 6.96–2.69 mg quercetin equivalent/g (Bhoyar et al. 2011); derivatives of quercetin and kaempferol identified in MeOH extracts of *C. spinosa* "flowering buds" along with hydroxycinnamic acids (caffeic acid, ferulic acid, p-cumaric acid, and cinnamic acid) (Bonina et al. 2002)
Acacetin 7-rutinoside 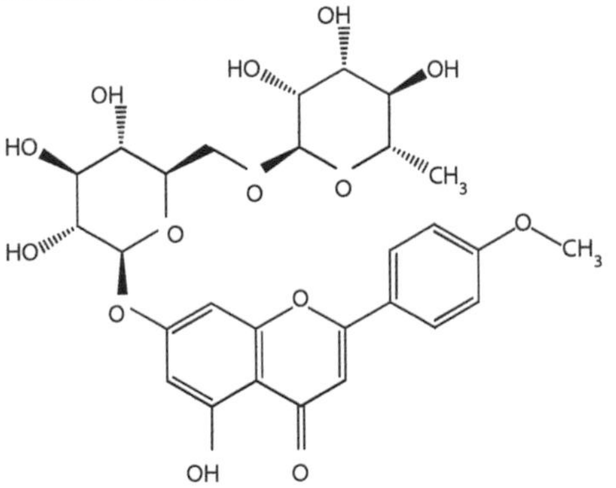	Sep. from *C. tenera* roots w/chromatog., structure confirmed w/NMR (Su et al. 2008)
Apigenin 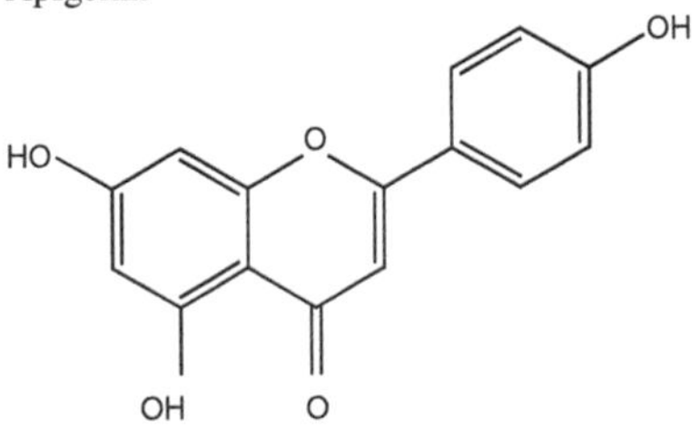	Ext. from whole plant of *C. himalayensis* (Li et al. 2008); isolated and purified from *C. spinosa* by solvent extn., silica gel, and Sephadex LH-20 gel column chromatog., recrystn., structures elucidated by ESI-MS, ^{1}H-NMR, and ^{13}C-NMR (Li et al. 2007)
Ginkgetin 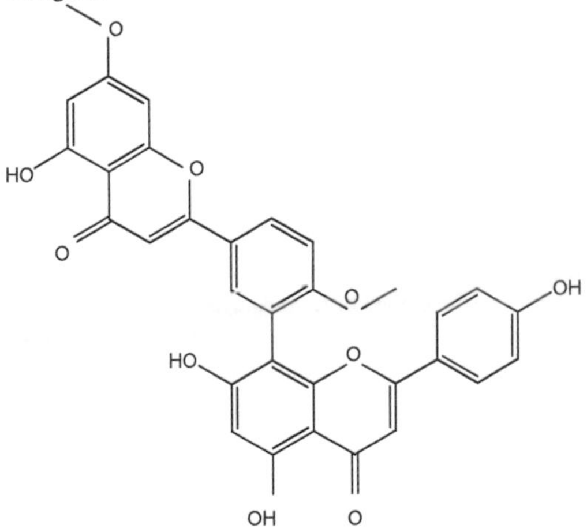	Biflavonoid isolated from fruit of *C. spinosa*, by secreted placental alk. phosphatase (SEAP) reporter assay inhibited NF-κB *in vitro* at 20 μ*M*, prev. isolated from *Gingko biloba* (Zhou et al. 2011)

(*Continued*)

TABLE 5.1 (*Continued*)
Flavonoids of *Capparis* spp.

Compound	Notes
Isoginkgetin 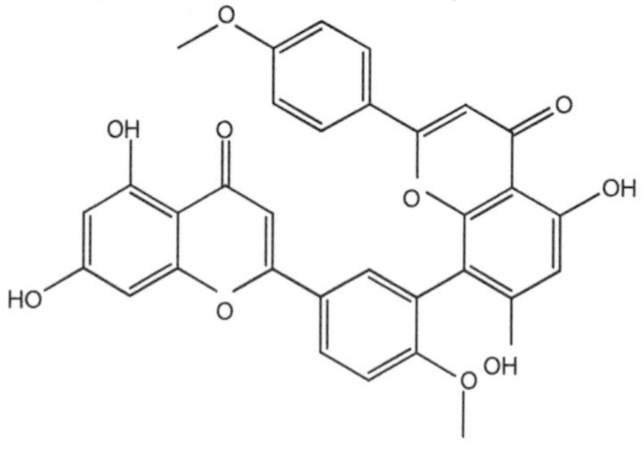	Biflavonoid isolated from fruit of *C. spinosa*, by secreted placental alk. phosphatase (SEAP) reporter assay inhibited NF-κB *in vitro* at 20 μ*M*, prev. isolated from *Gingko biloba* (Zhou et al. 2011)
Isoquercitin (Quercetin 3-O-glucoside) 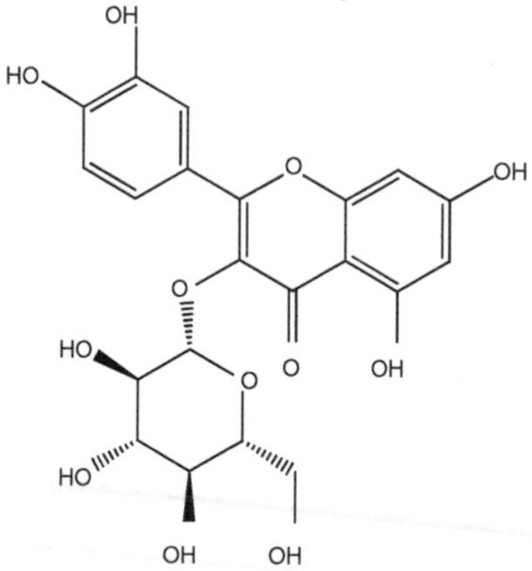	Isolated from methanol extract of aerial parts of *C. spinosa* (Sharaf et al. 2000)
Isorhamnetin 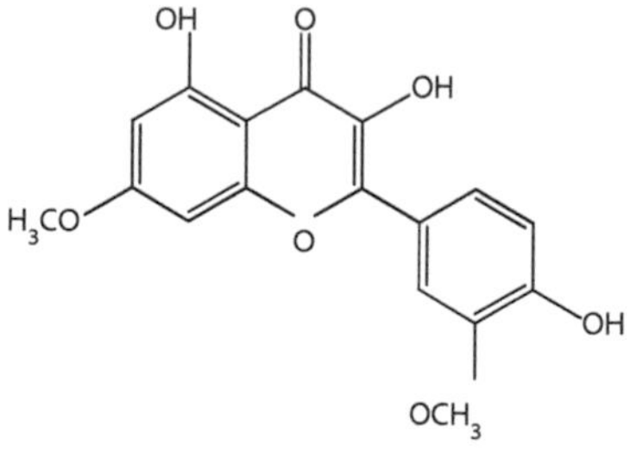	Isolated from *C. spinosa* fruits by solvent extn., column chromatog., structures identified w/physic-chem. consts., spectral data (Gan et al. 2009)
Kaempferol 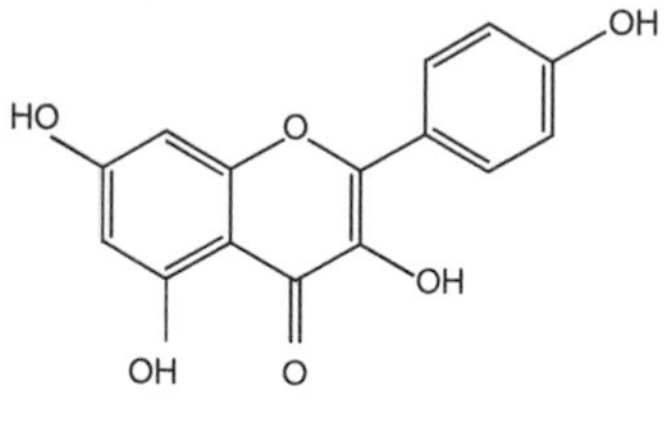	Inocencio et al. 2000, Conforti et al. 2011b, in 1 year 25% brine pickled *C. spinosa* buds (Tomas and Ferreres 1976a); isolated from whole plant of *C. himalayensis* (Li et al. 2008); isolated and purified from *C. spinosa* by solvent extn., silica gel and Sephadex LH-20 gel column chromatog., recrystn., structures elucidated by ESI-MS, ^{1}H-NMR, and ^{13}C-NMR (Li et al. 2007); in *C. spinosa* fruits (Gan et al. 2009)

TABLE 5.1 (*Continued*)
Flavonoids of *Capparis* spp.

Compound	Notes
Kaempferol 3,7-dirhamnoside 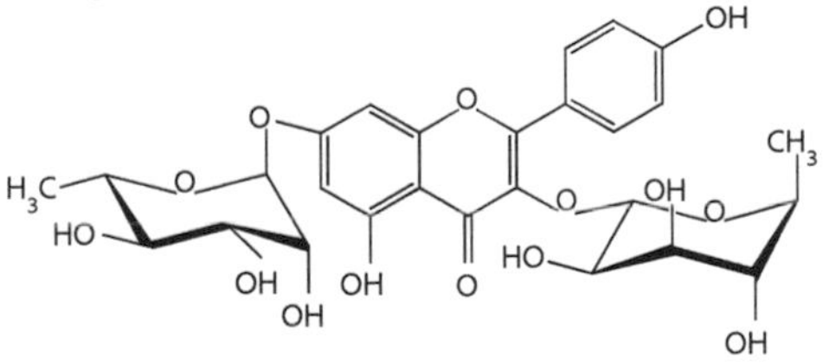	From methanol ext. *C. humilis*, detd. by comparison of NMR data with literature values (Sarragiotto et al. 2004)
Kaempferol 3-glucoside 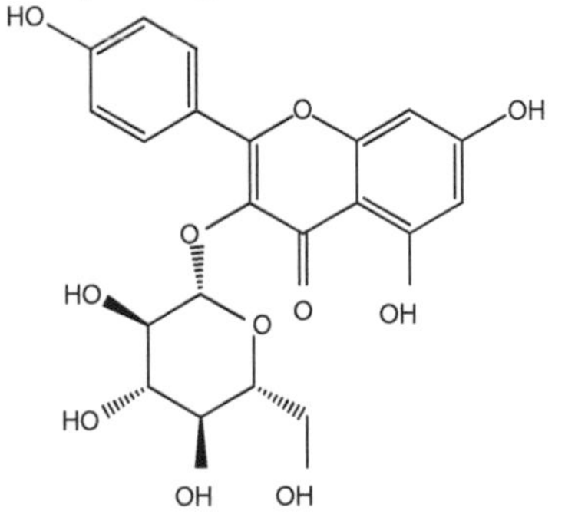	Rodrigo et al. 1992
Kaempferol-3-rhamnorutinoside	Rodrigo et al. 1992
Kaempferol 3-O-rhamnosyl-rutinoside	Conforti et al. 2011, in fresh buds of *C. spinosa* (Inocencio et al. 2000)
Kaempferol 3-O-rutinoside 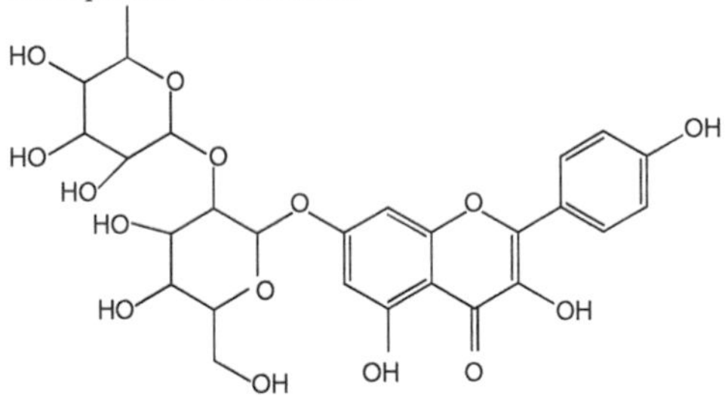	Isolated from *C. spinosa* buds by ion exclusion (Sephadex G-25) and paper chromatog., then extn. w/pet. ether, trichloroethylene, and MeOH, w/subsequent acid hydrolysis. Solvent systems for paper chromatog. were AcOH (15%); H_2O; BuOH-AcOH-H_2O (4:1:5, upper layer); PhOH-satd. H_2O; and *tert*-BuOH-AcOH-H_2O (3:1:1) (Tomas and Ferreres 1976b); Inocencio et al. 2000, Conforti et al. 2011, in fresh buds of *C. spinosa* (Inocencio et al. 2000); sepd. and purified by column (Rodrigo et al. 1992); with rutin, the most abundant flavonoid in *C. spinosa* (Giuffrida et al. 2002); sepd. and purified by column chromatog. from *C. spinosa* and structure analyzed by spectroscopic methods in fruits (Yang et al. 2010); in *C. spinosa* (Wang et al. 2009)
Luteolin 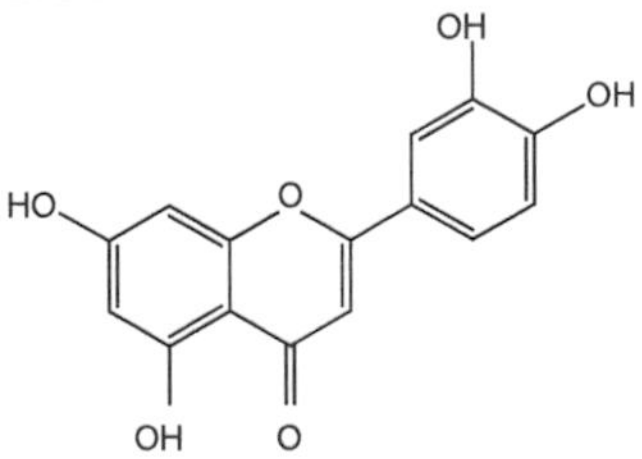	Ext. from whole plant of *C. himalayensis* (Li et al. 2008)

(*Continued*)

TABLE 5.1 (*Continued*)
Flavonoids of *Capparis* spp.

Compound	Notes
Luteolin-8-O-β-D-glucopyranoside 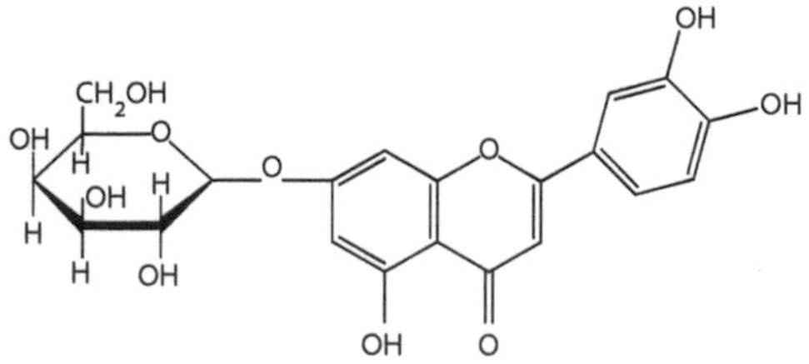	Ext. from seeds of *C. decidua* (in Hindi, *karer*) (Saxena and Goutam 2008)
Oroxylin A 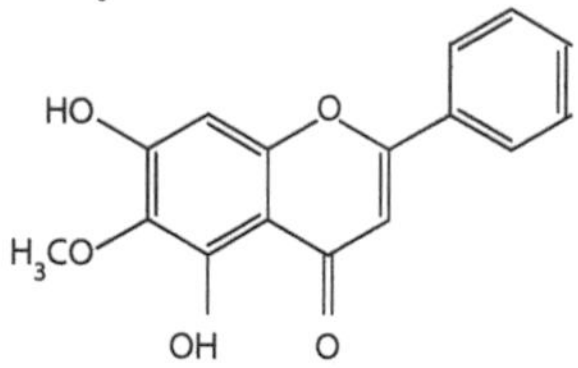	Isolated and purified from *C. spinosa* by solvent extn., silica gel and Sephadex LH-20 gel column chromatog., recrystn., structures elucidated by ESI-MS, ^{1}H-NMR and ^{13}C-NMR (Li et al. 2007)
Quercetin 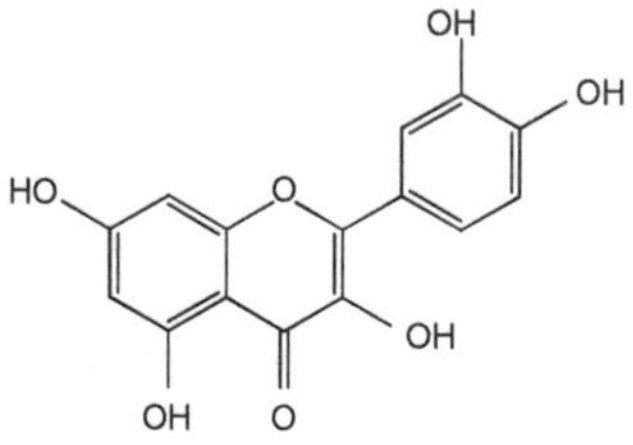	Charaux 1924, in 1 year 25% brine pickled *C. spinosa* buds (Tomas and Ferreres 1976a); isolated from whole plant of *C. himalayensis* (Li et al. 2008); HPLC detn. of quercetin in *C. spinosa* L. via Choromato bar: Phenmenex Lnna 5N C18 100 choromato bar (250 × 4.60 mm), mobile phase: methanol-H_3PO_4 (50:50), flow rate: 0.07 mL/min, column temp.: 30°, detection wavelength: 360 nm, quercitin ref. of std. curve f(x) = 6.1539e-007x + 0.0019 (r = 0.999), av. recovery 95.0%, RSD = 0.3% (Meiliwan et al. 2009); from *C. spinosa* detd. by RP-HPLC, Kromasil C18 column (250 mm × 4.6 mm, 5 μm), mobile phase CH_3OH-0.4% H_3PO_4 (47:53), flow rate 1.0 mL/min^{-1}, detection wavelength 370 nm, good linear relation between quercetin (2.0–20 μg/L^{-1}) and peak area value (r = 0.9993), av. recovery 98.8%, RSD 1.75% (n = 6) (Mamat et al. 2009)
Quercetin 7-O-β-D-glucopyranoside-β-L-rhamnopyranoside	From concd. EtOH ext. of aerial parts of *C. spinosa* treated w/n-BuOH, new flavonol glycoside, m. 178°–179°, $C_{27}H_{30}O_{16}$, identified by UV spectral anal., other methods (Artem'eva et al. 1981)
Quercetin 3-O-glucoside-7-O-rhamnoside 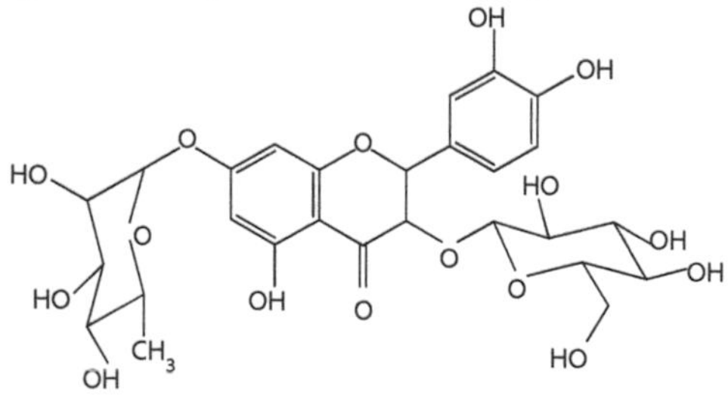	Isolated from methanol extract of aerial parts of *C. spinosa* (Sharaf et al. 2000)

TABLE 5.1 (*Continued*)
Flavonoids of *Capparis* spp.

Compound	Notes
Quercetin 3-O-[6‴-α-L-rhamnosyl-6″-β-D-glucosyl]-β-D-glucoside 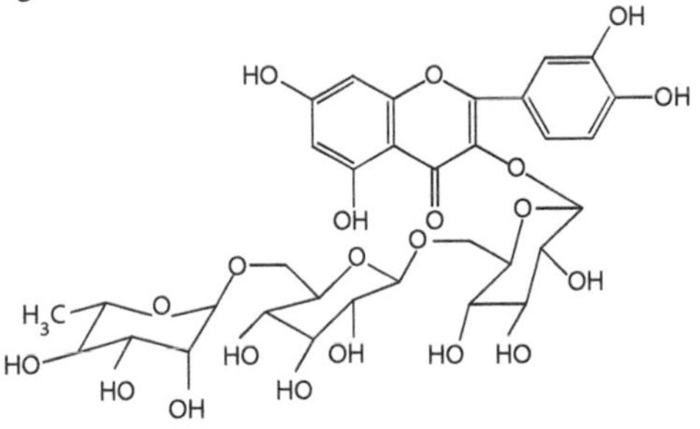	Isolated from methanol extract of aerial parts of *C. spinosa* (Sharaf et al. 2000)
Quercetin-3-rutinoside (Rutin) 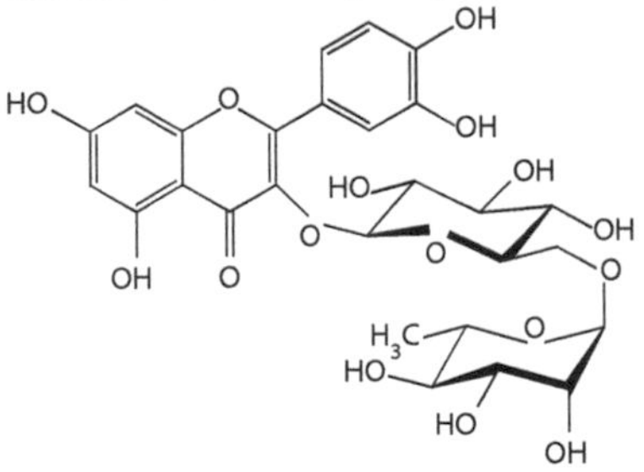	In MeOH ext. *C. spinosa* buds by HPLC (Germano et al. 2002); 13.76 mg in 8.6 g salt-stabilized *C. spinosa* buds (Tesoriere et al. 2007); *C. sicula, C. spinosa* leaf, fruit, flower, bud HPLC; fats extracted in diethyl ether, rutin ext. w/50% EtOH maceration, parts sep. by prep. TLC on silica gel plate w/butanol: acet. Ac. (4:1 vol), spots vis. w/UV (254 nm), rutin qual. comp. R(f) val. w/stand. UV/Vis spect. of sep. rutin comp. to standards w/charact. wavelengths at 260, 360 nm, purified rutin quant. by UV/Vis at 360 nm, (Ramezani et al. 2008); fruits of *C. moonii*, 0.01% of aq. ext. (Kanthamani et al. 1960); Egyptian *C. cartilaginea* and *C. deserti* (Hamed et al. 2007); dried leaves and bud of *C. spinosa* powdered and extd. w/boiling alc., and to the ext. boiling H_2O added (2x plant wt.), hot liquid filtered, aq. soln. shaken, while sl. warm, w/2x vol. Et_2O, aq. soln. sepd. by decantation—rutin sep. immediately on cooling (Charaux 1924; earlier, Brauns 1904, Schmidt et al. 1904, Wachs 1894, Wunderlich 1909); 0.10% of unsaponifiable matter in *C. spinosa* var. *aegyptia* to 1.00% in *C. ovata* var. *palaestina* (Ahmed et al. 1972); most abundant flavonoid in *C. spinosa* (Rodrigo et al. 1992); 2.1% in *C. spinosa* detd. by NMR, measuring doublet centered at 1.03 ppm (rhamnose secondary Me group); and singlets at 7.50 ppm (C-2 and C-6' arom. protons) (Khalifa et al. 1983); isolated from *C. spinosa* buds by ion exclusion (Sephadex G-25) and paper chromatog., then extn. w/pet. ether, trichloroethylene, and MeOH, w/subsequent acid hydrolysis, solvent systems for paper chromatog: AcOH (15%), H_2O, BuOH-AcOH-H_2O (4:1:5, upper layer), PhOH-satd. H_2O, and *tert*-BuOH-AcOH-H2O (3:1:1) (Tomas and Ferreres 1976b); comprises in methanol exts. of *C. spinosa* leaves 3.90% and of *C. ovata* leaves 2.32% (Tuerkoez et al. 1995); isolated from methanol extract of aerial parts of *C. spinosa* (Sharaf et al. 2000); in fresh buds of *C. spinosa* (Inocencio et al. 2000); along with kaempferol-3-rutinoside,

(*Continued*)

TABLE 5.1 (*Continued*)
Flavonoids of *Capparis* spp.

Compound	Notes
	most abundant flavonol in *C. spinosa* (Giuffrida et al. 2002); from leaf, fruit, and flower of *C. spinosa* growing wild in Khuzestan—after soxhlet extn. of fats in di-Et ether, extd. by maceration in 50% EtOH, sepd. by preparative TLC on silica gel precoated plate w/butanol: ac. Ac. (4:1, by vol.) as developing solvent, spots vis. under UV light (254 nm), qualified by compar. of its Rf value with std., UV/Vis spectrum of sepd. rutin also compared w/stds. w/characteristic wavelengths at 260 and 360 nm, calibration curve linear 0.156–2.5 μg mL^{-1} w/detection limit 0.0731 μg mL^{-1}, purity of extd. rutin from leaf, flower, and fruit detd. by HPLC, 90.41, 87.25, and 64.56%, resp., amt. in leaves, fruits, and flowers 61.09, 6.03 and 43.72 mg/100 g dried powder, resp. (Ramezani et al. 2008); by HPLC averaged on dry basis 2.76% for leaves, 1.8% for flowering buds, and 0.28% for fruits, highly variable among 24 populations in Jordan (Musallam et al. 2012); sepd. and purified from *C. spinosa* by column chromatog. and structure analyzed by spectroscopic methods (Yang et al. 2011); in *C. spinosa* fruits (Gan et al. 2009, ethanolic extract, Yang et al. 2010); in *C. spinosa* (Wang et al. 2009)
3-O-Rhamnorutinosyl kaempferol	Isolated from buds of *C. spinosa* and identified by column, paper, thin-layer, and gas chromatog. and visible, UV, and mass spectra (Tomas and Ferreres 1978)
Sakuranetin	Flavonone isolated from fruit of *C. spinosa* (Zhou et al. 2011)
Wogonin	Isolated and purified from *C. spinosa* by solvent extn., silica gel and Sephadex LH-20 gel column chromatog., recrystn., structures elucidated by ESI-MS, ^{1}H-NMR and ^{13}C-NMR (Li et al. 2007)

REFERENCES

Ahmed, Z.F., A.M. Rizk, F.M. Hammouda et al. 1972. Phytochemical investigation of Egyptian *Capparis* species. *Planta Medica* 21(2): 156–160.

Artem'eva, M.V., M.O. Karryev, A.A. Meshcheryakov, and V.P. Gordienko. 1981. [A new flavonol glycoside, quercetin-7-O-glucorhamnoside, from *Capparis spinosa* L.] *Izv. Akad. Nauk Turk. SSSR, Ser. Fiz. Tekh.* 3: 123–125.

Bao, X., R. Cao, H. Han, R. Li, and Y. Wang. 2009. Optimization of the extraction and separation of total flavonoid with macroporous resin from *Capparis spinosa* L. by orthogonal design. *Xinjiang Nongye Daxue Xuebao* 32(6): 38–41.

Bhoyar, M.S. G.P. Mishra, P.K. Naik, and R.B. Srivastava. 2011. Estimation of antioxidant activity and total phenolics among natural populations of caper (*Capparis spinosa*) leaves collected from cold arid desert of trans-Himalayas. *Australian Journal of Crop Science* 5(7): 912–919.

Bonina, F., C. Puglia, D. Ventura et al. 2002. *In vitro* antioxidant and *in vivo* photoprotective effects of a lyophilized extract of *Capparis spinosa* L buds. *Journal of Cosmetic Science* 53: 321–335.

Brauns, D.H. 1904. [The Cappernrutin, the rhamnoside of the Bluetenknospen of *Capparis spinosa*.] *Archiv der Pharmazie* 242: 556–60.

Charaux, C. 1924. [Presence of rutin in certain plants. Preparation and identification of this glucoside and of its decomposition products.] *Bulletin de la Societe de Chimie Biologique* 6: 641–647.

Chen, Y., G. Ma, and H. Bai. 2010. Antioxidative activities in vitro of total flavonoids extracted solution from *Capparis spinosa* L. stem. *Xinjiang Nongye Kexue* 47(12): 2489–2495.

Conforti, F., M.C. Marcotullio, F. Menichini, et al. 2011. The influence of collection zone on glucosinolates, polyphenols and flavonoids contents and biological profiles of *Capparis sicula* ssp. *sicula*. *Food Science and Technology International*. 17(2): 87–97.

Gan, Y., W. Chen, X. Wang, B. Han, C. Mu, H. Zhang, and Y. Zhuo. 2009. Chemical constituents of fruit of *Capparis spinosa* L. *Shihezi Daxue Xuebao, Ziran Kexueban* 27(3): 334–336.

Germano, M.P., R. De Pasquale, V. D'Angelo, S. Catania, V. Silvari, and C. Costa. 2002. Evaluation of extracts and isolated fraction from *Capparis spinosa* L. buds as an antioxidant source. *Journal of Agricultural and Food Chemistry* 50(5): 1168–1171.

Giuffrida, D., F. Salvo, M. Ziino, G. Toscano, and G. Dugo. 2002. Initial investigation on some chemical constituents of capers (*Capparis spinosa* L.) from the island of Salina. *Italian Journal of Food Science* 14(1): 25–33.

Hamed, A.R., K.A. Abdel-Shafeek, N.S. Abdel-Azim et al. 2007. Chemical Investigation of some Capparis species growing in Egypt and their antioxidant activity. *Evidence- Based and Complementary Alternative Medicine* 4(1): 25–28.

Inocencio, C., D. Rivera, F. Alcaraz, F.A. Tomás-Barberán. 2000. Flavonoid content of commercial capers (*Capparis spinosa*, *C. sicula* and *C. orientalis*) produced in Mediterranean countries. *European Food Research and Technology* 212: 70–74.

Jäger, A.K. and L. Saaby. 2011. Flavonoids and the CNS. *Molecules* 16(2): 1471–1485.

Jiang, X., J. Tang, and H. Bai. 2010. Visible spectrophotometric determination of total flavonoids in *Capparis spinosa* L. buds. *Shipin Kexue* (Beijing, China) 31(18): 252–254.

Kanthamani, S., C.R. Narayanan, and K. Venkataraman. 1960. Isolation of l-stachydrine and rutin from the fruits of *Capparis moonii*. *Journal of Scientific & Industrial Research (B)* 19B: 409–410.

Khalifa, T.I., J. Muhtadi Farid et al. 1983. PMR assay of rutin in drug plants and pharmaceuticals. *Zentralblatt für Pharmazie, Pharmakotherapie und Laboratoriumsdiagnostik* 122(8): 809–13.

Lee, E.R., G.H. Kang, and S.G. Cho. 2007. Effect of flavonoids on human health: old subjects but new challenges. *Recent Patents on Biotechnology* 1(2): 139–150.

Li, Y., Y. Feng, S. Yang, and L. Xu. 2007. Chemical components of *Capparis spinosa* L. *Zhongcaoyao* 38(4): 510–12.

Li, Y.Q., S.L. Yang, H.R. Li, and L.Z. Xu. 2008. Two new alkaloids from *Capparis himalayensis*. *Chemical and Pharmaceutical Bulletin*. 56(2): 189–191.

Magrone, T., G. Candore, C. Caruso, E. Jirillo, and V. Covelli. 2008. Polyphenols from red wine modulate immune responsiveness: biological and clinical significance. *Current Pharmaceutical Design*. 14(26): 2733–2748.

Mamat, T., Y. Nawurekez, and X. Ding. 2009. Determination of quercetin in Capparis spinosa L. for Uighur medicine by RP-HPLC. *Zhongguo Xiandai Zhongyao* 11(8): 30–31, 63.

Meiliwan, A., A. Ainiwaer, N. Haji, and Y. Wang. 2009. Content measurement of quercetin and analysis on amino acids in *Capparis spinosa* L. *Xinjiang Yike Daxue Xuebao* 32(10): 1422–1423, 1426.

Mennen, L.I., R. Walker, C. Bennetau-Pelissero, and A. Scalbert. 2005. Risks and safety of polyphenol consumption. *American Journal of Clinical Nutrition*. 81(1 Suppl): 326S–9S.

Musallam, I., M. Duwayri, R. Shibli, and F. Alali. 2012. Investigation of rutin content in different plant parts of wild caper (*Capparis spinosa* L.) populations from Jordan. *Research Journal of Medicinal Plant* 6(1): 27–36.

Ramezany, Z., N. Aghel, and H. Keyghobadi. 2008. Rutin from different parts of *Capparis spinosa* growing wild in Khuzestan/Iran. *Pakistan Journal of Biological Sciences* 11(5): 768–72.

Rodrigo, M., M.J. Lazaro, A. Alvarruiz et al. 1992. Composition of capers (*Capparis spinosa*): influence of cultivar, size and harvest date. *Journal of Food Science* 57(5): 1152–1154.

Sarragiotto, M.H., A.S. Nazari, M. Lins de Oliveira, W. Ferreira da Costa, and M. Conceicao de Souza. 2004. Proline betaine, N-methylproline, 3-carbomethyoxy-N-methylpyridinium and kaempferol 3,7-dirhamnoside from *Capparis humilis*. *Biochemical Systematics and Ecology* 32(5): 505–507.

Sawadogo, M., A.M. Tessier, and P. Delaveau. 1981. Chemical study of *Capparis corymbosa* Lam. Roots. *Plantes Medicinales et Phytotherapie* 15(4): 234–239.

Saxena, V.K. and A. Goutam. 2008. Isolation and study of the flavone glycoside luteolin-7-O-β-D-glucopyranoside from the seeds of the *Capparis decidua* (Forsk). *International Journal of Chemical Sciences* 6(1): 7–10.

Schmidt, E., H. Brauns, and N. Waliaschko. 1904. Rhamnosides. *Archiv der Pharmazie* 242: 219–224.

Sharaf, M., M.A. El-Ansari, and N.A.M. Saleh. 2000. Quercetin triglycoside from *Capparis spinosa*. *Fitoterapia* 71(1): 46–49.

Spatafora, C. and C. Tringali. 2012. Natural-derived polyphenols as potential anticancer agents. *Anticancer Agents in Medical Chemistry.* 12(8): 902–918.

Su, D.M., W.Z. Tang, S. Yu, Y. Liu, J. Qu, and D. Yu. 2008. Water-soluble constituents from roots of *Capparis tenera*. *Zhongguo Zhongyao Zazhi* 33(9): 1021–1023.

Tesoriere, L., D. Butera, C. Gentile, and M.A. Livrea. 2007. Bioactive components of caper (*Capparis spinosa* L.) from Sicily and antioxidant effects in a red meat simulated gastric digestion. *Journal of Agricultural and Food Chemistry* 55(21): 8465–8471.

Tomas, F. and F. Ferreres. 1976a. [Study of *Capparis spinosa* flavonoid fraction.] *Revista de Agroquimica y Tecnologia de Alimentos* 16(2): 252–256.

Tomas, F. and F. Ferreres. 1976b. [Flavonoid glycosides in floral buttons of *Capparis spinosa*. I.] *Revista de Agroquimica y Tecnologia de Alimentos* 16(4): 568–571.

Tomas, F. and F. Ferreres. 1978. [3-O-Rhamnorutinosyl kaempferol from floral buttons of *Capparis spinosa*. (Capparidaceae). *Revista de Agroquimica y Tecnologia de Alimentos* 18(2): 232–235.

Tuerkoez, S., G. Toker, and B. Sener. 1995. Investigation of some Turkish plants regarding of rutin. *Journal of Faculty of Pharmacy of Gazi University* 12(1): 17–21.

van Elswijk, D.A., U.P. Schobel, E.P. Lansky, H. Irth, and J. van der Greef. 2004. Rapid dereplication of estrogenic compounds in pomegranate (*Punica granatum*) using on-line biochemical detection coupled to mass spectrometry. *Phytochemistry* 65(2): 233–241.

Wachs, R. 1894. [Quercitrin and allied compounds.] *Pharmazeutische Post* 26: 529–530.

Wang, C., Z. Wang, T. Yang, X. Cheng, and J. Zhou. 2009. Extracts from Chinese medicine Capparis and its application as antiinflammatory and analgesic agents. *Chin. Pat.* Nov 26, 2008, CN 2008-10203382.

Wang, H., X. Zhao, Y. Wang, and S. Yin. 2007. [Potential toxicities of flavonoids]. *Wei Sheng Yan Jiu* 36(5): 640–2.

Wunderlich, A. 1909. [Rhamnosides of *Capparis spinosa* and *Globularia alypum*.] *Archiv der Pharmazie* 246: 256–9.

Yang, T., X. Cheng, F. Yu, G. Chou, C. Wang, and Z. Wang. 2010. The chemical constituents of *Capparis spinosa* L. fruits. *Xibei Yaoxue Zazhi* 25(4): 260–3.

Yang, T., H. Liu, X. Cheng, F. Yu, G. Chou, C. Wang, and Z. Wang. 2011. The chemical constituents from stems and leaves of *Capparis spinosa* L. *Xibei Yaoxue Zazhi* 26(1): 16–18.

Zhang, H., X. Zhu, Y. Gan, Y. Zhuo, X. Wang, and W. Chen. 2010. Effects of different extraction methods on indicator constituents and main pharmacological effects of *Capparis spinosa*. *Zhongchengyao* 32(2): 298–300.

Zhang, Y., W. Chen, X. Peng, F. Wang, B. Han, and F. Jiang. 2007. [Preliminary experiment on the chemical constituents of *Capparis spinosa* L. fruits.] *Shihezi Daxue Xuebao, Ziran Kexueban* 25(4): 481–483.

Zhou, F., A. Meiliwan, and A. Hajiakber. 2010. Preliminary study on chemical components of leaves of *Capparis spinosa* L. *Guoyi Guoyao* 21(3): 515–516.

Zhou, H.F., C. Xie, R. Jian, J. Kang, Y. Li, C.L. Zhuang, F. Yang, L.L. Zhang, L. Lai, T. Wu, and X. Wu. 2011. Biflavonoids from Caper (*Capparis spinosa* L.) fruits and their effects in inhibiting NF-κB activation. *Journal of Agricultural and Food Chemistry* 59(7): 3060–3065.

6 Glucosinolates and Isothiocyanates

Glucosinolates are small organic compounds containing sulfur (S), nitrogen (N), carbons (C), and hydrogens (H), also incorporating molecules of glucose, resembling fusion between glucose molecules and sulfur-containing amino acids. Glucosinolates are ubiquitous throughout the order Brassicales and constitute what is colloquially called "mustard oils," as in a defense against herbivores. Such "mustard oil bombs" (Hall et al. 2002) occur only in plants of Brassicales (which includes the mustard and cabbage families, Brassicaceae, and the caper family, Capparaceae) with a single exception, the genus *Drypetes* in the family Euphorbiaceae of order Malpighiales (Figure 6.1) (Johnson et al. 2009). Curiously, *Drypetes* bears a superficial resemblance to *Capparis* in the appearance of its leaves.

An alternative name for the order Brassicales is in fact Capparales, which still is sometimes used. This entire clade, which contains glucosinolates, can claim a common ancestor, one with the genetically transmitted secret of the mustard oil bomb. In a sense, then, the secret is more caper than cabbage, but the two are obviously closely related, and that aspect of the common phylogeny so well recognized is the smell of volatile sulfur compounds that give cooking and fermenting cabbage its distinctive aroma. These volatile compounds are known as isothiocyanates and are liberated when glucosinolates are separated from their sugars during cooking and fermentation. For this reason, isothiocyanates are included in this chapter, although the complex bouquet of caper involves numerous other volatile compounds, most notably the terpenoids, and these are discussed separately in Chapter 10; of course, they are all together in the caper scent and taste.

The medical and nutritional benefits of glucosinolates and isothiocyanates are now well recognized in animal models of carcinogenesis and cardiovascular and neurological diseases

FIGURE 6.1 *Drypetes* sp. (From http://en.wikipedia.org/wiki/File:Drypetes_deplanchei_Greybark_Mt_Eliza_track_Lord_Howe_Island_6June2011.jpg.)

(carcinogenesis particularly concerning the Keapl-Nrf2-ARE pathway and its cytoprotective and antioxidant cellular defense system) (Dinkova-Kostova and Kostov 2012, Boddupalli et al. 2012). Similarly, the importance of sulfur-containing organic compounds such as glucosinolates and isothiocyanates as a potent wall of plant defense against pathogens is increasingly recognized (Nwachukwu et al. 2012), highlighting another example of plant defense producing pharmaceutically important benefits for the treatment of mammals by upregulating a global antioxidant defense system, including mild agonism at the CB2 mammalian cannabinoid receptor (Gertsch et al. 2010) and potential as targeted anticancer agents in leukemia (Prashar et al. 2012).

Table 6.1 summarizes details regarding known glucosinolates and their isothiocyanates found in the *Capparis* genus. As can be seen, the table includes both widely known compounds and a few that are probably unique to *Capparis* (Figures 6.2 and 6.3).

FIGURE 6.2 All aerial parts of *Capparis spinosa* are rich in glucosinolates and isothiocyanates. (14.5.2010, Jerusalem, Israel: by Helena Paavilainen.)

FIGURE 6.3 Leaves of *Capparis flexuosa* contain several glucosinolates and isothiocyanates. (From near Grand Ton, Guadeloupe, Lesser Antilles. © William Hawthorne.)

TABLE 6.1
Glucosinolates and Isothiocyanates of *Capparis* spp.

Glucosinolates of *Capparis*

Compound	Notes
Glucosinolates in general 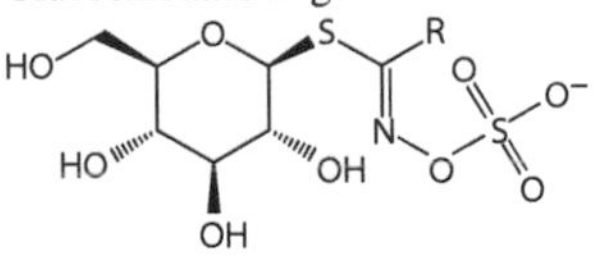 R varies	34.5–84.6 μmol/g for *C. ovata* and 42.6–88.9 μmol/g for *C. spinosa*, resp., as dry wt. (Matthäus and Özcan 2005); reviewed in *Capparis* genus (Gaind et al. 1975); vary according to size in young shoots and raw flower buds of *C. spinosa* var. *spinosa* and *C. ovata* Desf. var. *canescens* at three different sizes ($x \le 8$ mm, $8 < x \le 13$ mm, and $x > 13$ mm) by HPLC w/UV detection, harvested 8/2001 in Turkey, 12 identified in young shoots and buds of both species, range of total 6.55 μmol/g in large *C. spinosa* buds to 45.56 μmol/g in *C. ovata* young shoots (Matthäus and Özcan 2002); in *C. sicula* ssp. *sicula*, higher content in plants growing on a castle wall than in clay "calanchi," the plant's natural habitat, even though the opposite was true for flavonoids (Conforti et al. 2011)
3-Butenyl-glucosinolate 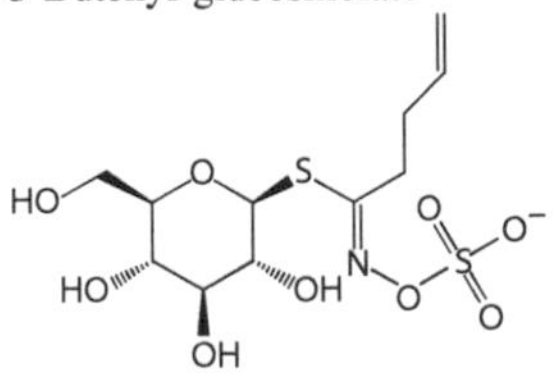	In dried leaves of *C. flexuosa* by identifying derivs. of the isothiocyanates produced by their enzymic hydrolysis (Kjaer and Schuster 1971)
Butyl-glucosinolate 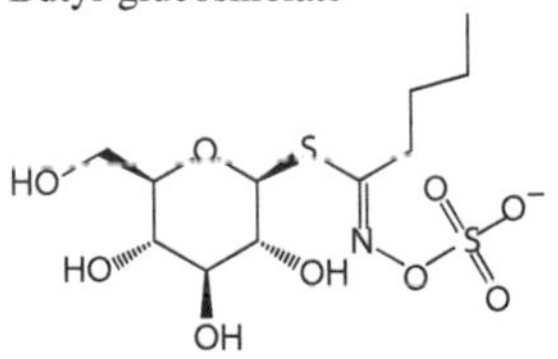	In dried leaves of *C. flexuosa* by identifying derivs. of the isothiocyanates produced by their enzymic hydrolysis (Kjaer and Schuster 1971)
Epiprogoitrin 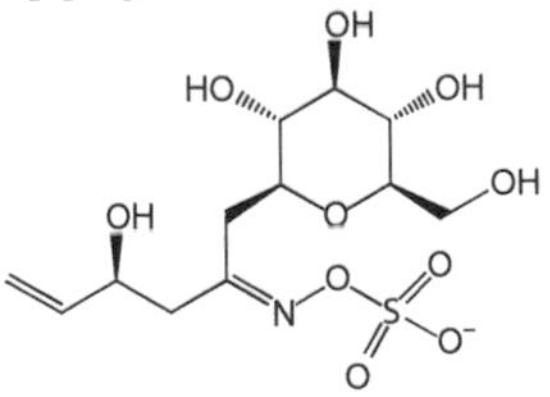	From *C. ovata*, desulfoglucosinolate, extd. and quantified from leaves, seeds, flowers, flower buds, and young shoots, 39.35 ± 0.09 and 25.56 ± 0.11 μmol g^{-1} dry wt. in seed and leaf exts. (Bor et al. 2009)
Glucobrassicin 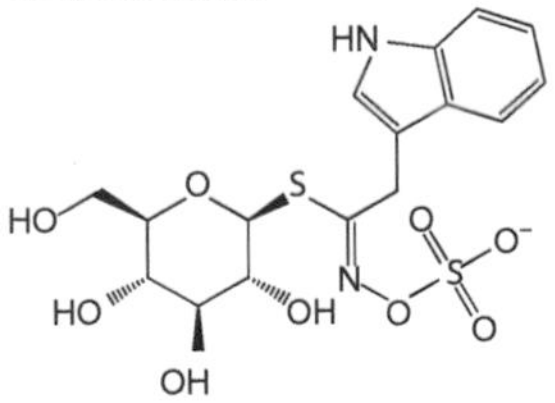	From *C. ovata*, desulfoglucosinolate, extd. and quantified from leaves, seeds, flowers, flower buds, and young shoots, 39.35 ± 0.09 and 25.56 ± 0.11 μmol g^{-1} dry wt. in seed and leaf exts. (Bor et al. 2009); in roots of *C. spinosa* detd. by HPLC and mass spec., w/respect to a 4-methoxyoxindole in roots of *C. tomentosa* (Schraudolf 1988); in varying amounts in *C. ovata* var. *palaestina*, *C. spinosa* var. *aegyptia*, and *C. spinosa* var. *deserti* (Ahmed et al. 1972)
Glucocapangalin	In varying amounts in *C. ovata* var. *palaestina*, *C. spinosa* var. *aegyptia*, and *C. spinosa* var. *deserti* (Ahmed et al. 1972)

(*Continued*)

TABLE 6.1 (*Continued*)
Glucosinolates and Isothiocyanates of *Capparis* spp.

Compound	Notes
Glucosinolates of *Capparis*	
Glucocapangulin 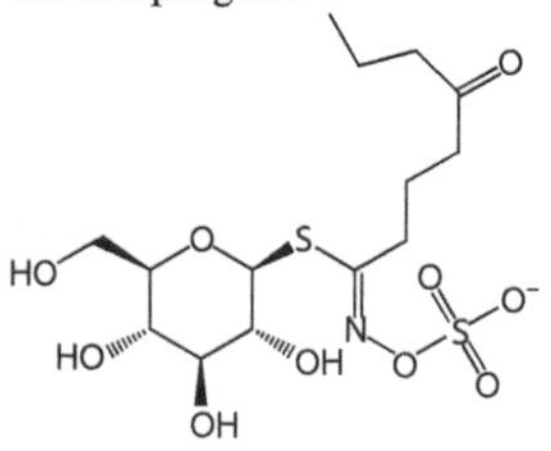	Revealed by paper chromatography as glycoside in MeOH ext. of seed embryos of *C. angulata*, purification w/acid-washed alumina and acetylation → tetraacetate as cryst. K salt, m. 168–9°, [α]22D −190.0° (H_2O), −16.5° (MeOH), hydrolysis of the cryst. acetate w/20% HCl, 50° 2.5 h → 50% 5-oxooctanoic acid, m. 32°; S-benzylthiuronium salt m. 132°; p-bromophenacyl ester m. 89°; semicarbazone m. 186°, which detd. the side-chain structure (Kjaer et al. 1960)
Glucocapparin 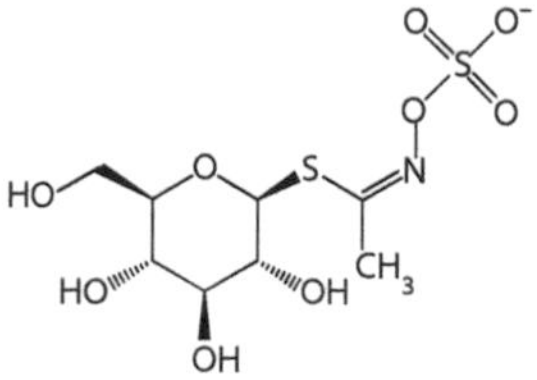	From *C. ovata*, most prominent desulfoglucosinolate, extd. and quantified from leaves, seeds, flowers, flower buds, and young shoots, 39.35 ± 0.09 and 25.56 ± 0.11 μmol g^{-1} dry wt. in seed and leaf exts. (Bor et al. 2009); quantified in diff. parts of *C. spinosa* w/HPLC, Phnomenex ODS (4.6 × 250 mm, 5 μm) column, mobile phase: methanol-0.002 mol/L^{-1} KCl soln. (15:85), flow rate 0.8 mL/min^{-1}, detection wavelength 224 nm, linear range 1.96–58.8 μg/mL^{-1}, linear relation (r = 0.9999), Y = 13685.4X - 109.1, av. recovery 96.3% (RSD = 2.0%), column temp. 25° (Meiliwan et al. 2008a,b); > 95% of total glucosinolates in *C. ovata* and *C. spinosa* (Matthäus and Özcan 2005); from seeds of *C. decidua*, sepd. from pericarp, defatted by refluxing w/petrol. ether, then extd. 3x w/70% aq. MeOH, ext. concd. and chromatographed using various solvent systems (Juneja et al. 1970); in varying amounts in *C. ovata* var. *palaestina*, *C. spinosa* var. *aegyptia*, and *C. spinosa* var. *deserti* (Ahmed et al. 1972); ~ 90% of total glucosinolates, varied according to bud size, in *C. spinosa* var. *spinosa* and *C. ovata* Desf. var. *canescens*, most variability in *C. spinosa* buds (Matthäus and Özcan 2002)
Glucocappasalin 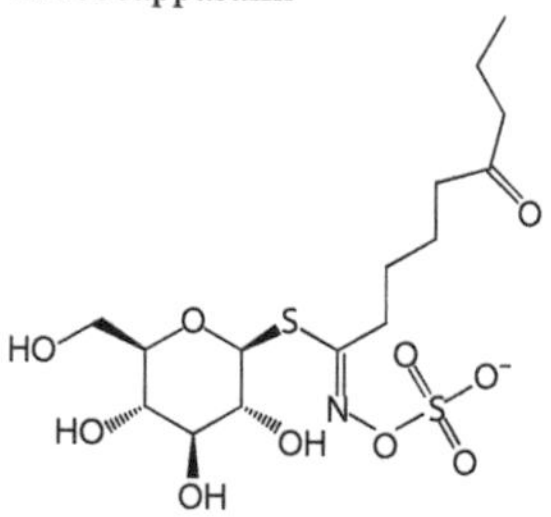	Paper chromatography of *C. salicifolia* seed, tree indigenous to Chaco region of Argentina, ion-exchange purification and acetylation yielded the tetraacetate as cryst. K salt, m. 133–6° (decompn.), IR → oxo group, analysis → $C_{23}H_{34}O_{14}NS_2K$, 0.5 H_2O for the acetate w/(C_7H_{15})(CO) as side chain, NMR → unbranched C8-chain contg. CO– in γ position vis-à-vis terminal Me, therefore higher homolog of glucocapangulin (Kjaer and Thomsen 1962)
Glucocleomin 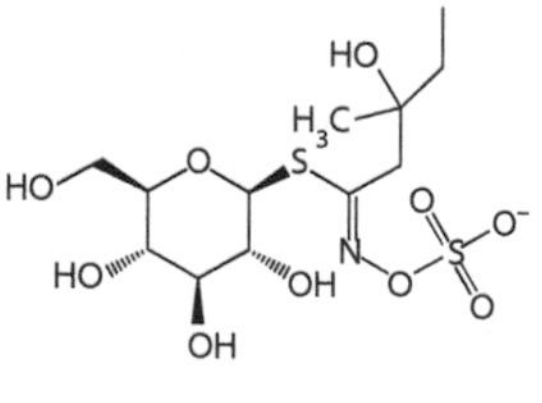	In varying amounts in *C. ovata* var. *palaestina*, *C. spinosa* var. *aegyptia*, and *C. spinosa* var. *deserti* (Ahmed et al. 1972)

TABLE 6.1 (*Continued*)
Glucosinolates and Isothiocyanates of *Capparis* spp.

Glucosinolates of *Capparis*	
Compound	**Notes**
Glucoiberin 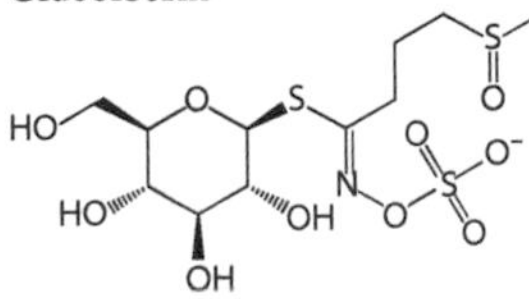	From *C. ovata*, desulfoglucosinolate, extd. and quantified from leaves, seeds, flowers, flower buds, and young shoots, 39.35 ± 0.09 and 25.56 ± 0.11 μmol g^{-1} dry wt. in seed and leaf exts. (Bor et al. 2009); in roots of *C. spinosa* detd. by HPLC and mass spec., w/ respect to a 4-methoxyoxindole in roots of *C. tomentosa* (Schraudolf 1988); in varying amounts in *C. ovata* var. *palaestina*, *C. spinosa* var. *aegyptia*, and *C. spinosa* var. *deserti* (Ahmed et al. 1972)
Gluconapin	From *C. ovata*, desulfoglucosinolate, extd. and quantified from leaves, seeds, flowers, flower buds, and young shoots, 39.35 ± 0.09 and 25.56 ± 0.11 μmol g^{-1} dry wt. in seed and leaf exts. (Bor et al. 2009)
Glucosinalbin	From *C. ovata*, desulfoglucosinolate, extd. and quantified from leaves, seeds, flowers, flower buds, and young shoots, 39.35 ± 0.09 and 25.56 ± 0.11 μmol g^{-1} dry wt. in seed and leaf exts. (Bor et al. 2009)
2-Hydroxy-3-butenylglucosinolate	In dried leaves of *C. flexuosa* by identifying derivs. of the isothiocyanates produced by their enzymic hydrolysis (Kjaer and Schuster 1971)
3-Hydroxybutyl-glucosinolate	In dried leaves of *C. flexuosa* by identifying derivs. of the isothiocyanates produced by their enzymic hydrolysis (Kjaer and Schuster 1971)
4-Hydroxybutyl-glucosinolate 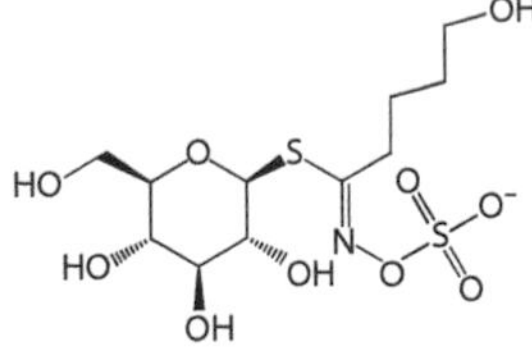	In dried leaves of *C. flexuosa* by identifying derivs. of the isothiocyanates produced by their enzymic hydrolysis (Kjaer and Schuster 1971)

(*Continued*)

TABLE 6.1 (*Continued*)
Glucosinolates and Isothiocyanates of *Capparis* spp.

Glucosinolates of *Capparis*	
Compound	**Notes**
2-Hydroxyethyl-glucosinolate 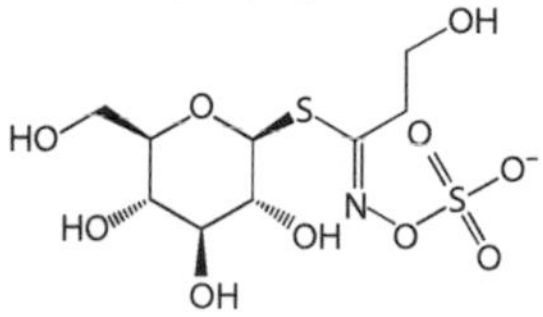	From *C. masaikai* seeds, chem. degradative, spectroscopic studies → isolation and characterization, simplest glucosinolate to spontaneously cyclize on rx. w/thioglucoside glucohydrolase (EC 3.2.3.1, myrosinase), → oxazolidine-2-thione at pH 6.0, 3% of dry seed meal (Hu 1988, Hu et al. 1989)
4-Methoxyglucobrassicin	In roots of *C. spinosa* detd. by HPLC and mass spec., w/respect to a 4-methoxyoxindole in roots of *C. tomentosa* (Schraudolf 1988)
3-Methyl-3-butenylglucosinolate 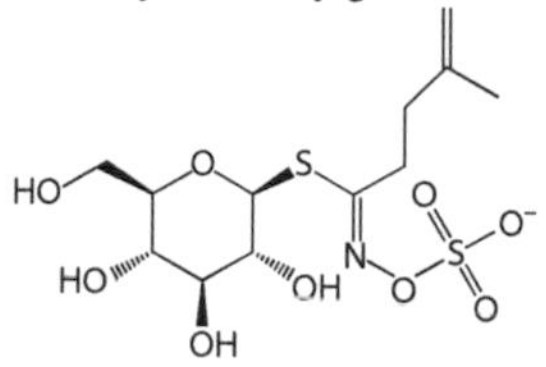	By paper chromatography of 70% MeOH ext. of dried leaf material of *C. linearis* → only one thioglucoside, 3-methyl-3-butenylglucosinolate, acetylation of the K salt → tetraacetate, which crystd. as the tetramethylammonium salt hemihydrate m. 168–71° decompn., [α]24D −22° (c 0.3, H_2O), confirmed by IR, NMR, MS (Kjaer and Wagnieres 1965)
Neoglucobrassicin 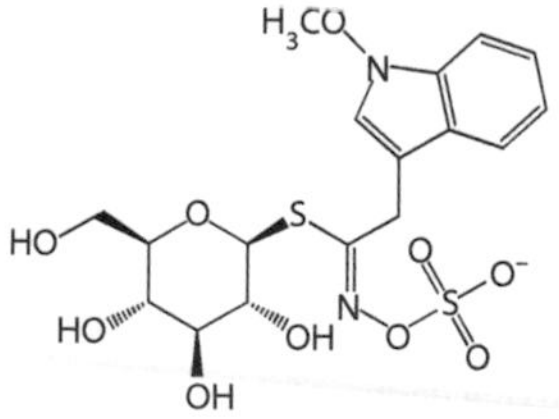	In roots of *C. spinosa* detd. by HPLC and mass spec., w/respect to a 4-methoxyoxindole in roots of *C. tomentosa* (Schraudolf 1988), in varying amounts in *C. ovata* var. *palaestina*, *C. spinosa* var. *aegyptia*, and *C. spinosa* var. *deserti* (Ahmed et al. 1972)
Progoitrin 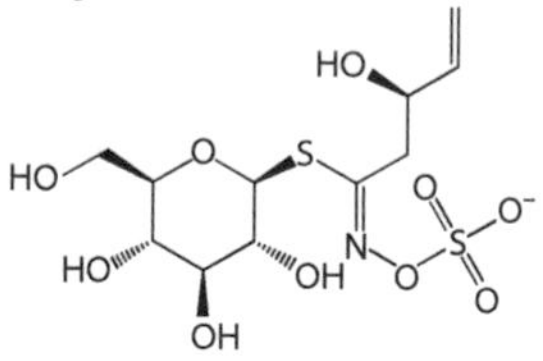	From *C. ovata*, desulfoglucosinolate, extd. and quantified from leaves, seeds, flowers, flower buds, and young shoots, 39.35 ± 0.09 and 25.56 ± 0.11 μmol g^{-1} dry wt. in seed and leaf exts. (Bor et al. 2009)
Sinigrin 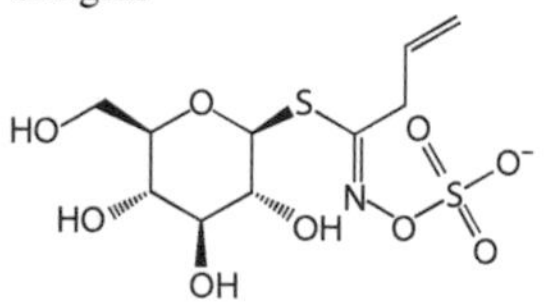	From *C. ovata*, desulfoglucosinolate, extd. and quantified from leaves, seeds, flowers, flower buds, and young shoots, 39.35 ± 0.09 and 25.56 ± 0.11 μmol g^{-1} dry wt. in seed and leaf exts. (Bor et al. 2009); in roots of *C. spinosa* detd. by HPLC and mass spec., w/respect to a 4-methoxyoxindole in roots of *C. tomentosa* (Schraudolf 1988); in varying amounts in *C. ovata* var. *palaestina*, *C. spinosa* var. *aegyptia*, and *C. spinosa* var. *deserti* (Ahmed et al. 1972)

TABLE 6.1 (*Continued*)
Glucosinolates and Isothiocyanates of *Capparis* spp.

Isothiocyanates of *Capparis*

Compound	Notes
Isothiocyanates in general R–N=C=S R varies	Comprise 83.14% of compounds, by GC-MS, in naptha of *C. spinosa* fruits (Xie et al. 2007)
Butyl-isothiocyanate	From essential oil of *C. flexuosa* from Brazil, structure by IR, MS, NMR spectroscopy, chem. derivatization (Gramosa et al. 1997); 6% of *C. spinosa* leaf oil and a main component in *C. spinosa* fresh fruit and root oils (Afsharypuor et al. 1998)
Isopropyl isothiocyanate	11% of *C. spinosa* leaf oil and a principal component in fresh fruits and roots (Afsharypuor et al. 1998)
3-Methyl-3-butenenyl-isothiocyanate	From essential oil of *C. flexuosa* from Brazil, structure by IR, MS, NMR spectroscopy, chem. derivatization (Gramosa et al. 1997)
3-Methyl-3-butenylglucosinolate	From myrosinase-catalyzed hydrolysis of 3-methyl-3-butenylglucosinolate in *C. linearis* leaf, n22D 1.519, confirmed by IR, NMR, MS (Kjaer and Wagnieres 1965)
Methyl-isothiocyanate H_3C–N=C=S	Comprises 4.5% of volatile compounds in caper bud oil, obtained by steam distillation followed by solvent extraction, and 20% in leaves of *C. ovata* Desf. var. *canescens* (El-Ghorab et al. 2007); obtained by enzymatic hydrolysis of MeOH extract of *C. decidua* seeds, potently antibacterial (Juneja et al. 1970), a main component in *C. spinosa* fresh fruits and roots (Afsharypuor et al. 1998)
4,5,6,7-Tetrahydroxydecyl isothiocyanate	Isol. from *C. grandis* roots, breakdown product of parent glucosinolate (Gaind et al. 1975)

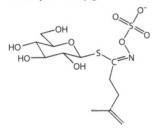

REFERENCES

Afsharypuor, S., K. Jeiran, and A.A. Jazy. 1998. First investigation of the flavor profiles of the leaf, ripe fruit and root of *Capparis spinosa* var. *mucronifolia* from Iran. *Pharmaceutica Acta Helvetiae* 72(5): 307–309.

Ahmed, Z.F., A.M. Rizk, F.M. Hammouda, and M.M.S. El-Nasr. 1972. Glucosinolates of Egyptian *Capparis* species. *Phytochemistry* 11(1): 251–256.

Boddupalli, S., J.R. Mein, S. Lakkanna, and D.R. James. 2012. Induction of phase 2 antioxidant enzymes by broccoli sulforaphane: perspectives in maintaining the antioxidant activity of vitamins A, C, and E. *Front Genet* 3: 7.

Bor, M., O. Ozkur, F. Ozdemir, and I. Turkan. 2009. Identification and characterization of the glucosinolate-myrosinase system in caper (*Capparis ovata* Desf.). *Plant Molecular Biology Reporter* 27(4): 518–525.

Conforti, F., M.C. Marcotullio, F. Menichini, G.A. Statti, L. Vannutelli, G. Burini, F. Menichini, and M. Curini. 2011. The influence of collection zone on glucosinolates, polyphenols and flavonoids contents and biological profiles of *Capparis sicula* ssp. *sicula*. *Food Science and Technology International* 17(2): 87–97.

Dinkova-Kostova, A.T. and R.V. Kostov. 2012. Glucosinolates and isothiocyanates in health and disease. *Trends in Molecular Medicine* 18(6): 337–347.

El-Ghorab, A., T. Shibamoto, and M. Özcan. 2007. Chemical composition and antioxidant activities of buds and leaves of capers (*Capparis ovata* Desf. var. *canescens*) cultivated in Turkey. *Journal of Essential Oil Research* 19(1): 72–77.

Gaind, K.N., K.S. Gandhi, T.R. Juneja, A. Kjaer, and B.J. Nielsen. 1975. 4,5,6,7-Tetrahydroxydecyl isothiocyanate derived from a glucosinolate in *Capparis grandis*. *Phytochemistry* 14(5–6): 1415–8.

Gertsch, J., R.G. Pertwee, and V. Di Marzo. 2010. Phytocannabinoids beyond the *Cannabis* plant—do they exist? *Br J Pharmacol* 160(3): 523–9.

Gramosa, N.V., T.L.G. Lemos, and R. Braz-Filho. 1997. Volatile constituents isolated from *Capparis flexuosa* of Brazil. *Journal of Essential Oil Research* 9(6): 709–712.

Hall, J.C., K.J. Sytsma, and H.H. Iltis. 2002. Phylogeny of Capparaceae and Brassicaceae based on chloroplast sequence data. *American Journal of Botany* 89(11): 1826–1842.

Hu, Z. 1988. [The glucosinolates and thioglucosidase of *Capparis masaikai* seeds.] *Yunnan Zhi Wu Yan Jiu* 10(2): 167–174.

Hu, Z., J.A. Lewis, A.B. Hanley, and G.R. Fenwick. 1989. 2-Hydroxyethylglucosinolate from *Capparis masaikai* of Chinese origin. *Phytochemistry* 28(4): 252–254.

Johnson, S.D., M.E. Griffiths, C.I. Peter, and M.J. Lawes. 2009. Pollinators, "mustard oil" volatiles, and fruit production in flowers of the dioecious tree *Drypetes natalensis* (Putranjivaceae). *American Journal of Botany* 96(11): 2080–2086.

Juneja, T.R., K.N. Gaind, and A.S. Panesar. 1970. *Capparis decidua*: Study of isothiocyanate glucoside. *Research Bulletin of the Panjab University: Science* 21(3–4): 519–521.

Kjaer, A. and A. Schuster. 1971. Glucosinolates in *Capparis flexuosa* of Jamaican origin. *Phytochemistry* 10(12): 3155–3160.

Kjaer, A. and H. Thomsen. 1962. Isothiocyanates. XLVI. Glucocappasalin, a new naturally occurring 1-thioglucoside. *Acta Chemica Scandinavica* 16: 2065–2066.

Kjaer, A., H. Thomsen, and S.E. Hansen. 1960. Isothiocyanates. XXXVIII. Glucocapangulin, a novel isothiocyanate-producing glucoside. *Acta Chemica Scandinavica* 14(5): 1226–1227.

Kjaer, A. and W. Wagnieres. 1965. Isothiocyanates. LIII. 3-Methyl-3-butenylglucosinolate, a new isothiocyanate-producing thioglucoside. *Acta Chemica Scandinavica* 19(8): 1989–1991.

Matthäus, B. and M. Özcan. 2002. Glucosinolate composition of young shoots and flower buds of capers (*Capparis* species) growing wild in Turkey. *Journal of Agricultural and Food Chemistry* 50(25): 7323–7325.

Matthäus, B. and M. Özcan. 2005. Glucosinolates and fatty acid, sterol, and tocopherol composition of seed oils from *Capparis spinosa* var. *spinosa* and *Capparis ovata* Desf. var. *canescens* (Coss.) Heywood. *Journal of Agricultural and Food Chemistry* 53(18): 7136–7141.

Meiliwan, A., L. Jiang, and H. Aisa. 2008a. Determination of glucocapparin in *Capparis spinosa* L. *Shizhen Guoyi Guoyao* 19(9): 2084–2085.

Meiliwan, A., L. Jiang, and A. Ajiaikebaier. 2008b. HPLC method for assaying glucocapparin in *Capparis spinosa*. *Zhongguo Zhong Yao Za Zhi* 33(9): 1092–1093.

Nwachukwu, I.D., A.J. Slusarenko, and M.C. Gruhlke. 2012. Sulfur and sulfur compounds in plant defence. *Natural Product Communications* 7(3): 395–400.

Prashar, A., F. Siddiqui, and A.K. Singh. 2012. Synthetic and green vegetable isothiocyanates target red blood leukemia cancers. *Fitoterapia* 83(2): 255–265.

Schraudolf, H. 1988. Indole glucosinolates of *Capparis spinosa*. *Phytochemistry* 28(1): 259–60.

Xie, L., D. Ma, S. Xue, and C. Tian. 2007. Study on naphtha and fatty acid by GC-MS in *Capparis spinosa* L. fruit. *Shipin Kexue* 28(5): 262–264.

7 Minerals

In ancient and medieval systems of plant alchemy, three aspects of plants are considered. The first is the plant "sulfur," defined as the masculine solar principle, manifest in the active extract of the specific plant material. The plant "mercury" is common to all plants and is considered interchangeable among them; this is the passive female lunar alcohol, which is achieved by fermentation (*rubedo*) and distillation (sublimation). The third aspect contains both Martian and Venusian elements and, unlike the mercurial lunar aspect, is specific to the plant, although not with the same specificity as the sulfur solar aspect. This third aspect of a plant's nature, which in a sense represents the "core" of a plant, is the plant's minerals, defined as what is left of the plant after all the extractions and fermentations have been completed, all the alcohol drained off, and the remaining skeleton of the plant consumed by fire and its ashes placed in a 1000°C calcination furnace for a prescribed period of multiple hours. The ensuing powder constitutes the plant's minerals and is typically recombined in plant alchemy with the plant's or other plants' spirits and their plant mercury and consumed together as containing the essence of the plant, with the minerals considered more sublime and "elemental" than the active sulfur aspect coveted in medicine and in medicinal chemistry.

We now have extremely accurate means for identifying different minerals in these calcinates or in other preparations from plants. We also have information concerning the different elements found in the plant world and in the soil in which the plants grow and the importance of many of these elements to human, animal, and plant metabolism. Chromium, for example, is known to reduce insulin resistance and thus benefit diabetes and metabolic syndrome (Hua et al. 2012). Selenium (Figure 7.1) and zinc may benefit immunity (Siegfried et al. 2012). Selenium is incorporated into selenoproteins having far-reaching pleiotropic, antioxidant, and anti-inflammatory actions and promoting production of active thyroid hormone (Rayman 2012); both elements may be included along with antioxidant compounds as micronutrient therapy for victims of trauma (Reddell and

FIGURE 7.1 Buds of *Capparis spinosa* are rich in selenium. (20.5.2010, Ein Karem, Jerusalem, Israel: by Helena Paavilainen.)

Cotton 2012). The modern version of the ancient alchemists' preoccupation with inorganic elements as a means of treating human disease is borne out by explorations into the creation of complexes between pharmacologically active organic molecules and zinc, copper, or manganese to create a new generation of clinically useful metallopharmaceutics (Yoshikawa and Yasui 2012) for treating "incurable" diseases, including diabetes mellitus.

In Table 7.1., certain elements such as carbon, chlorine, and oxygen, not usually considered minerals, are also included. This anomaly owes to the inclusion of these compounds in the exposition of research of particular groups studying minerals in *Capparis*. Also, note that the *Capparis* species, like all plants, draw minerals from the soil. Thus, the mineral content of plants from different areas

TABLE 7.1
Minerals of *Capparis* spp.

Mineral	Notes
Ag Silver	In *C. spinosa*, in higher concentration than in underlying soil (Lezhneva 1974)
Al Aluminum	Found in *C. zeylanica* using energy dispersive X-ray spectroscopy (EDX) (Padhan et al. 2010); 14.91–118.81 mg/kg detd. in seed oils of *C. ovata* Desf. var. *canescens* (Coss.) Heywood and *C. spinosa* var. *spinosa* by inductively coupled plasma atomic emission spectrometry (ICP-AES) (Özcan 2008)
Ba Barium	Traces found in organs of *C. ovata* Desf. var. *canescens* (Coss.) Heywood, detd. by ICP-AES (Özcan 2005)
C Carbon	Found in *C. zeylanica* using EDX (Padhan et al. 2010)
Ca Calcium	Found in *C. zeylanica* using EDX (Padhan et al. 2010); high amount in all *C. ovata* parts by ICP-AES (Ogut and Er 2010); 1.04–76.39 mg/kg detd. in seed oils of *C. ovata* Desf. var. *canescens* (Coss.) Heywood and *C. spinosa* var. *spinosa* by ICP-AESD (Özcan 2008); highest levels, 598.34–16,947.1 ppm, in buds > of young shoots, flowers, caper berries (fruit), and seeds of *C. ovata* Desf. var. *canescens* (Coss.) Heywood, detd. by IVP-AES (Özcan 2005); 90 mg/100 g in unripe fruit (*ker*) of *C. decidua* (Chauhan et al. 1986)
Cd Cadmium	Very low amount in all *C. ovata* parts by ICP-AES (Ogut and Er 2010); traces in organs of *C. ovata* Desf. var. *canescens* (Coss.) Heywood, detd. by ICP-AES (Özcan 2005)
Cl Chlorine	Found in *C. zeylanica* using EDX (Padhan et al. 2010)
Cr Chromium	Very low amount in all *C. ovata* parts by ICP-AES (Ogut and Er 2010); among the most plentiful trace elements found in *C. spinosa* seeds by ICP-AES (Meiliwan et al. 2009); traces in organs of *C. ovata* Desf. var. *canescens* (Coss.) Heywood, detd. by ICP-AES (Özcan 2005)
Cu Copper	Found in *C. zeylanica* using EDX (Padhan et al. 2010); among the most plentiful trace elements found in *C. spinosa* seeds by ICP-AES (Meiliwan et al. 2009); traces in organs of *C. ovata* Desf. var. *canescens* (Coss.) Heywood, detd. by ICP-AES (Özcan 2005); 1.1 mg/100 g in unripe fruit (*ker*) of *C. decidua* (Chauhan et al. 1986); in *C. spinosa*, in higher concentration than in underlying soil (Lezhneva 1974)
Fe Iron	Found in *C. zeylanica* using EDX (Padhan et al. 2010); among the most plentiful trace elements found in *C. spinosa* seeds by ICP-AES (Meiliwan et al. 2009); 78.83–298.14 mg/kg detd. in seed oils of *C. ovata* Desf. var. *canescens* (Coss.) Heywood and *C. spinosa* var. *spinosa* by ICP-AES (Özcan 2008); 3.5 mg/100 g in unripe fruit (*ker*) of *C. decidua* (Chauhan et al. 1986)
K Potassium	Found in *C. zeylanica* using EDX (Padhan et al. 2010); 4,057.1 to 37,368.1 ppm in all *C. ovata* parts by ICP-AES (Ogut and Er 2010); highest levels, 3,093.1–28,163.9 ppm, in buds > of young shoots, flowers, caper berries (fruit), and seeds (low of 3,093.1 ppm) of *C. ovata* Desf. var. *canescens* (Coss.) Heywood, detd. by ICP-AES (Özcan 2005)
Li Lithium	Traces found in organs of *C. ovata* Desf. var. *canescens* (Coss.) Heywood, detd. by ICP-AES (Özcan 2005)
Mg Magnesium	Found in *C. zeylanica* using EDX (Padhan et al. 2010); 19.4 ppm (middle bud) to 1,559.6 ppm (large bud) in all *C. ovata* parts by ICP-AES (Ogut and Er 2010); found in *C. spinosa* seeds by ICP-AES (Meiliwan et al. 2009); 102.15–1,655.33 mg/kg detd. in seed oils of *C. ovata* Desf. var. *canescens* (Coss.) Heywood and *C. spinosa* var. *spinosa* by ICP-AES (Özcan 2008)

TABLE 7.1 (*Continued*)
Minerals of *Capparis* spp.

Mineral	Notes
Mn Manganese	Found in *C. spinosa* seeds by ICP-AES (Meiliwan et al. 2009); and 1.9 mg/100 g in unripe fruit (*ker*) of *C. decidua* (Chauhan et al. 1986)
Mo Molybdenum	In *C. spinosa*, in higher concentration than in underlying soil (Lezhneva 1974)
Na Sodium	High amount in all *C. ovata* parts by ICP-AES (Ogut and Er 2010); 505.78–4,489.51 mg/kg detd. in seed oils of *C. ovata* Desf. var. *canescens* (Coss.) Heywood and *C. spinosa* var. *spinosa* by ICP-AES (Özcan 2008); highest levels, 57.9–444.3 ppm, in flower buds > of young shoots, flowers, caper berries (fruit), and seeds of *C. ovata* Desf. var. *canescens* (Coss.) Heywood, detd. by ICP-AES (Özcan 2005)
Ni Nickel	Traces found in organs of *C. ovata* Desf. var. *canescens* (Coss.) Heywood, detd. by ICP-AES (Özcan 2005)
O Oxygen	Found in *C. zeylanica* using EDX (Padhan et al. 2010)
P Phosphorus	High amount in all *C. ovata* parts by ICP-AES (Ogut and Er 2010); 1,489–11,524 mg/kg detd. in seed oils of *C. ovata* Desf. var. *canescens* (Coss.) Heywood and *C. spinosa* var. *spinosa* by ICP-AES (Özcan 2008); highest levels, 1,690.5–4,153.9 ppm, in flower buds > of young shoots, flowers, caper berries (fruit), and seeds of *C. ovata* Desf. var. *canescens* (Coss.) Heywood, detd. by ICP-AES (Özcan 2005); 179 mg/100 g in unripe fruit (*ker*) of *C. decidua* (Chauhan et al. 1986)
Pb Lead	Very low amount in all *C. ovata* parts by ICP-AES (Ogut and Er 2010); traces in organs of *C. ovata* Desf. var. *canescens* (Coss.) Heywood, detd. by ICP-AES (Özcan 2005); in *C. spinosa*, in higher concentration than in underlying soil (Lezhneva 1974)
S Sulfur	In S8 form, first notice of elemental S in plant source, by GC-MS in *C. spinosa* bud (Brevard et al. 1992)
Se Selenium	Traces found in organs of *C. ovata* Desf. var. *canescens* (Coss.) Heywood, detd. by ICP-AES (Özcan 2005); highest amount in *C. spinosa* buds (0.8 µg/g dry wt) relative to cress (*Lepidium sativum*) 0.25 and dandelion (*Taraxacum dens-leonis*) 0.15 (Herrero Latorre et al. 1987)
Si Silicon	Found in *C. zeylanica* using EDX (Padhan et al. 2010)
Sr Strontium	In *C. spinosa*, plant reacted with some anatomical and physiological changes to its uptake (Koval'skii and Zasorina 1965)
Zn Zinc	Found in *C. zeylanica* using EDX (Padhan et al. 2010); among the most plentiful trace elements found in *C. spinosa* seeds by ICP-AES (Meiliwan et al. 2009); highest levels, 21.1–35.6 ppm, in buds > of young shoots, flowers, caper berries (fruit), and seeds of *C. ovata* Desf. var. *canescens* (Coss.) Heywood, detd. by ICP-AES (Özcan 2005); 1.6 mg/100 g in unripe fruit (*ker*) of *C. decidua* (Chauhan et al. 1986)

will vary depending on the soils from those areas. Still, *Capparis* may concentrate certain elements, as it does for silver (Ag). Other minerals, such as cadmium, may be more representative of an uptake of contaminants in the soil than the plant itself.

Minerals clearly influence human physiology and influence organic processes in multiple ways, creating synergistic interactions, for example, between selenium and tocopherols (Zu and Ip 2003) for inducing apoptosis in cancer cells. In short, minerals should be a conscious consideration in the creation of all complex medical products. The plant itself, following extraction of the sulfur actives, may also prove to be the best source of these minerals, particularly from such a mineral-rich genus as *Capparis*.

REFERENCES

Brevard, H., M. Brambilla, A. Chaintreau, J.P. Marion, and H. Diserens. 1992. Occurrence of elemental sulfur in capers (*Capparis spinosa* L.) and first investigation of the flavor profile. *Flavour and Fragrance Journal* 7(6): 313–321.

Chauhan, E.M., A. Duhan, and C.M. Bhat. 1986. Nutritional value of ker (*Capparis decidua*) fruit. *Journal of Food Science and Technology* 23(2): 106–108.

Herrero Latorre, C., M. Mejuto Marti, M. Bollain Rodriguez, and F. Bermejo Martínez. 1987. Fluorimetric determination of selenium in foods. II. Vegetables. *Anales de Bromatologia* 39(1): 133–137.

Hua, Y., S. Clark, and N. Sreejayan. 2012. Molecular mechanisms of chromium in alleviating insulin resistance. *Journal of Nutritional Biochemistry* 23(4): 313–319.

Koval'skii, V.V. and E.F. Zasorina. 1965. Biogeochemical aspect of strontium. *Agrokhimiya* 4: 78–88.

Lezhneva, N.D. 1974. Distribution of some ore elements in plants in the Almalyk ore field. *Zapiski Uzbekistanskogo Otdeleniya Vsesoyuznogo Mineralogicheskogo Obshchestva* 27: 101–102.

Meiliwan, A., A. Ainiwaer, Hajinisha, and Y. Wang. 2009. The study of trace elements in *Capparis spinosa* L. *Guangpuxue Yu Guangpu Fenxi* 29(8): 2266–2267.

Ogut, M. and F. Er. 2010. Mineral contents of different parts of capers (*Capparis ovata* Desf.). *Journal of Food, Agriculture & Environment* 8(2, Pt. 1): 216–217.

Özcan, M. 2005. Mineral composition of different parts of *Capparis ovata* Desf var. *canescens* (Coss.) Heywood growing wild in Turkey. *Journal of Medicinal Food* 8(3): 405–407.

Özcan, M.M. 2008. Investigation on the mineral contents of capers (*Capparis* spp.) seed oils growing wild in Turkey. *Journal of Medicinal Food* 11(3): 596–599.

Padhan, A.R., A.K. Agrahari, A. Meher, and M.R. Mishra. 2010. Elemental analysis by energy dispersive X-ray spectroscopy (EDX) of *Capparis zeylanica* Linn. plant. *Journal of Pharmacy Research* 3(4): 669–670.

Rayman, M.P. 2012. Selenium and human health. *Lancet* 379(9822): 1256–1268.

Reddell, L. and B.A. Cotton. 2012. Antioxidants and micronutrient supplementation in trauma patients. *Current Opinion in Clinical Nutrition & Metabolic Care* 15(2): 181–187.

Siegfried, N., J.H. Irlam, M.E. Visser, and N.N. Rollins. 2012. Micronutrient supplementation in pregnant women with HIV infection. *Cochrane Database Systematic Reviews* 3: CD009755.

Yoshikawa, Y. and H. Yasui. 2012. Zinc complexes developed as metallopharmaceutics for treating diabetes mellitus based on the bio-medicinal inorganic chemistry. *Current Topics In Medicinal Chemistry* 12(3): 210–218.

Zu, K. and C. Ip. 2003. Synergy between selenium and vitamin E in apoptosis induction is associated with activation of distinctive initiator caspases in human prostate cancer cells. *Cancer Research.* 63(20): 6988–6995.

8 Proteins and Amino Acids

Depending on point of view, proteins, like any other plant fraction, may be desirable or undesirable, a coveted value or an impurity of which to be cleansed. The most coveted is the specific protein mabinlin II (Hu et al. 1998), very sweet and possessing greater thermostability than any other sweet protein (Guan et al. 2000). It was originally extracted from the seeds of *C. masakai* Levl., indigenous to southern China. There is a vast literature on this specific protein from the point of view of mass production by biofermentation/biotechnology (Faus 2000, Masuda and Kitabatake 2006); commercial potential as a natural nonsugar sweetener (Kant 2005); identification of the protein for standardization (Gnanavel and Serva Peddha 2011); regulation of the sweet protein-creating gene in situ (Hu et al. 2009); crystal structure of the protein (Li et al. 2006, 2008); cloning and sequencing of DNA (Hu et al. 1998); and chemical synthesis (Kohmura and Ariyoshi 1998). Yet, in wine making, proteins are undesirable and may be removed from wine by "fining"; the same is often true in other types of plant fermentations and extracts relevant for pharmaceutics. "Cleaning up" many fermented natural products often means eliminating proteins.

The greater part of the proteins have been found in seeds, and the details of the work on protein content and specific proteins characterized in *Capparis* are summarized in Table 8.1. There are known proteins globulin, glutelin, and albumin; a novel lectin with pronounced biological activity; a possibly genus-unique cysteine protease, proline betaine; and the amino acids aspartic and glutamic acid.

TABLE 8.1
Proteins of *Capparis* spp.

Compound	Notes
Proteins in general	18.31% of *C. spinosa* determined by semimicro Kjeldahl and grading method, 18 amino acids (AAs) identified, ~ 26% of the AAs were essential AAs (Nurmamat and Korbanjhon 2011); 27% of total storage in Tunisian *C. spinosa* seeds (Tlili et al. 2011); eval. in *C. spinosa* L. var. *spinosa* and *C. ovata* Desf. var. *canescens* seeds (Akgul and Özcan 1999); ~ 25% in seeds of *C. flexuosa* of Venezuela (Grunwald 1946); 14.88% of dry wt. of unripe *C. decidua* fruit (Chauhan et al. 1986)
Albumin	Minor component of total protein in *C. spinosa* (Nurmamat and Korbanjhon 2011); 53% of total protein in unripe *C. decidua* fruit (Chauhan et al. 1986)
Aspartic acid (structure: HO, O, OH, NH_2)	Most prevalent AA, ~ 18% of total, in *C. spinosa* (Nurmamat and Korbanjhon 2011)

(*Continued*)

TABLE 8.1 (*Continued*)
Proteins of *Capparis* spp.

Compound	Notes
Capparin (cysteine protease)	Purified w/ammonium sulfate fractionation and CM Sephadex column, optimum pH = 5.0, optimum temp. 60°, v_{max} and K_m values detd. by Lineweaver-Burk graphics 1.38 μg/(L/min) and 0.88 μg/L, resp., purifn. degree and the mol. mass (46 kDa) detd. by sodium dodecyl sulfate-polyacrylamide gel electrophoresis (SDS-PAGE) and gel filtration chromatog. (Demir et al. 2008)
3-Carbomethoxy-N-methylpyridinium 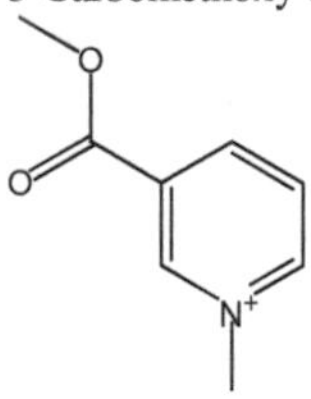	In methanol extd. dried *C. humilis*, sep. by TLC, confirmed by NMR (Sarragiotto et al. 2004)
Dimeric 62-kDa lectin w/novel N-terminal amino acid sequence	Purified from *C. spinosa* seeds, anion-exchange chromatog., cation-exchange chromatog., and finally gel filtration by FPLC on Superdex 75, ~ 100-fold purifn., stable, hemagglutinating activity at pH 1–12, up to 40°C, ↓ by D(+)-galactose, α-lactose, raffinose, and rhamnose at 1 m*M*, by 25 m*M* L(+)-arabinose, and by 100 m*M* D(+) GlcN (glucosamine), ↓ HIV-1 reverse transcriptase w/IC_{50} of 0.28 μ*M* and proliferation of hepatoma HepG2 and breast cancer MCF-7 cells w/IC_{50} of 2 μ*M*, → apoptosis in HepG2 and MCF-7 cells, weaker mitogenic activity on mouse splenocytes than ConA, ↓ mycelial growth in *Valsa mali* w/IC_{50} of 18 μ*M* (Lam et al. 2009)
Gliadin	Minor component of total protein in *C. spinosa* (Nurmamat and Korbanjhon 2011)
Globulin	42.2% of total protein in *C. spinosa* (Nurmamat and Korbanjhon 2011); 16% of total protein in unripe fruit of *C. decidua* (Chauhan et al. 1986)
Glutamic acid 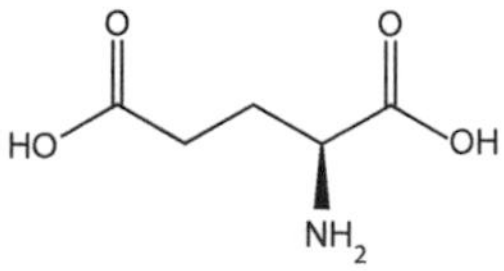	Second most prevalent AA, ~ 12% of total, in *C. spinosa* (Nurmamat and Korbanjhon 2011)
Glutelin	37.2% of total protein in *C. spinosa* (Nurmamat and Korbanjhon 2011); 12% of total protein in unripe fruit of *C. decidua* (Chauhan et al. 1986)

TABLE 8.1 (*Continued*)
Proteins of *Capparis* spp.

Compound	Notes
3-Hydroxyprolinebetaine 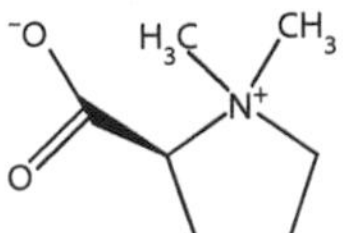	McLean et al. 1996
Mabinlin II	Sweet protein w/highest known thermostability, isol. from seeds of *C. masaikai* Levl. grown in south China, 2 crystal forms obtnd. w/hanging-drop vapor-diffusion method: 1st diffracted to 2.8 Å resoln. in space group P2, unit-cell parameters a = 50.16, b = 50.17, c = 76.60 Å, and $\beta = 99.6°$, 4 mols. per asym. unit, solvent content 35.3% (Guan et al. 2000); reviewed and compared to miraculin from *Richardella dulcifica* seeds and curculin from *Curculigo latifolia* seeds, and *Aeromonas caviae* T-64 aminopeptidase (debittering protein) (Nirasawa 2000); primers synthesized according to the 185 bp of cDNA encoding B subunit of mabinlin II, cloned per AA sequence, total mRNA extd. from *C. masaikai* seeds and purified, full-length cDNA cloned per rapid amplification of cDNA ends (RACE), amplified DNA 583 bp, coding sequence 465 bp, and encodes 155 AAs (Hu et al. 1998)
N-Methyl-proline	In methanol extd. dried *C. humilis*, sep. by TLC, confirmed by NMR (Sarragiotto et al. 2004)
Prolamin	11% of total protein in unripe fruit of *C. decidua* (Chauhan et al. 1986)
Proline betaine	McLean et al. 1996 in methanol extd. dried *C. humilis*, sep. by TLC, confirmed by NMR (Sarragiotto et al. 2004)

(*Continued*)

TABLE 8.1 (*Continued*)
Proteins of *Capparis* spp.

Compound	Notes
Protein w/N-terminal amino acid sequence similar to imidazoleglycerol phosphate synthase	From fresh *C. spinosa* seeds, anion-exchange chromatography on DEAE-cellulose, cation-exchange chromatography on SP-Sepharose, and finally gel filtration by fast protein liquid chromatography on Superdex 75, adsorbed using 20 m*M* Tris-HCl buffer (pH 7.4) and desorbed using 1 *M* NaCl in the starting buffer from the DEAE-cellulose column and SP-Sepharose column, molecular mass of 38 kDa in gel filtration and SDS-PAGE, indicating that it was monomeric, ↓ proliferation of hepatoma HepG2 cells, colon cancer HT29 cells, and breast cancer MCF-7 cells w/IC(50) ~ 1, 40, and 60 μ*M*, resp., ↓ HIV-1 reverse transcriptase w/IC_{50} of 0.23 μ*M*, ↓ mycelial growth in fungus *Valsa mali*, no hemagglutinating, ribonuclease, mitogenic, or protease inhibitory activities (Lam and Ng 2009)

REFERENCES

Akgul, A. and M. Özcan. 1999. Some compositional characteristics of capers (*Capparis* spp.) seed and oil. *Grasas y Aceites (Sevilla)* 50(1): 49–52.

Chauhan, E.M., A. Duhan, and C.M. Bhat. 1986. Nutritional value of ker (*Capparis decidua*) fruit. *Journal of Food Science and Technology* 23(2): 106–108.

Demir, Y., A.A. Gungor, E.D. Duran, and N. Demir. 2008. Cysteine protease (capparin) from capsules of caper (*Capparis spinosa*). *Food Technology and Biotechnology* 46(3): 286–291.

Faus, I. 2000. Recent developments in the characterization and biotechnological production of sweet-tasting proteins. *Applied Microbiology and Biotechnology* 53(2):145–151.

Gnanavel, M. and M. Serva Peddha. 2011. Identification of novel sweet protein for nutritional applications. *Bioinformation* 7(3): 112–114.

Grunwald, O. 1946. The oleaginous fruits of a Venezuelan Capparidacea. *Venezuela, Ministerio agr y cria, Inst exptl agr y zootec, El Valle, Bot* 3: 33.

Guan, R.J., J.M. Zheng, Z. Hu, and D.C. Wang. 2000. Crystallization and preliminary X-ray analysis of the thermostable sweet protein mabinlin II. *Acta Crystallographica*, Section D: *Biological Crystallography* D56(7): 918–919.

Hu, X., J. Guo, and X. Zheng. 1998. Cloning and sequencing of cDNA encoding sweet-tasting plant protein mabinlin II from C. masaikai. *Shengming Kexue Yanjiu* 2(3): 189–193.

Hu, X.W., S.X. Liu, J.C. Guo, J.T. Li, R.J. Duan, and S.P. Fu. 2009. Embryo and anther regulation of the mabinlin II sweet protein gene in *Capparis masaikai* Lév I. *Functional & Integrative Genomics* 9(3): 351–361.

Kant, R. 2005. Sweet proteins—potential replacement for artificial low calorie sweeteners. *Nutrition Journal* 4:5.

Kohmura, M. and Y. Ariyoshi. 1998. Chemical synthesis and characterization of the sweet protein mabinlin II. *Biopolymers* 46(4):215–223.

Lam, S.K., Q.F. Han, and T.B. Ng. 2009. Isolation and characterization of a lectin with potentially exploitable activities from caper (*Capparis spinosa*) seeds. *Bioscience Reports* 29(5): 293–299.

Lam, S.K. and T.B. Ng. 2009. A protein with antiproliferative, antifungal and HIV-1 reverse transcriptase inhibitory activities from caper (*Capparis spinosa*) seeds. *Phytomedicine* 16(5): 444–450.

Li, D.F., P. Jiang, D.Y. Zhu, Y. Hu, M. Max, and D.C. Wang. 2008. Crystal structure of mabinlin II: a novel structural type of sweet proteins and the main structural basis for its sweetness. *Journal of Structural Biology* 162(1): 50–62.

Li, D.F., D.Y. Zhu, Z. Hu, and D.C. Wang. 2006. Crystallization and preliminary X-ray analysis of the highly thermostable sweet protein mabinlin II. *Protein and Peptide Letters* 3: 319–321.

Masuda, T. and N. Kitabatake. 2006. Developments in biotechnological production of sweet proteins. *Journal of Bioscience and Bioengineering* 102(5): 375–389.

McLean, W.F.H., G. Blunden, and K. Jewers. 1996. Quaternary ammonium compounds in the Capparaceae. *Biochemical Systematics and Ecology* 24(5): 427–434.

Nirasawa, S. 2000. Sweet proteins and taste-modifying proteins. *Baiosaiensu to Indasutori* 58(4): 268–9.

Nurmamat, E. and B. Korbanjhon. 2011. Study on fat and protein components of Uygur folk medicine *Capparis spinosa* L. *Medicinal Plant: The Journal Board of Medicinal Plant, Cranston, USA* 2(3), 30–32.

Sarragiotto, M.H., A.S. Nazari, M. Lins de Oliveira, W. Ferreira da Costa, and M. Conceicao de Souza. 2004. Proline betaine, N-methylproline, 3-carbomethyoxy-N-methylpyridinium and kaempferol 3,7-dirhamnoside from *Capparis humilis*. *Biochemical Systematics and Ecology* 32(5): 505–507.

Tlili, N., T. El Guizani, N. Nasri, A. Khaldi, and S. Triki. 2011. Protein, lipid, aliphatic and triterpenic alcohol content of caper seeds "Capparis spinosa." *Journal of the American Oil Chemists' Society* 88(2): 265–270.

9 Sterols

Sterols—including cholesterol—are precursors to steroid hormones and vitamins A and D and are a subset of steroids in general (i.e., steroid alcohols). They comprise a small portion of seed oils and appear in the unsaponified fraction if the oil contacts lye. The rich fat seeds of *Capparis* fruits (Figures 9.1 and 9.2) yield as much as 15% oil or higher, and 1 or 2% of this may be sterols, which contain a polar OH element in contradistinction to the nonpolar element. Sterols have estrogenic activity documented for some of the major phytosterols, such as β-sitosterol and stigmasterol. Sterols also exert cholesterol-lowering effects in hypercholesterolemic individuals. In addition to β-sitosterol and stigmasterol, other known phytosterols, such as cholesterol, campesterol, and daucosterol, are present in seed oil; other lesser-known or even genus-unique forms may occur in other plant parts, such as the roots. A possible species-specific sterol for *C. moonii* fruits, moonisterol, has also been described.

Sterols other than cholesterol have recently gained attention as surrogate serum markers of cholesterol metabolism, and the sum of these compounds in serum depends partly on oral intake of phytosterols from plants (Mackay and Jones 2012). The nutritional value of sterols in olive oil was considered in a recent review (Ghanbari et al. 2012). Sterols may also serve as pharmaceutical matrices for transdermal preparations. In a recent study, cholesterol was found to enhance activity of anti-inflammatory moisturizers when used topically (Byun et al. 2012). Sex steroids, such as progesterone, may also be produced in plants from sterol intermediates, such as cholesterol (Janeczko 2012). Sterols β-sitosterol and stigmasterol, like other plant compounds, also are not fixed sums but vary according to the situation; for example, these compounds are increased in plants' response to stress, such as from proximity to other plants, or to microbes or viruses, suggesting a possible role in mediation of plants' intrinsic immunity (Srivastava et al. 2012). Similarly, β-sitosterol and stigmasterol may exert neuroprotective (Brimson et al. 2012) and anticancer effects in mammals (Baskar et al. 2012) by preventing lipid peroxidation and other effects for up-regulating antioxidant

FIGURE 9.1 Seed of *Capparis sola*. (9.1.2007: by C. E. Timothy Paine.)

status, as well as effects on signal transduction, immune response, and cholesterol metabolism of the host (Awad and Fink 2000). Moreover, these actions in different settings may rely on common mechanisms; further, sterol deprivation in the nematode *Chaenorhabditis elegans* results in shortening of lifespan (Cheong et al. 2011).

A summary of the sterols in *Capparis* is given in Table 9.1.

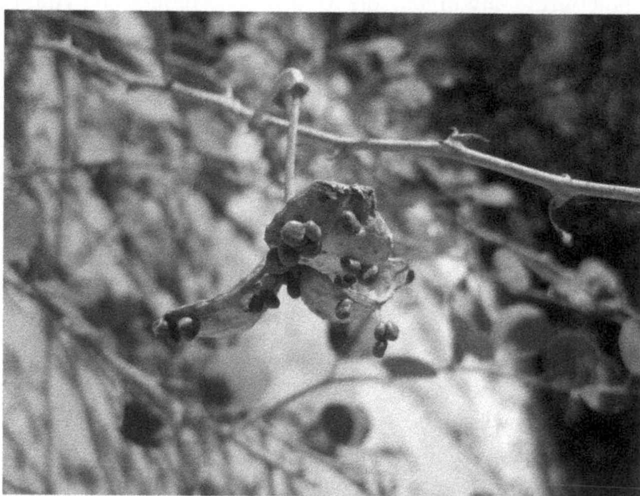

FIGURE 9.2 Seeds of *Capparis spinosa* contain interesting sterols. Ripe, dehisced fruit on which the seeds still cling. (16.8.2013, French Hill, Jerusalem, Israel: by Helena Paavilainen.)

TABLE 9.1
Sterols of *Capparis* spp.

Compound	Notes
Sterols in general (sterol nucleus)	Detection, isolation, and capillary GC-GC/MS quant. evaluation as chem. markers for distinguishing *C. orientalis* and *C. sicula* ssp. *sicula* (Conforti et al. 2011a); from *C. spinosa* seed oil 2,240.4 mg/kg of total extd. lipids (Tlili et al. 2010); total 4,875.5 to 12,189.1 mg/kg (*C. ovata*), 4,961.8 to 10,009.1 mg/kg (*C. spinosa*) (Matthäus and Özcan 2005); from aerial parts of *C. decidua* (Abdel-Mogib et al. 2000); ident. in C. *decidua* (Rai 1987a)
Δ5-Avenasterol	Comprises 6% of total sterols in *C. spinosa* seed oil (Tlili et al. 2010); between 138.8 and 599.4 mg/kg seed oil of *C. spinosa* and *C. ovata* (Matthäus and Özcan 2005)

TABLE 9.1 (*Continued*)
Sterols of *Capparis* spp.

Compound	Notes
Brassicasterol 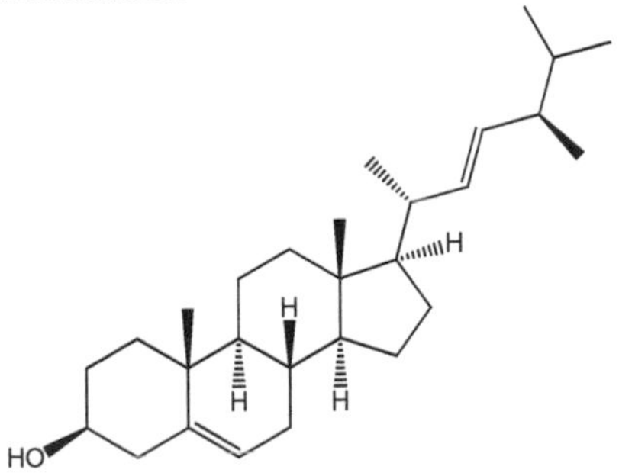	In *C. spinosa* seed, 3.39 mg/kg total extracted lipids (Tlili et al. 2010)
Campestanol 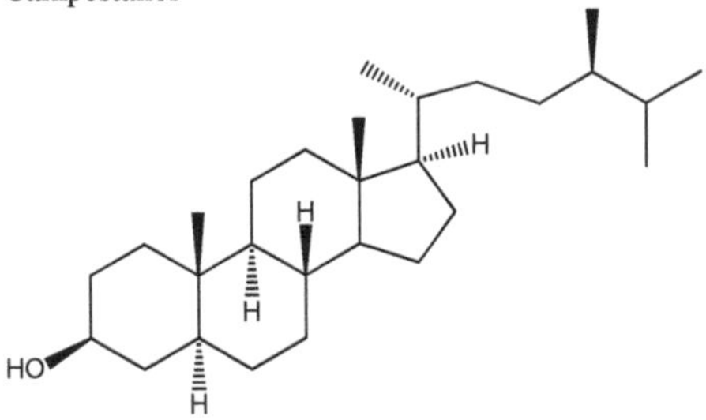	In *C. spinosa* seed oil (Tlili et al. 2010)
Campesterol 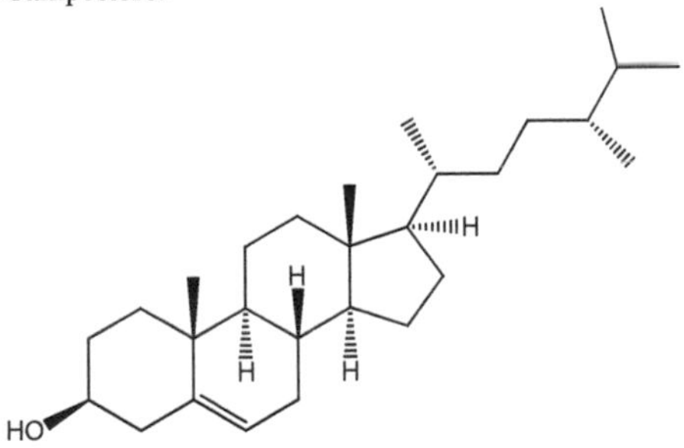	Comprises 17.05% of total sterols in *C. spinosa* seed oil, 382 mg/kg of total extd. lipids (Tlili et al. 2010); 16% of total sterols in *C. spinosa* and *C. ovata* (Matthäus and Özcan 2005); in n-hexane fraction of *C. formosana* stems ether ext. (Liu et al. 1977)
Cholesterol 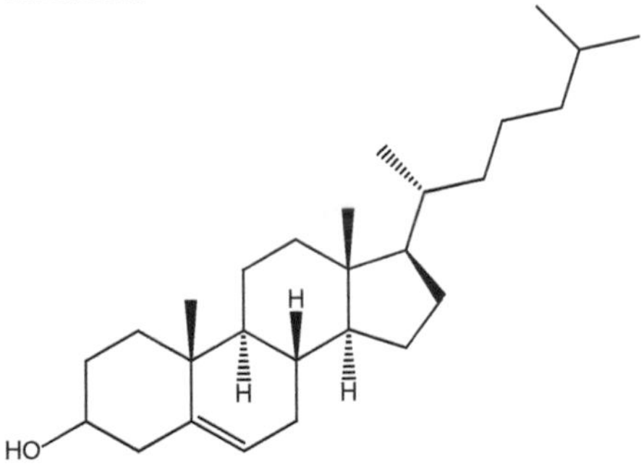	Sepd. from the root bark of *C. corymbosa* (Sawadogo et al. 1981); minor component of *C. spinosa* seed oil (Tlili et al. 2010)

(*Continued*)

TABLE 9.1 (*Continued*)
Sterols of *Capparis* spp.

Compound	Notes
β-Daucosterol 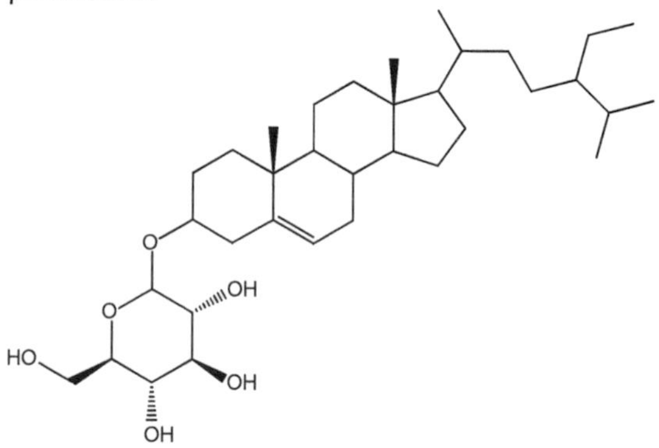	From *C. spinosa* isol. by solvent extn., silica gel, and Sephadex LH-20 gel column chromatog. and recrystn., structure elucidated by ESI-MS, ^{1}H-NMR and ^{13}C-NMR (Li et al. 2007); from stems and leaves of *C. spinosa* sepd. and purified by column chromatog., structure analyzed by spectroscopic methods (Yang et al. 2011); from fruit of *C. spinosa* (Yang et al. 2010, Gan et al. 2009); isolated from *C. spinosa* pericarp by column chromatog. and identified via physicochem. consts. and spectral anal. (Xiao et al. 2008); from *C. spinosa* fruits (Yu et al. 2006); from *C. spinosa* (Jiang et al. 2005); ~ 60% of the total amt. of sterols in *C. spinosa* and *C. ovata* (Matthäus and Özcan 2005)
Gramisterol 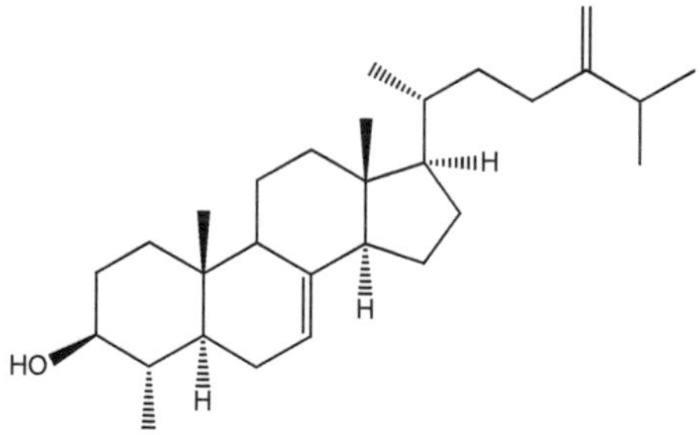	Identified in *C. spinosa* seeds (Tlili et al. 2011)
24-β-Methylcholest-7-ene-22-one-3β-ol	From alc. ext. of *C. decidua* root bark (Gupta and Ali 1998)
24-β-Methylcholest-9(11)-ene-22-one-3α-ol	From alc. ext. of *C. decidua* root bark (Gupta and Ali 1998)
Moonisterol (24 S-Stigmasta-5,7-dien-3β-ol) 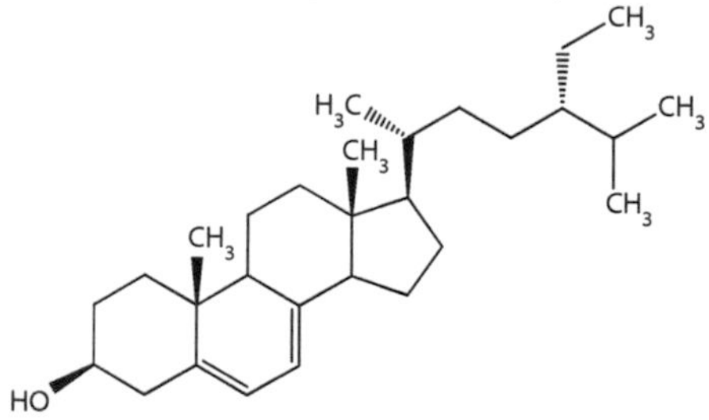	In alc. ext. of *C. moonii* fruits (Ramachandram et al. 2004)

TABLE 9.1 (*Continued*)
Sterols of *Capparis* spp.

Compound	Notes
β-Sitosterol 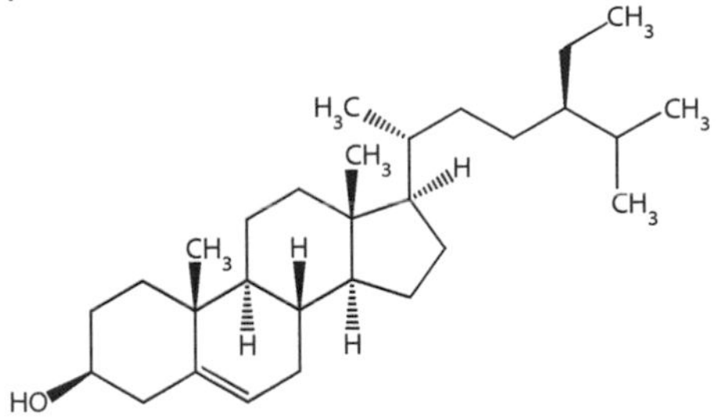	Sepd. from *C. corymbosa* root bark (Sawadogo et al. 1981); from *C. spinosa* isol. by solvent extn., silica gel, and Sephadex LH-20 gel column chromatog. and recrystn., structure elucidated by ESI-MS, ^{1}H-NMR and ^{13}C-NMR (Li et al. 2007); from stems and leaves of *C. spinosa* sepd. and purified by column chromatog., structure analyzed by spectroscopic methods (Yang et al. 2011); from fruit of *C. spinosa* (Yang et al. 2010); quantified from *C. decidua* by novel TLC densitometric method from methanolic ext. using solvent sys. toluene:methanol (9:1, vol./vol.), validated by ICH guidelines for precision, repeatability, and accuracy, linearity range 80–480 ng/spot, content 0.0244% wt./wt. (Rathee et al. 2010); most abundant (57.53%) sterol in *C. spinosa* seed oil, 1,390 mg/kg (Tlili et al. 2010); isolated from *C. spinosa* pericarp by column chromatog. and identified via physicochem. consts. and spectral anal. (Xiao et al. 2008); isolated from *C. spinosa*, 100, 150. and 200 mg/kg → statistically significant and dose-dependent anti-inflammatory act. in carrageenan-induced rat paw edema, sim. to indomethacin (10 mg/kg), known anti-inflammatory agent, 150 mg/kg → ↓ arachidonic acid-elicited rat paw edema 1 h after arachidonic acid challenge, topical application signif. ↓ ear inflammation 2° to multiple applications of tetradecanoyl-phorbol 13-acetate, ↓ exudate vol., tot. leukocytes and neutrophil migration into rat pleural cavity, suggesting anti-inflammatory activity, at least in part via ↓ of cyclooxygenase (COX) and 5-lipoxygenase (LOX) (Perianayagam et al. 2008); from *C. spinosa* fruits (Yu et al. 2006); from *C. spinosa* (Jiang et al. 2005); in alc. ext. of *C. moonii* fruits (Ramachandram et al. 2004); unidentified isomer noted in *C. decidua* (Rai 1987b); in n-hexane fraction of *C. formosana* stems ether ext. (Liu et al. 1977), "liberal" occurrence in *C. sepiaria* leaves (Chaudhury and Ghosh 1970a,b); from *C. aphylla* roots (3.85 kg, 32% moisture) in 7 l. EtOH 3 days, repeated 2x (6 L EtOH), combined exts. concd. to 1.5 L, extd. w/300 mL AcOEt, dried over Na_2SO_4 → 28 g semisolid grease-leached w/ benzene → 11.5 g. benzene-sol. matter, chromatographed on 215 g neutral alumina, eluents benzene, 30% and 60% ether in benzene, 25, 50, and 75% $CHCl_3$ in ether, $CHCl_3$, and 1, 5, 10, 15, 20, 25% MeOH in $CHCl_3$, benzene eluent collected in 18 fractions, fractions 11–18 combined → 11.5 g crude cryst. (m. 105–120°), crystd. from MeOH 10 x → β-sitosterol (m. 136–137°) (Manzoor-i-Khuda and Kapadia 1968)

(*Continued*)

TABLE 9.1 (*Continued*)
Sterols of *Capparis* spp.

Compound	Notes
β-Sitosterol glucoside 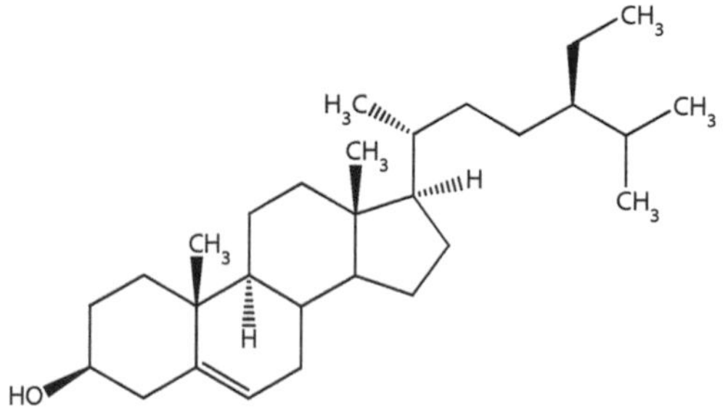	From root of *C. corymbosa* (Sawadogo et al. 1981)
β-Sitosteryl-glucoside-6′-octadecanoate	From *C. spinosa* (Khanfar et al. 2003); from *C. spinosa* fruits ethanol ext., column chrom., spectroscop. (Yang et al. 2010)
α-Spinasterol-3-O-β-D-glucopyranoside	Isolated from *C. spinosa* pericarp by column chromatog. and identified via physicochem. consts. and spectral anal. (Xiao et al. 2008)
Stigmasterol 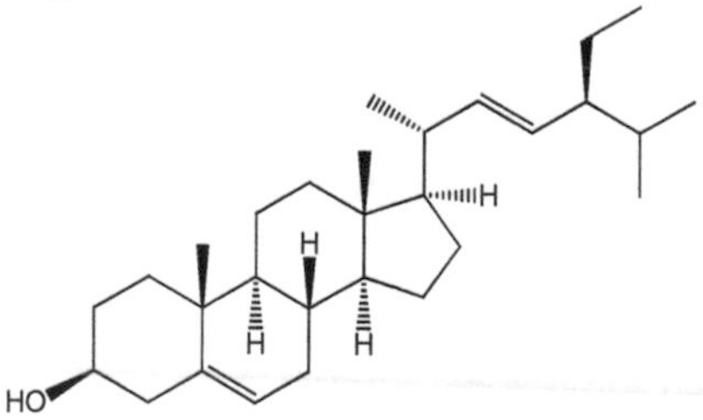	Comprises 265 mg/kg of total extracted lipids from *C. spinosa* seed, 11.85% of total sterols (Tlili et al. 2010); 10% of total sterols in *C. spinosa* and *C. ovata* (Matthäus and Özcan 2005); in n-hexane fraction of *C. formosana* ether stems ext. (Liu et al. 1977)
Taraxasterol 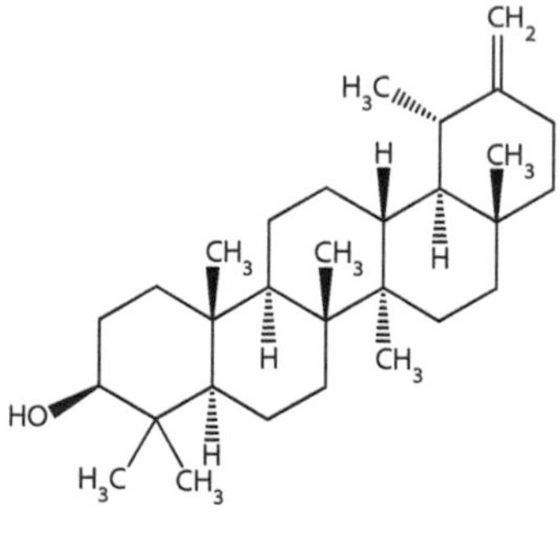	In *C. sepiaria* leaves (Chaudhury and Ghosh 1970a,b)

REFERENCES

Abdel-Mogib, M., S.T. Ezmirly, and S.A. Basaif. 2000. Phytochemistry of *Dipterygium glaucum* and *Capparis decidua*. *Journal of Saudi Chemical Society* 4(1): 103–108.

Awad, A.B. and C.S. Fink. 2000. Phytosterols as anticancer dietary components: evidence and mechanism of action. *Journal of Nutrition* 130(9): 2127–2130.

Baskar, A.A., K.S. Al Numair, M. Gabriel Paulraj, M.A. Alsaif, M.A. Muamar, and S. Ignacimuthu. 2012. β-sitosterol prevents lipid peroxidation and improves antioxidant status and histoarchitecture in rats with 1,2-dimethylhydrazine-induced colon cancer. *Journal of Medicinal Food* 15(4): 335–343.

Brimson, J.M., S.J. Brimson, C.A, Brimson, V. Rakkhitawatthana, and T. Tencomnao. 2012. *Rhinacanthus nasutus* extracts prevent glutamate and amyloid-β neurotoxicity in HT-22 mouse hippocampal cells: possible active compounds include lupeol, stigmasterol and β-sitosterol. *International Journal of Molecular Science* 13(4): 5074–5097.

Byun, H.J., K.H. Cho, H.C. Eun, M.J. Lee, Y. Lee, S. Lee, and J.H. Chung. 2012. Lipid ingredients in moisturizers can modulate skin responses to UV in barrier-disrupted human skin in vivo. *Journal of Dermatological Science* 65(2): 110–117.

Chaudhury, N.A. and D. Ghosh. 1970a. Insecticidal plants: chemical examination of the leaves of *Capparis sepiaria. Journal of the Indian Chemical Society* 47(8): 751–754.

Chaudhury, N.A. and D. Ghosh. 1970b. Taraxasterol and other triterpenoids in *Capparis sepiaria* leaves. *Phytochemistry* 9(8): 1885.

Cheong, M.C., K. Na, H. Kim, S.K. Jeong, H.J. Joo, D.J. Chitwood, and Y.K. Paik. 2011. A potential biochemical mechanism underlying the influence of sterol deprivation stress on *Caenorhabditis elegans* longevity. *Journal of Biological Chemistry* 286(9):7248–7256.

Conforti, F., S. Modesto, F. Menichini, G.A. Statti, D. Uzunov, U. Solimene, P. Duez, and F. Menichini. 2011a. Correlation between environmental factors, chemical composition, and antioxidative properties of caper species growing wild in Calabria (South Italy). *Chemistry & Biodiversity* 8(3): 518–531.

Gan, Y., W. Chen, X. Wang, B. Han, C. Mu, H. Zhang, and Y. Zhuo. 2009. Chemical constituents of fruit of *Capparis spinosa* L. *Shihezi Daxue Xuebao, Ziran Kexueban* 27(3): 334–336.

Ghanbari, R., F. Anwar, K.M. Alkharfy, A.H. Gilani, and N. Saari. 2012. Valuable nutrients and functional bioactives in different parts of olive (*Olea europaea* L.)—a review. *International Journal of Molecular Science* 13(3): 3291–3340.

Gupta, J. and M. Ali. 1998. Phytoconstituents of *Capparis decidua* root barks. *Journal of Medicinal and Aromatic Plant Sciences* 20(3): 683–689.

Janeczko, A. 2012. The presence and activity of progesterone in the plant kingdom. *Steroids* 77(3): 169–173.

Jiang, W., W.H. Lin, and S.D. Guo. 2005. Study on chemical constituents of *Capparis spinosa. Harbin Shangye Daxue Xuebao, Ziran Kexueban* 21(6): 684–686, 689.

Khanfar, M.A., S.S. Sabri, M.A. Zarga, and K.P. Zeller. 2003. The chemical constituents of *Capparis spinosa* of Jordanian origin. *Natural Product Research* 17(1): 9–14.

Li, Y., Y. Feng, S. Yang, and L. Xu. 2007. Chemical components of *Capparis spinosa* L. *Zhongcaoyao* 38(4): 510–512.

Liu, K.C., C.J. Chou, and W.C. Pan. 1977. Studies on the constituents of the stems of *Capparis formosana* Hemsl. *Taiwan Yaoxue Zazhi* 28(1–2): 2–5.

Mackay, D.S. and P.J. Jones. 2012. Plasma noncholesterol sterols: current uses, potential and need for standardization. *Current Opinion in Lipidology* 23(3): 241–247.

Manzoor-i-Khuda, M. and Z. Kapadia. 1968. Isolation of β-sitosterol from *Capparis aphylla. Pakistan Journal of Scientific and Industrial Research* 11(1): 108.

Matthäus, B. and M. Özcan. 2005. Glucosinolates and fatty acid, sterol, and tocopherol composition of seed oils from *Capparis spinosa* var. *spinosa* and *Capparis ovata* Desf. var. *canescens* (Coss.) Heywood. *Journal of Agricultural and Food Chemistry* 53(18): 7136–7141.

Perianayagam, J.B., S.K. Sharma, and K.K. Pillai. 2008. Anti-inflammatory potential of β-sitosterol on acute and chronic inflammation models. *Recent Progress in Medicinal Plants* 22: 25–35.

Rai, S. 1987a. Chemical examination of edible plants of Rajasthan desert with special reference to Capparidaceae. *Current Agriculture* 11(1–2): 15–23.

Rai, S. 1987b. Oils and fats in arid plants with particular reference to *Capparis decidua* L. *Transactions of Indian Society of Desert Technology* 12(2): 99–105.

Ramachandram, R., M. Ali, and S.R. Mir. 2004. Phytoconstituents from *Capparis moonii* fruits. *Indian Journal of Natural Products* 20(1): 40–42.

Rathee, S., O.P. Mogla, P. Rathee, and D. Rathee. 2010. Quantification of β-sitosterol using HPTLC from *Capparis decidua* (Forsk.) Edgew. *Pharma Chemica* 2(4): 86–92.

Sawadogo, M. A.M. Tessier, and P. Delaveau. 1981. Chemical study of *Capparis corymbosa* Lam. roots. *Plantes Medicinales et Phytotherapie* 15(4): 234–239.

Srivastava, S., H. Bisht, O.P. Sidhu, A. Srivastava, P.C. Singh. R.M. Pandey, S.K. Raj, R. Roy, and C.S. Nautiyal. 2012. Changes in the metabolome and histopathology of *Amaranthus hypochondriacus* L. in response to Ageratum enation virus infection. *Phytochemistry* 80: 8–16.

Tlili, N., T. El Guizani, N. Nasri, A. Khaldi, and S. Triki. 2011. Protein, lipid, aliphatic and triterpenic alcohol content of caper seeds "Capparis spinosa." *Journal of the American Oil Chemists' Society* 88(2): 265–270.

Tlili, N., N. Nasri, E. Saadaoui, A. Khaldi, and S. Triki. 2010. Sterol composition of caper (*Capparis spinosa*) seeds. *African Journal of Biotechnology* 9(22): 3328–3333.

Xiao, W., N. Li, and X. Li. 2008. Isolation and identification of organic acids from pericarp of *Capparis spinosa* L. *Shenyang Yaoke Daxue Xuebao* 25(10): 790–792.

Yang, T., X. Cheng, F. Yu, G. Chou, C. Wang, and Z. Wang. 2010. The chemical constituents of *Capparis spinosa* L. fruits. *Xibei Yaoxue Zazhi* 25(4): 260–263.

Yang, T., H. Liu, X. Cheng, F. Yu, G. Chou, C. Wang, and Z. Wang. 2011. The chemical constituents from stems and leaves of *Capparis spinosa* L. *Xibei Yaoxue Zazhi* 26(1): 16–18.

Yu, Y., H. Gao, Z. Tang, X. Song, and L. Wu. 2006. Several phenolic acids from the fruit of *Capparis spinosa*. *Asian J Trad Med* 1(2): 101–4.

10 Triterpenoids and Other Volatile Compounds Excepting Isothiocyanates

The distinctive flavor and aroma of pickled caper buds is comprised of hundreds of compounds whose volatility allows these compounds to travel through the air, constituting an olfactory "air space" around the plant itself (Figure 10.1). In some cases, the movement of these molecules may extend over many miles, reaching out directly to pollinating insects or hummingbirds and, in other cases, to mammals with other means of helping the plants to propagate by eating their fruits and excreting their seeds. The complexity of this volatile segment of the *Capparis* chemistry correlates with the great complexity of the *Capparis* flavor and delicate bouquet.

One segment of this chemistry was considered in the previous chapter on glucosinolates. When the glucosinolates are relieved of their sugar moieties, the liberated isothiocyanates are free to roam and impart their distinctive but subtle sulfur-like aroma. Volatile compounds of all stripes may also exert medicinal effects either, and rarely, as stand-alone compounds or more usually as part of the chemical complexes that constitute the pharmacological totality of plants. In the latter cases, the volatiles may be part of the plants' "entourage" system, exerting complementary or facilitating actions to those of the heavier compounds that remain more deeply ensconced within the plant's tissues. One large and important class of such volatile compounds is derived from repeating isoprene units, the class known as terpenes or terpenoids. These compounds are of enormous pharmacological and medical importance owing to the ability of multiple natural systems to be able to gently release these compounds in a very delicate and tightly controlled manner in response to exceptionally minor alterations in chemical conditions (Herrmann 2007). Such changes may be in response to myriad signals, a not insignificant number of which are also volatile compounds.

Exogenous volatile compounds, both natural and especially human made, may exert noxious effects in human beings. The most pervasive example of this phenomenon is air pollution, which includes a large number of volatile compounds, the waste products of industrialization, in the formlessness of "fumes" (Crinnion 2012) that synergize unfortunately with particulate matter, also often related to waste products of industrial processes, to create or exacerbate numerous diseases, ranging from lung cancer to allergy, asthma, and dermatological atopia (Boutin-Forzano et al. 2005). Noxious volatile air pollution occurs not only outside our dwellings but also inside, where, for example, formaldehyde still contained in home fittings and furnishings may continue to be emitted for months or years after installation and result in considerable negative impact on health in sensitive individuals (Sorg et al. 2001).

Such compounds can also remain in the body and its tissues for many hours, days, weeks, or longer and may be detected even in their original form in the exhaled breath (Beauchamp 2011). Furthermore, when such noxious volatile compounds or other external or internal stimuli do finally result in a serious illness, the illness itself may produce its own metabolic derangements, resulting in still other volatile compounds as disease-specific "signatures," for example, in lung cancer (Goldstraw et al. 2011, Horváth et al. 2009) and asthma (Vijverberg et al. 2011). In addition to such diseases that may befall the unwitting individual exposed to toxic volatile compounds, volatiles may cause diseases, disorders, and even death when the individual inhales them intentionally in the form of industrial cleaning fluids, glue, or other similar preparations for the purpose of finding a legal

FIGURE 10.1 The strong aroma of caper buds has its origin in volatile compounds. Commercial pickled capers. (By Helena Paavilainen.)

(although definitely not safe) means of altering the consciousness or sensorium for the purpose of self-transcendence, relief of existential pain, or "getting high" (Pauluhn 2004, Wille and Lambert 2004).

Volatile compounds also figure prominently in medical therapy, not to mention perfumery. In both, these therapeutic agents are contained in the so-called essential or ethereal "oils" of plants, also known as "volatile oils." Such include the most volatile compounds, the "high notes" of perfumes; the intermediately volatile compounds, "the middle notes" of perfumes; and the only modestly volatile compounds, the "low notes" of perfumes. While the high notes usually derive from the most delicate plant parts, such as flowers, the low notes typically are sourced from "harder" plant parts, the resins from myrrh being a classic example possessing far-reaching therapeutic power (Nomicos 2007).

Volatile essential oils evolved to answer strategic requirements of the plants, for example, as protection against bacterial or fungal attack (Kalemba and Kunicka 2003, Pauli 2006, van Vuuren 2008), that may readily translate into antibacterial or antifungal agents for human medicine; other compounds may protect the plants against insects (Chaudhury and Ghosh 1970) and similarly translate into insecticides or insect repellants for human use. However, not in all cases is the jump from plant physiology to human medicine so facile. Thus, essential oils and their volatile compounds affect mammals in many complex ways, most notably through complex psychosomatic interactions (Pisseri et al. 2008), resulting in analgesia (de Sousa 2011, Sakurada et al. 2009) and anesthesia, involving modulation of two pore domain potassium channels (Bayliss and Barrett 2008) that are characteristically associated with human health- and organ-preserving antioxidant function (Landoni et al. 2009, Li Volti et al. 2008, Soukup et al. 2009).

Halitosis, so-called bad breath, is a superb example of a medical condition for which the "medium is the message" (McLuhan 1964) or, in homeopathic parlance, for which "like cures like" (Tedesco and Cicchetti 2001). To wit, halitosis, caused by chronic disturbances in oral flora, gastrointestinal dysfunction, or dietary indiscretions (Loesche and Kazor 2002, Porter 2011), is the expression of volatile waste products of digestion, which may or may not include sulfur-containing compounds, and is experienced by another's olfactory brain as unpleasant. The cure is essential oils' volatile compounds from plants with a "minty" aroma and a power extending into complex realms in the mammal on both the neuro-olfactory and deeper neural circuits stemming from the primacy of olfaction in evoking human consciousness, which may feed back through neurolymphic circuits to help modify immunity and the disease process itself, as well as its symptoms (Mimica-Dukic and Bozin 2008), while killing microorganisms (Quirynen et al. 2002).

Other medicinal plants in which volatile compounds figure prominently in their key medicinal actions include garlic (*Allium sativa*) (Swiderski et al. 2007); black seed (*Nigella sativa*) (Gali-Muhtasib et al. 2006); passion flowers (*Passiflora* sp.) (Dhawan et al. 2004); lavender

(*Lavendula* sp.) (Woronuk et al. 2011); *liǎo gē wáng* (*Wikstroemeia indica*) (Y.M. Li et al. 2009b); juniper berries (*Juniperus* sp.) (Anon., 2001); the Native American "devil's club" (*Oplopanax horridus*) (Calway et al. 2012); and the Ayurvedic "Indian cactus" or "famine food," *yugmaphallottama* (*Caralluma fimbriata*) (Dutt et al. 2012). All of these plants' actions share the singular quality of producing volatile oils that "go to the head" as well as the internal organs to effect complex somatopsychic reverberations owing to the connectedness of the rhinencephalon, the "smell brain," to brain architecture governing both somatic responses and complex memory.

Such plant drugs may also be particularly suitable for use in children, at least in part in certain cases through the use of pleasant tastes (e.g., mint and fruit essences) that are due to volatile compounds (Mennella and Beauchamp 2008). Volatile compounds are responsible for all of olfaction and most of taste.

Cutting edge pharmaceutical technology will likely make increasing utilization of volatile compounds in its formulations to improve taste and smell of delivery, enhance penetration into hard-to-reach fatty places, and to influence the central nervous system. Electronic noses aim to provide computer-controlled means of smell and of recording such in quantifiable impressions amenable to quality control of food products and medical diagnosis (Wilson and Baietto 2011), while nanoencapsulation may provide a means for containing these volatile principles (Lertsutthiwong and Rojsitthisak 2011). Also to be explored will be the role of such compounds in enhancing synergy of complex plant mixtures (Efferth and Koch 2011).

Capparis genus and its capers have invested considerable evolutionary energy to create the very impressive volatile resources outlined in Table 10.1. Most of this work has focused on edible capers, most especially *C. spinosa* capers. Many workers contributed to this base, but a special note of appreciation is due Romeo et al. (2007) for their innovative use of a closed-head system of analysis for unraveling the highly complex volatile components of capers that make them so highly prized a culinary feature, likely also a medicine or a component of one, and why the plant evolved them. Was it a means of defense (certainly), to attract pollinators (also) and perhaps also propagators such as ourselves and other mammals who collect and excrete viable seeds?

TABLE 10.1
Volatile Oils of *Capparis* spp.

Compound	Notes
Volatile oils in general	From supercritical CO_2 extracted *C. spinosa* seeds, by GC-MS, 14 kinds of volatile oils sepd., 13 kinds of compds. (Ren et al. 2009); using supercritical CO_2 extraction and GC-MS, w/*C. spinosa* best extn. conditions incl. pressure 30 mPa, temp. 55°C, time 2 h, yield of volatile oil 10% (Liu et al. 2009); essential oil of *C. spinosa* leaves extd. by steam distn., by GC-MS 50 peaks isol., 34 compds. identified, accounting for 94.93% of total volatile oil, sulfur compds., 80.72% of tot. volatile oil, ketones 9.19%, aldehydes 2.36%, terpenes 2.26%, hydrocarbons 0.32%, alcs. 1.09% (G.Q. Li et al. 2009); from *C. spinosa* seeds by steam distillation, 1.12%, sepd. and identified w/GC-MS, 7 compounds ident. (Meiliwan et al. 2009); in *C. ovata* Desf. var. *canescens* 86 volatile compounds in bud oil, 100 in leaf oil by steam distn., solvent extn., GC/MS, ↓ hexanal oxidn. by 80% over 40 days at 200 μg/mL, ↓ oxidn. of 1,1-diphenyl-2-picrylhydrazyl hydrate by > 70% at 500 μg/mL, but lower antioxidant act. than dichlomethane and methanol extracts of same plant parts (El-Ghorab et al. 2007)

(Continued)

TABLE 10.1 (*Continued*)
Volatile Oils of *Capparis* spp.

Compound	Notes
	Terpenoids of *Capparis*
Acyclic terpenoids	Determined by chromatographic separation from aerial parts of *C. decidua* (Abdel-Mogib et al. 2000)
α-Amyrin	Pentacyclic triterpene isolated as its crystal acetate from leaves of *C. sepiaria* (Chaudhury and Ghosh 1970)
β-Amyrin 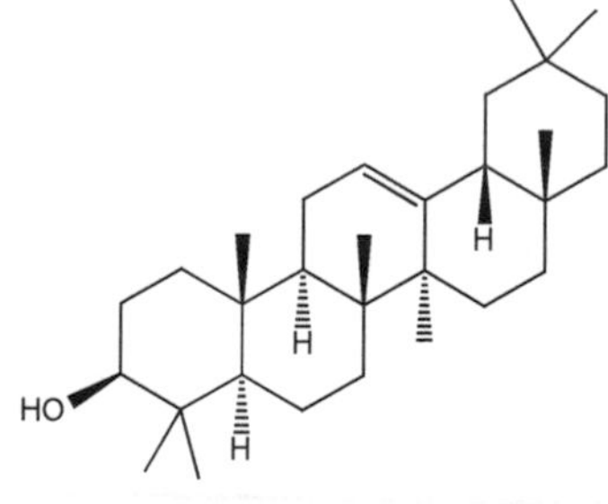	Pentacyclic triterpene identified in *C. spinosa* seeds (Tlili et al. 2011); as its crystal acetate from leaves of *C. sepiaria* (Chaudhury and Ghosh 1970)
δ-3-Carene	11 PPM (mg/kg) by headspace solid-phase microextraction (HS-SPME)/GC–MS in lactic-fermented *C. spinosa* buds collected on the island of Salina (Eolian Archipelago, Sicily) in August 2004 (Romeo et al. 2007)
Carvone	81 PPM (mg/kg) by HS-SPME/GC–MS in lactic-fermented *C. spinosa* buds collected on the island of Salina (Eolian Archipelago, Sicily) in August 2004 (Romeo et al. 2007)
β-Cimene	23 PPM (mg/kg) by HS-SPME/GC–MS in lactic-fermented *C. spinosa* buds collected on the island of Salina (Eolian Archipelago, Sicily) in August 2004 (Romeo et al. 2007)

TABLE 10.1 (*Continued*)
Volatile Oils of *Capparis* spp.

Compound	Notes
Citrostadienol	Major compd. (ca. 170 mg kg^{-1}) in *C. spinosa* seeds (Tlili et al. 2011)
α-Curcumene	86 PPM (mg/kg) by HS-SPME/GC–MS in lactic-fermented *C. spinosa* buds collected on the island of Salina (Eolian Archipelago, Sicily) in August 2004 (Romeo et al. 2007)
Cycloartanol 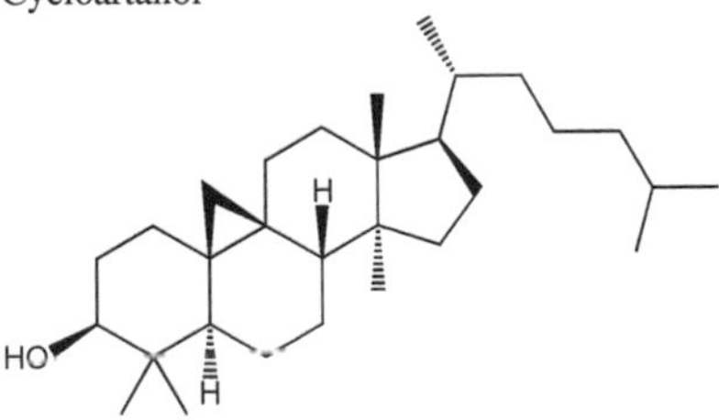	Identified in *C. spinosa* seeds (Tlili et al. 2011)
Erythrodiol	Pentacylic triterpene isolated as its crystal acetate from leaves of *C. sepiaria* (Chaudhury and Ghosh 1970)
(E,E)-α-Farnesene	84 PPM (mg/kg) by HS-SPME/GC–MS in lactic-fermented *C. spinosa* buds collected on the island of Salina (Eolian Archipelago, Sicily) in August 2004 (Romeo et al. 2007)
(E,Z)-α-Farnesene 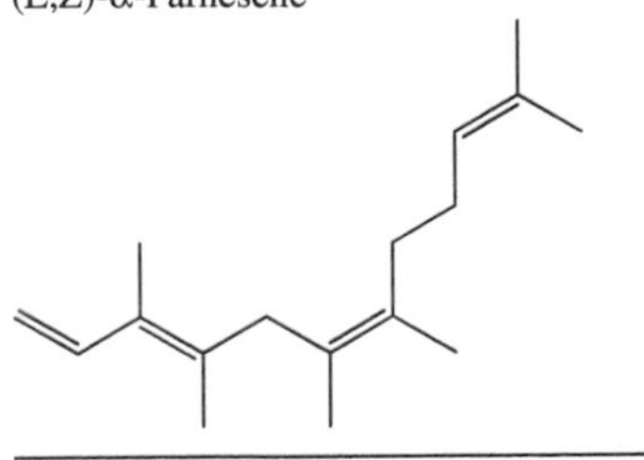	79 PPM (mg/kg) by HS-SPME/GC–MS in lactic-fermented *C. spinosa* buds collected on the island of Salina (Eolian Archipelago, Sicily) in August 2004 (Romeo et al. 2007)

(*Continued*)

TABLE 10.1 (*Continued*)
Volatile Oils of *Capparis* spp.

Compound	Notes
β-Farnesene 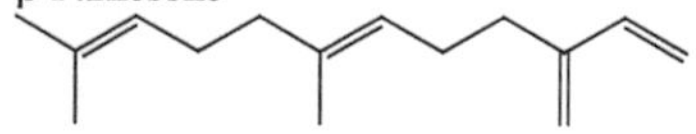	71 PPM (mg/kg) by HS-SPME/GC–MS in lactic-fermented *C. spinosa* buds collected on the island of Salina (Eolian Archipelago, Sicily) in August 2004 (Romeo et al. 2007)
D-Fenchyl alcohol 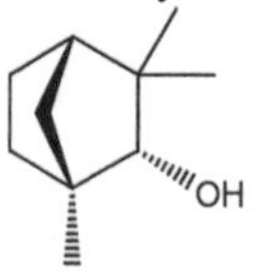	57 PPM (mg/kg) by HS-SPME/GC–MS in lactic-fermented *C. spinosa* buds collected on the island of Salina (Eolian Archipelago, Sicily) in August 2004 (Romeo et al. 2007)
Gramisterol 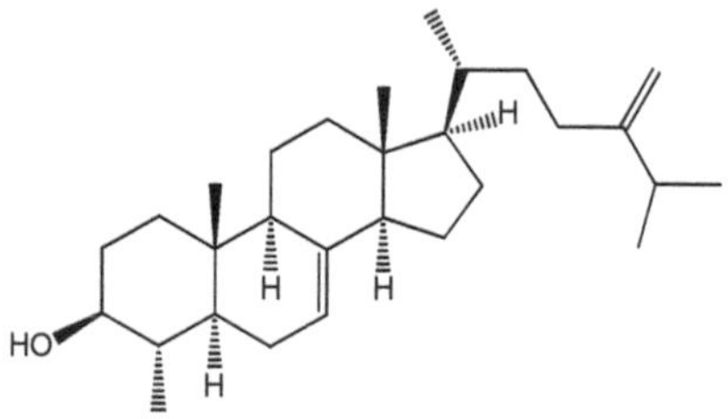	Identified in *C. spinosa* seeds (Tlili et al. 2011)
Hexahydrofarnesyl acetate 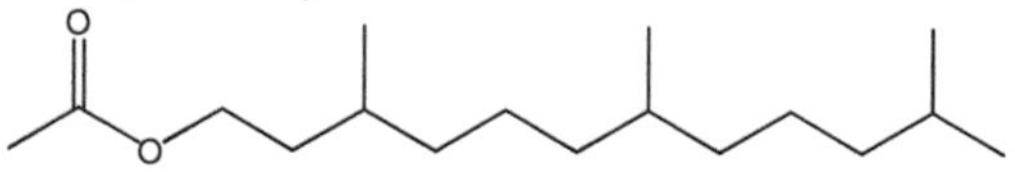	119 PPM (mg/kg) by HS-SPME/GC–MS in lactic-fermented *C. spinosa* buds collected on the island of Salina (Eolian Archipelago, Sicily) in August 2004 (Romeo et al. 2007)
(E)-β-Ionone 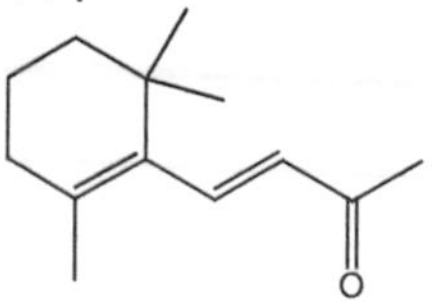	103 PPM (mg/kg) by HS-SPME/GC–MS in lactic-fermented *C. spinosa* buds collected on the island of Salina (Eolian Archipelago, Sicily) in August 2004 (Romeo et al. 2007)
Isomenthol 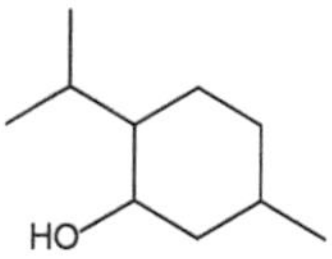	65 PPM (mg/kg) by HS-SPME/GC–MS in lactic-fermented *C. spinosa* buds collected on the island of Salina (Eolian Archipelago, Sicily) in August 2004 (Romeo et al. 2007)
Limonene 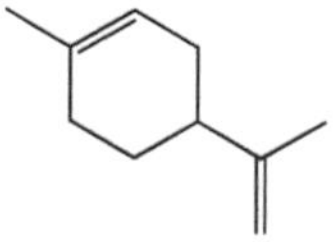	16 PPM (mg/kg) by HS-SPME/GC–MS in lactic-fermented *C. spinosa* buds collected on the island of Salina (Eolian Archipelago, Sicily) in August 2004 (Romeo et al. 2007)
β-Linalool 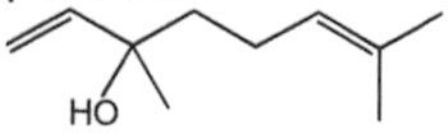	49 PPM (mg/kg) by HS-SPME/GC–MS in lactic-fermented *C. spinosa* buds collected on the island of Salina (Eolian Archipelago, Sicily) in August 2004 (Romeo et al. 2007)
Lupine triterpenoids	Determined by chromatographic separation from aerial parts of *C. decidua* (Abdel-Mogib et al. 2000)

TABLE 10.1 (*Continued*)
Volatile Oils of *Capparis* spp.

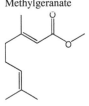

Compound	Notes
Menthol	59 PPM (mg/kg) by HS-SPME/GC–MS in lactic-fermented *C. spinosa* buds collected on the island of Salina (Eolian Archipelago, Sicily) in August 2004 (Romeo et al. 2007)
p-Menthone 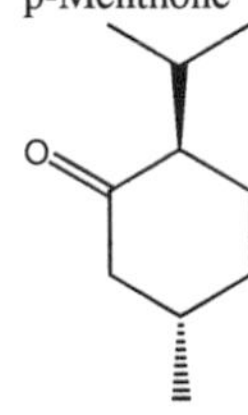	43 PPM (mg/kg) by HS-SPME/GC–MS in lactic-fermented *C. spinosa* buds collected on the island of Salina (Eolian Archipelago, Sicily) in August 2004 (Romeo et al. 2007)
1-Methoxy-4-(1-propenyl) benzene 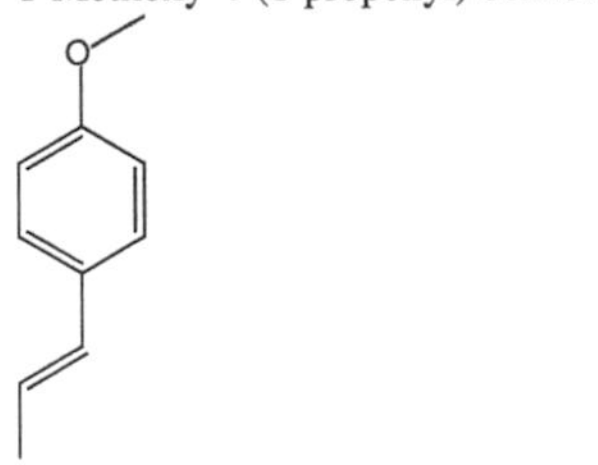	12.5% of volatile oil extd. and purified from *C. spinosa* seeds by steam dist., GC-MS (Meiliwan et al. 2009)
2,4 Methylcycloartenol	Identified in *C. spinosa* seeds (Tlili et al. 2011)
Methylgeranate	75 PPM (mg/kg) by HS-SPME/GC–MS in lactic-fermented *C. spinosa* buds collected on the island of Salina (Eolian Archipelago, Sicily) in August 2004 (Romeo et al. 2007)
Monogynol A (3β,20-dihydroxylupane)	Lupine triterpenoid confirmed by ^{1}H- and ^{13}C-NMR in aerial parts of *C. decidua* (Abdel-Mogib et al. 2000)
D-Nerolidol 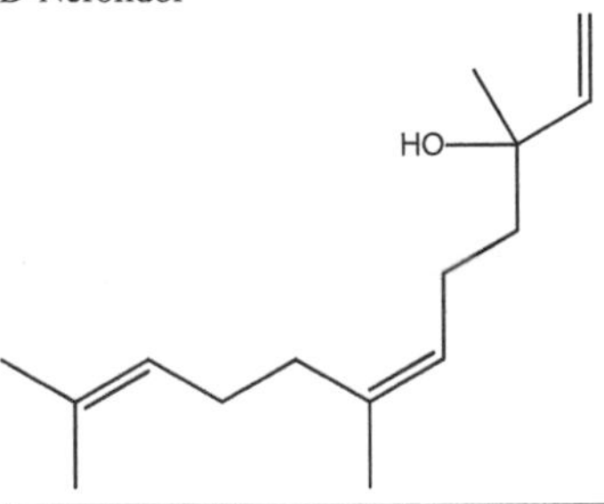	108 PPM (mg/kg) by HS-SPME/GC–MS in lactic-fermented *C. spinosa* buds collected on the island of Salina (Eolian Archipelago, Sicily) in August 2004 (Romeo et al. 2007)

(*Continued*)

TABLE 10.1 (*Continued*)
Volatile Oils of *Capparis* spp.

Compound	Notes
trans-Nerolidol 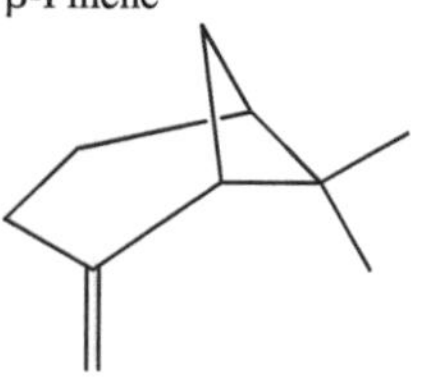	112 PPM (mg/kg) by HS-SPME/GC–MS in lactic-fermented *C. spinosa* buds collected on the island of Salina (Eolian Archipelago, Sicily) in August 2004 (Romeo et al. 2007)
β-Pinene	7 PPM (mg/kg) by HS-SPME/GC–MS in lactic-fermented *C. spinosa* buds collected on the island of Salina (Eolian Archipelago, Sicily) in August 2004 (Romeo et al. 2007)
Taxasterol	Pentacylic triterpene isolated as its crystal acetate from leaves of *C. sepiaria* (Chaudhury and Ghosh 1970)
4-Terpineol	61 PPM (mg/kg) by HS-SPME/GC–MS in lactic-fermented *C. spinosa* buds collected on the island of Salina (Eolian Archipelago, Sicily) in August 2004 (Romeo et al. 2007)
α-Terpineol	74 PPM (mg/kg) by HS-SPME/GC–MS in lactic-fermented *C. spinosa* buds collected on the island of Salina (Eolian Archipelago, Sicily) in August 2004 (Romeo et al. 2007)
Tetracosanol	Identified in *C. spinosa* seeds (Tlili et al. 2011)
trans-Theaspirane	2.6% of *C. ovata* leaf oil, steam dist., solvent ext., GC-MS (El-Ghorab et al. 2007)
Thymol	5.1% of *C. ovata* bud oil, steam dist., solvent ext., GC-MS, 15.5% of leaf oil (El-Ghorab et al. 2007)
Triterpenic alcohols in general	Identified in *C. spinosa* seeds 396.82 mg kg^{-1} of total extracted lipids (Tlili et al. 2011)

TABLE 10.1 (*Continued*)
Volatile Oils of *Capparis* spp.

Compound	Notes
4-Vinyl guaiacol 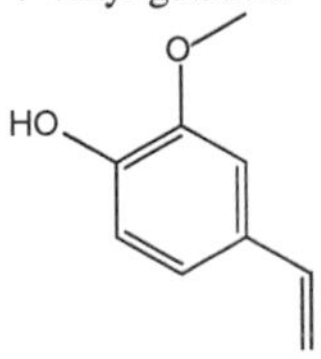	5.3% of *C. ovata* bud oil, steam dist., solvent ext., GC-MS, 4.3% in leaf oil (El-Ghorab et al. 2007)
Volatile Acids of *Capparis*	
Acetic acid 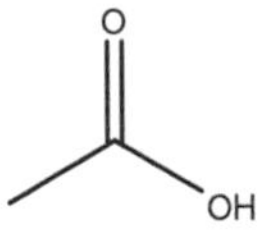	39 PPM (mg/kg) by HS-SPME/GC–MS in lactic-fermented *C. spinosa* buds collected on the island of Salina (Eolian Archipelago, Sicily) in August 2004 (Romeo et al. 2007)
Benzoic acid 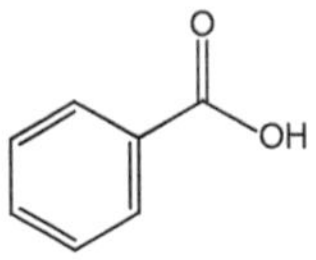	135 PPM (mg/kg) by HS-SPME/GC–MS in lactic-fermented *C. spinosa* buds collected on the island of Salina (Eolian Archipelago, Sicily) in August 2004 (Romeo et al. 2007); from *C. spinosa* leaves and stems, column chrom., spectroscop. (Yang et al. 2011)
N-Decanoic acid 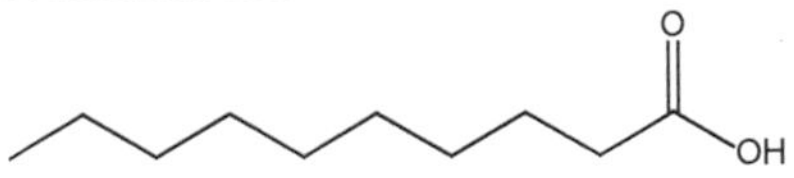	129 PPM (mg/kg) by HS-SPME/GC–MS in lactic-fermented *C. spinosa* buds collected on the island of Salina (Eolian Archipelago, Sicily) in August 2004 (Romeo et al. 2007)
Heptanoic acid 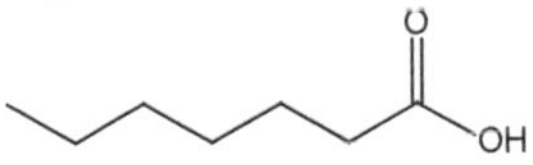	104 PPM (mg/kg) by HS-SPME/GC–MS in lactic-fermented *C. spinosa* buds collected on the island of Salina (Eolian Archipelago, Sicily) in August 2004 (Romeo et al. 2007); 4.8% of *C. ovata* bud oil, steam dist., solvent ext., GC-MS (El-Ghorab et al. 2007)
Hexanoic acid 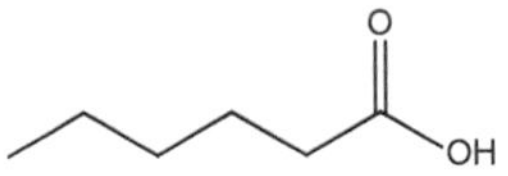	93 PPM (mg/kg) by HS-SPME/GC–MS in lactic-fermented *C. spinosa* buds collected on the island of Salina (Eolian Archipelago, Sicily) in August 2004 (Romeo et al. 2007)
Octanoic acid 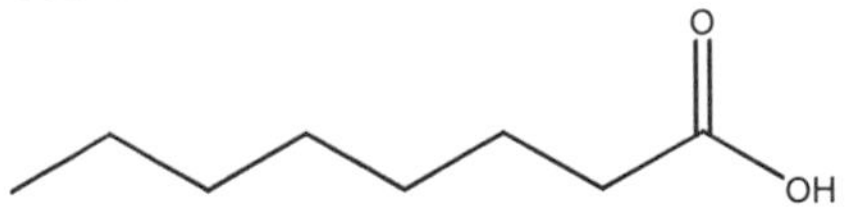	114 PPM (mg/kg) by HS-SPME/GC–MS in lactic-fermented *C. spinosa* buds collected on the island of Salina (Eolian Archipelago, Sicily) in August 2004 (Romeo et al. 2007); 4.8% of *C. ovata* bud oil, steam dist., solvent ext., GC-MS (El-Ghorab et al. 2007)
Tetradecanoic acid 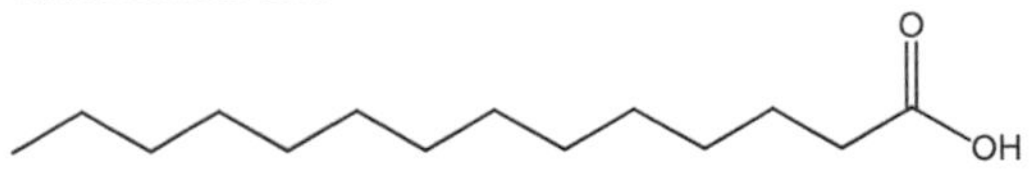	146 PPM (mg/kg) by HS-SPME/GC–MS in lactic-fermented *C. spinosa* buds collected on the island of Salina (Eolian Archipelago, Sicily) in August 2004 (Romeo et al. 2007)

(*Continued*)

TABLE 10.1 (*Continued*)
Volatile Oils of *Capparis* spp.

Compound	Notes
	Volatile Esters of *Capparis*
Benzyl acetate 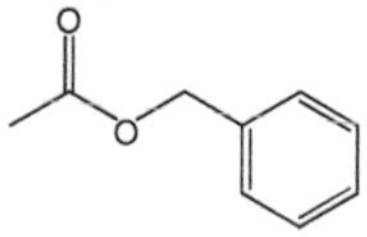	80 PPM (mg/kg) by HS-SPME/GC–MS in lactic-fermented *C. spinosa* buds collected on the island of Salina (Eolian Archipelago, Sicily) in August 2004 (Romeo et al. 2007)
Benzyl benzoate 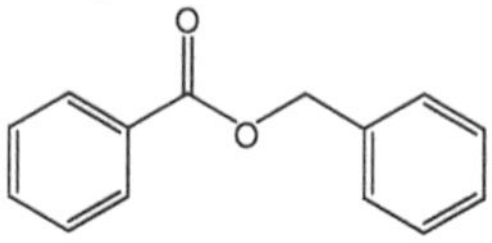	144 PPM (mg/kg) by HS-SPME/GC–MS in lactic-fermented *C. spinosa* buds collected on the island of Salina (Eolian Archipelago, Sicily) in August 2004 (Romeo et al. 2007)
Benzyl isovalerate 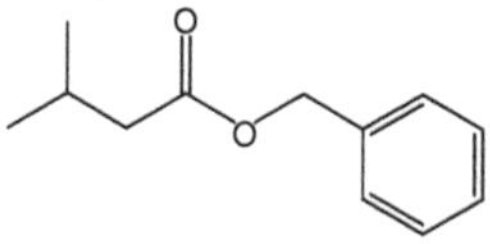	98 PPM (mg/kg) by HS-SPME/GC–MS in lactic-fermented *C. spinosa* buds collected on the island of Salina (Eolian Archipelago, Sicily) in August 2004 (Romeo et al. 2007)
Benzyl tiglate 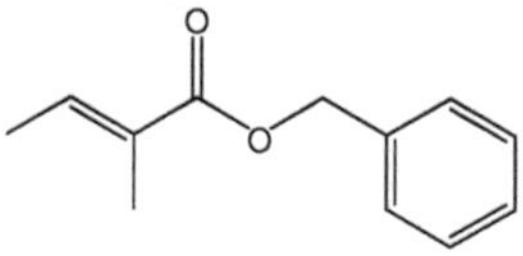	118 PPM (mg/kg) by HS-SPME/GC–MS in lactic-fermented *C. spinosa* buds collected on the island of Salina (Eolian Archipelago, Sicily) in August 2004 (Romeo et al. 2007)
1-Butanol 3-methyl benzoate 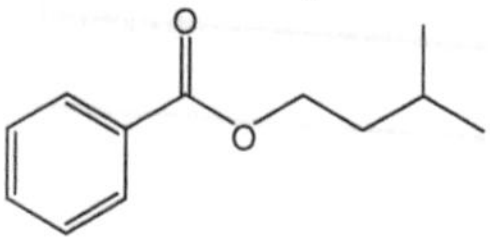	101 PPM (mg/kg) by HS-SPME/GC–MS in lactic-fermented *C. spinosa* buds collected on the island of Salina (Eolian Archipelago, Sicily) in August 2004 (Romeo et al. 2007)
Butyl butanoate 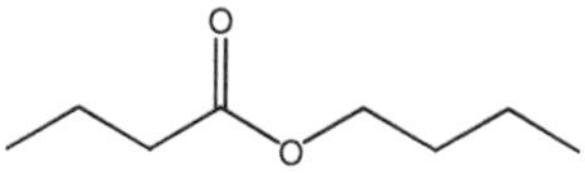	17 PPM (mg/kg) by HS-SPME/GC–MS in lactic-fermented *C. spinosa* buds collected on the island of Salina (Eolian Archipelago, Sicily) in August 2004 (Romeo et al. 2007)
Butyl hexanoate 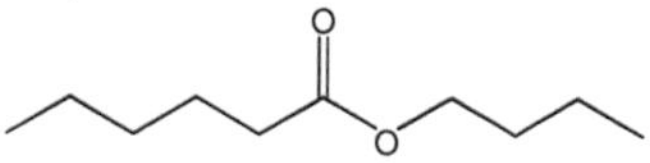	35 PPM (mg/kg) by HS-SPME/GC–MS in lactic-fermented *C. spinosa* buds collected on the island of Salina (Eolian Archipelago, Sicily) in August 2004 (Romeo et al. 2007)
Butyl octanoate 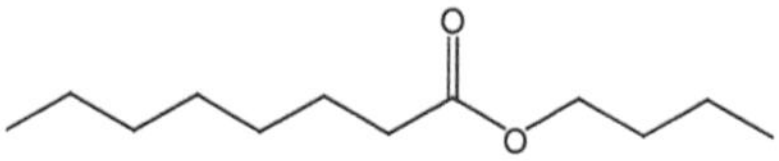	62 PPM (mg/kg) by HS-SPME/GC–MS in lactic-fermented *C. spinosa* buds collected on the island of Salina (Eolian Archipelago, Sicily) in August 2004 (Romeo et al. 2007)
Butyl tetradecanoate 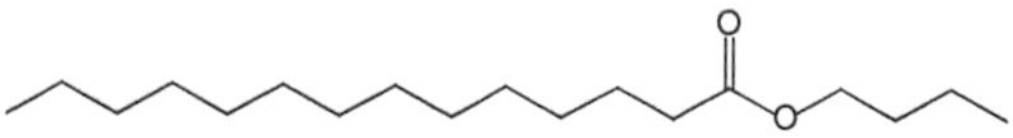	127 PPM (mg/kg) by HS-SPME/GC–MS in lactic-fermented *C. spinosa* buds collected on the island of Salina (Eolian Archipelago, Sicily) in August 2004 (Romeo et al. 2007)

TABLE 10.1 (*Continued*)
Volatile Oils of *Capparis* spp.

Compound	Notes
2,6-Cresotic acid methyl ester 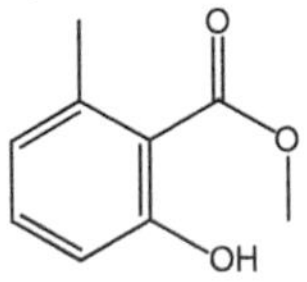	107 PPM (mg/kg) by HS-SPME/GC–MS in lactic-fermented *C. spinosa* buds collected on the island of Salina (Eolian Archipelago, Sicily) in August 2004 (Romeo et al. 2007)
Ethyl benzoate 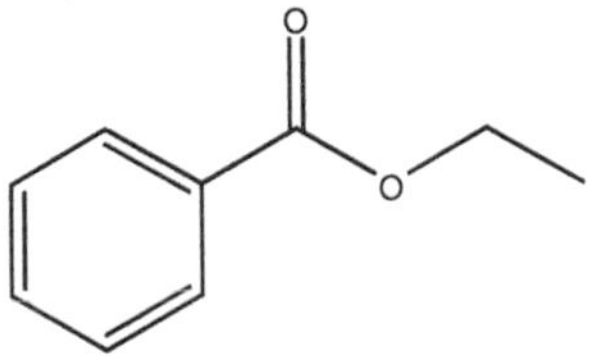	72 PPM (mg/kg) by HS-SPME/GC–MS in lactic-fermented *C. spinosa* buds collected on the island of Salina (Eolian Archipelago, Sicily) in August 2004 (Romeo et al. 2007)
Ethyl butanoate 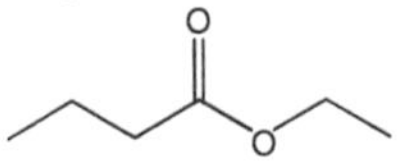	5 PPM (mg/kg) by HS-SPME/GC–MS in lactic-fermented *C. spinosa* buds collected on the island of Salina (Eolian Archipelago, Sicily) in August 2004 (Romeo et al. 2007)
3-Ethylbutyl hexanoate C6 isopentile	41 PPM (mg/kg) by HS-SPME/GC–MS in lactic-fermented *C. spinosa* buds collected on the island of Salina (Eolian Archipelago, Sicily) in August 2004 (Romeo et al. 2007)
Ethyl cinnamate 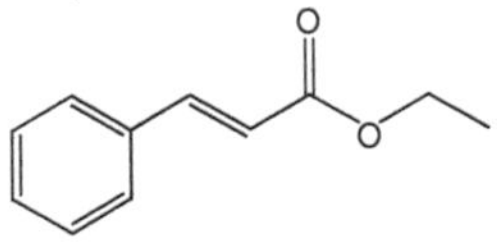	120 PPM (mg/kg) by HS-SPME/GC–MS in lactic-fermented *C. spinosa* buds collected on the island of Salina (Eolian Archipelago, Sicily) in August 2004 (Romeo et al. 2007)
Ethyl decanoate 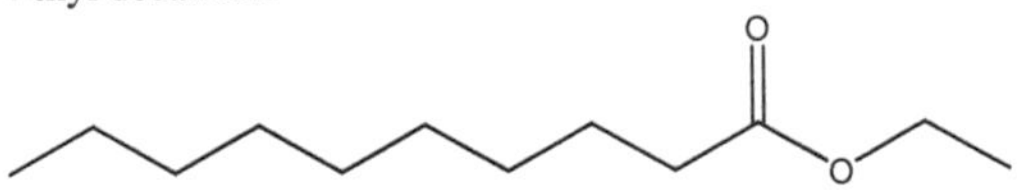	64 PPM (mg/kg) by HS-SPME/GC–MS in lactic-fermented *C. spinosa* buds collected on the island of Salina (Eolian Archipelago, Sicily) in August 2004 (Romeo et al. 2007)
Ethyl hexadecanoate 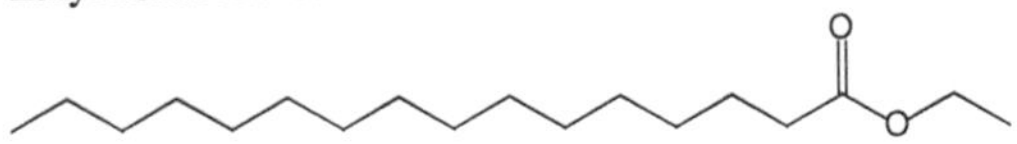	128 PPM (mg/kg) by HS-SPME/GC–MS in lactic-fermented *C. spinosa* buds collected on the island of Salina (Eolian Archipelago, Sicily) in August 2004 (Romeo et al. 2007)
Ethyl 9-hexadecenoate 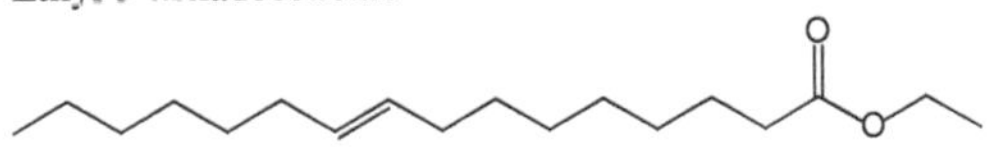	130 PPM (mg/kg) by HS-SPME/GC–MS in lactic-fermented *C. spinosa* buds collected on the island of Salina (Eolian Archipelago, Sicily) in August 2004 (Romeo et al. 2007)
2-Ethyl-hexyl-adipate	66.5% of volatile oil extd. and purified from *C. spinosa* seeds by steam dist., GC-MS (Meiliwan et al. 2009)

(*Continued*)

TABLE 10.1 (*Continued*)
Volatile Oils of *Capparis* spp.

Compound	Notes
Ethyl 9,12-octadecadienoate 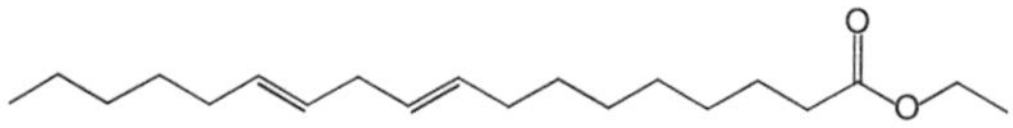	141 PPM (mg/kg) by HS-SPME/GC–MS in lactic-fermented *C. spinosa* buds collected on the island of Salina (Eolian Archipelago, Sicily) in August 2004 (Romeo et al. 2007)
Ethyl octadecanoate 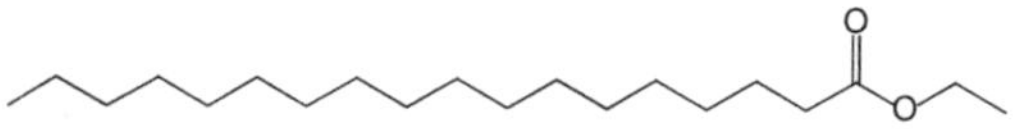	138 PPM (mg/kg) by HS-SPME/GC–MS in lactic-fermented *C. spinosa* buds collected on the island of Salina (Eolian Archipelago, Sicily) in August 2004 (Romeo et al. 2007)
Ethyl 9,12,15-octadecatrienoate 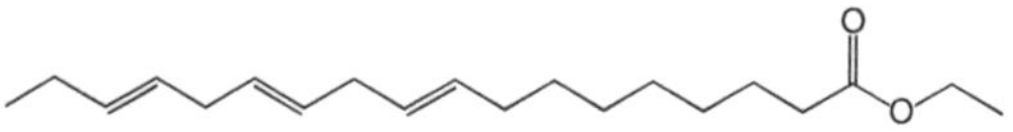	143 PPM (mg/kg) by HS-SPME/GC–MS in lactic-fermented *C. spinosa* buds collected on the island of Salina (Eolian Archipelago, Sicily) in August 2004 (Romeo et al. 2007)
Ethyl 9-octadecenoate 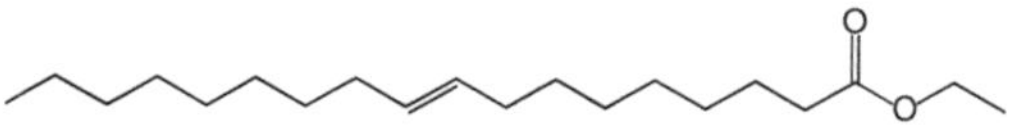	139 PPM (mg/kg) by HS-SPME/GC–MS in lactic-fermented *C. spinosa* buds collected on the island of Salina (Eolian Archipelago, Sicily) in August 2004 (Romeo et al. 2007)
Ethyl octanoate 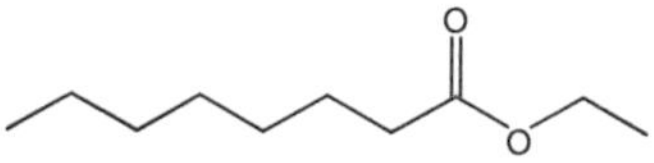	37 PPM (mg/kg) by HS-SPME/GC–MS in lactic-fermented *C. spinosa* buds collected on the island of Salina (Eolian Archipelago, Sicily) in August 2004 (Romeo et al. 2007)
Ethyl pentadecanoate 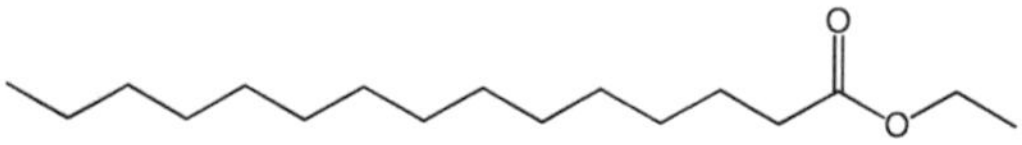	122 PPM (mg/kg) by HS-SPME/GC–MS in lactic-fermented *C. spinosa* buds collected on the island of Salina (Eolian Archipelago, Sicily) in August 2004 (Romeo et al. 2007)
Ethyl phenylacetate 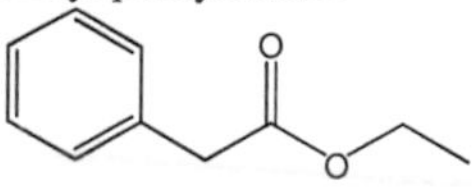	88 PPM (mg/kg) by HS-SPME/GC–MS in lactic-fermented *C. spinosa* buds collected on the island of Salina (Eolian Archipelago, Sicily) in August 2004 (Romeo et al. 2007)
Ethyl salicylate 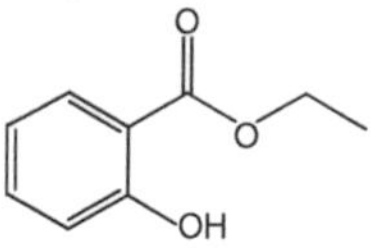	91 PPM (mg/kg) by HS-SPME/GC–MS in lactic-fermented *C. spinosa* buds collected on the island of Salina (Eolian Archipelago, Sicily) in August 2004 (Romeo et al. 2007)
Hexyl acetate 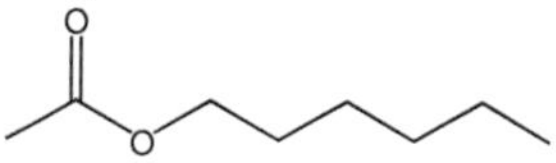	22 PPM (mg/kg) by HS-SPME/GC–MS in lactic-fermented *C. spinosa* buds collected on the island of Salina (Eolian Archipelago, Sicily) in August 2004 (Romeo et al. 2007); 3.6% of *C. ovata* leaf oil, steam dist., solvent ext., GC-MS (El-Ghorab et al. 2007)
Isopentyl octanoate 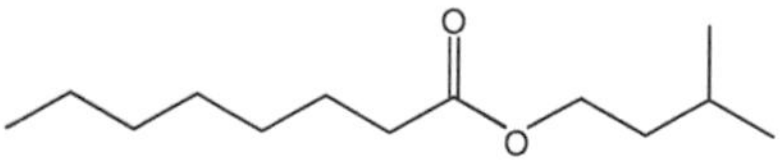	70 PPM (mg/kg) by HS-SPME/GC–MS in lactic-fermented *C. spinosa* buds collected on the island of Salina (Eolian Archipelago, Sicily) in August 2004 (Romeo et al. 2007)
Methyl benzoate 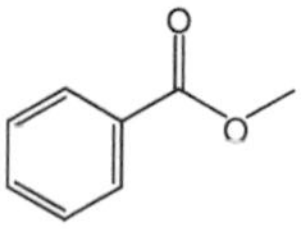	63 PPM (mg/kg) by HS-SPME/GC–MS in lactic-fermented *C. spinosa* buds collected on the island of Salina (Eolian Archipelago, Sicily) in August 2004 (Romeo et al. 2007)

TABLE 10.1 (*Continued*)
Volatile Oils of *Capparis* spp.

Compound	Notes
2-Methylbutyl-acetate	8 PPM (mg/kg) by HS-SPME/GC–MS in lactic-fermented *C. spinosa* buds collected on the island of Salina (Eolian Archipelago, Sicily) in August 2004 (Romeo et al. 2007)
Methyl cinnamate 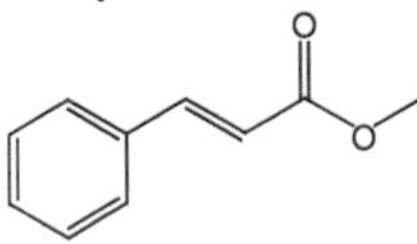	115 PPM (mg/kg) by HS-SPME/GC–MS in lactic-fermented *C. spinosa* buds collected on the island of Salina (Eolian Archipelago, Sicily) in August 2004 (Romeo et al. 2007)
Methyl heptanoate	26 PPM (mg/kg) by HS-SPME/GC–MS in lactic-fermented *C. spinosa* buds collected on the island of Salina (Eolian Archipelago, Sicily) in August 2004 (Romeo et al. 2007)
Methyl hexadecanoate	126 PPM (mg/kg) by HS-SPME/GC–MS in lactic-fermented *C. spinosa* buds collected on the island of Salina (Eolian Archipelago, Sicily) in August 2004 (Romeo et al. 2007)
Methyl isohexadecanoate	124 PPM (mg/kg) by HS-SPME/GC–MS in lactic-fermented *C. spinosa* buds collected on the island of Salina (Eolian Archipelago, Sicily) in August 2004 (Romeo et al. 2007)
Methyl octadecanoate	134 PPM (mg/kg) by HS-SPME/GC–MS in lactic-fermented *C. spinosa* buds collected on the island of Salina (Eolian Archipelago, Sicily) in August 2004 (Romeo et al. 2007)
Methyl octadecenoate	136 PPM (mg/kg) by HS-SPME/GC–MS in lactic-fermented *C. spinosa* buds collected on the island of Salina (Eolian Archipelago, Sicily) in August 2004 (Romeo et al. 2007)
Methyl octanoate	31 PPM (mg/kg) by HS-SPME/GC–MS in lactic-fermented *C. spinosa* buds collected on the island of Salina (Eolian Archipelago, Sicily) in August 2004 (Romeo et al. 2007)
Methyl salicylate	87 PPM (mg/kg) by HS-SPME/GC–MS in lactic-fermented *C. spinosa* buds collected on the island of Salina (Eolian Archipelago, Sicily) in August 2004 (Romeo et al. 2007)
Phenylmethyl pentanoate	97 PPM (mg/kg) by HS-SPME/GC–MS in lactic-fermented *C. spinosa* buds collected on the island of Salina (Eolian Archipelago, Sicily) in August 2004 (Romeo et al. 2007)

(*Continued*)

TABLE 10.1 (*Continued*)
Volatile Oils of *Capparis* spp.

Compound	Notes
	Volatile Aldehydes of *Capparis*
Benzaldehyde	48 PPM (mg/kg) by HS-SPME/GC–MS in lactic-fermented *C. spinosa* buds collected on the island of Salina (Eolian Archipelago, Sicily) in August 2004 (Romeo et al. 2007); 1.4% of volatile oil extd. and purified from *C. spinosa* seeds by steam dist., GC-MS (Meiliwan et al. 2009)
Benzylacetaldehyde	131 PPM (mg/kg) by HS-SPME/GC–MS in lactic-fermented *C. spinosa* buds collected on the island of Salina (Eolian Archipelago, Sicily) in August 2004 (Romeo et al. 2007)
(E)-2-Butenal	6 PPM (mg/kg) by HS-SPME/GC–MS in lactic-fermented *C. spinosa* buds collected on the island of Salina (Eolian Archipelago, Sicily) in August 2004 (Romeo et al. 2007)
(E)-Cinnamaldehyde 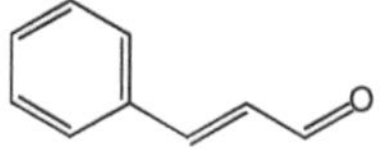	113 PPM (mg/kg) by HS-SPME/GC–MS in lactic-fermented *C. spinosa* buds collected on the island of Salina (Eolian Archipelago, Sicily) in August 2004 (Romeo et al. 2007)
(E,E)-2,4-Decadienal	90 PPM (mg/kg) by HS-SPME/GC–MS in lactic-fermented *C. spinosa* buds collected on the island of Salina (Eolian Archipelago, Sicily) in August 2004 (Romeo et al. 2007)
(E)-2-Decenal	67 PPM (mg/kg) by HS-SPME/GC–MS in lactic-fermented *C. spinosa* buds collected on the island of Salina (Eolian Archipelago, Sicily) in August 2004 (Romeo et al. 2007)
2,4-Dimethyl benzaldehyde	82 PPM (mg/kg) by HS-SPME/GC–MS in lactic-fermented *C. spinosa* buds collected on the island of Salina (Eolian Archipelago, Sicily) in August 2004 (Romeo et al. 2007)
2,2-Dimethyl 3,4-pentadienal	54 PPM (mg/kg) by HS-SPME/GC–MS in lactic-fermented *C. spinosa* buds collected on the island of Salina (Eolian Archipelago, Sicily) in August 2004 (Romeo et al. 2007)
4-Ethyl benzaldehyde	78 PPM (mg/kg) by HS-SPME/GC–MS in lactic-fermented *C. spinosa* buds collected on the island of Salina (Eolian Archipelago, Sicily) in August 2004 (Romeo et al. 2007)
5-Ethyl 2-furaldehyde	66 PPM (mg/kg) by HS-SPME/GC–MS in lactic-fermented *C. spinosa* buds collected on the island of Salina (Eolian Archipelago, Sicily) in August 2004 (Romeo et al. 2007)

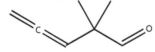

TABLE 10.1 (*Continued*)
Volatile Oils of *Capparis* spp.

Compound	Notes
Ethanal methyl pentyl acetal 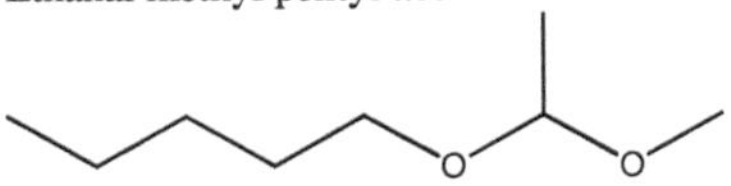	5.9% of *C. ovata* bud oil, steam dist., solvent ext., GC-MS (El-Ghorab et al. 2007)
Furfural 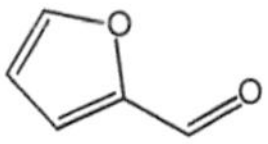	42 PPM (mg/kg) by HS-SPME/GC–MS in lactic-fermented *C. spinosa* buds collected on the island of Salina (Eolian Archipelago, Sicily) in August 2004 (Romeo et al. 2007); 7.4% of *C. ovata* bud oil, steam dist., solvent ext., GC-MS (El-Ghorab et al. 2007)
(E,E)-2,4-Heptadienal 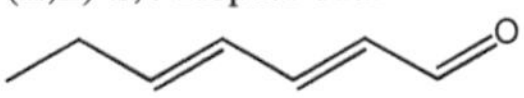	45 PPM (mg/kg) by HS-SPME/GC–MS in lactic-fermented *C. spinosa* buds collected on the island of Salina (Eolian Archipelago, Sicily) in August 2004 (Romeo et al. 2007)
Heptanal 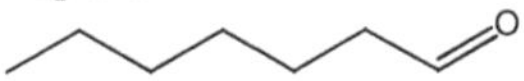	13 PPM (mg/kg) by HS-SPME/GC–MS in lactic-fermented *C. spinosa* buds collected on the island of Salina (Eolian Archipelago, Sicily) in August 2004 (Romeo et al. 2007)
(E)-2-Heptenal 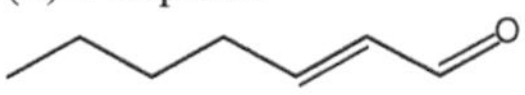	28 PPM (mg/kg) by HS-SPME/GC–MS in lactic-fermented *C. spinosa* buds collected on the island of Salina (Eolian Archipelago, Sicily) in August 2004 (Romeo et al. 2007)
(E,E)-2,4-Hexadienal 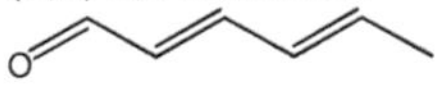	34 PPM (mg/kg) by HS-SPME/GC–MS in lactic-fermented *C. spinosa* buds collected on the island of Salina (Eolian Archipelago, Sicily) in August 2004 (Romeo et al. 2007)
(E)-2-Hexenal 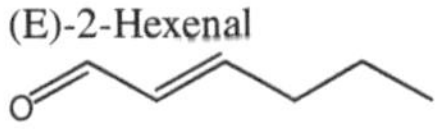	18 PPM (mg/kg) by HS-SPME/GC–MS in lactic-fermented *C. spinosa* buds collected on the island of Salina (Eolian Archipelago, Sicily) in August 2004 (Romeo et al. 2007)
3-Methoxycinnamaldehyde 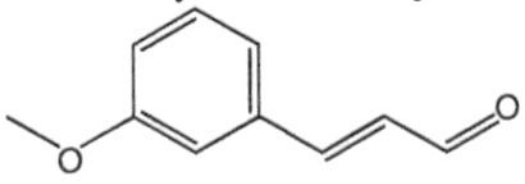	142 PPM (mg/kg) by HS-SPME/GC–MS in lactic-fermented *C. spinosa* buds collected on the island of Salina (Eolian Archipelago, Sicily) in August 2004 (Romeo et al. 2007)
3-Methyl butanal 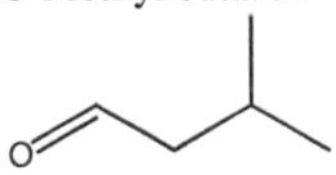	3 PPM (mg/kg) by HS-SPME/GC–MS in lactic-fermented *C. spinosa* buds collected on the island of Salina (Eolian Archipelago, Sicily) in August 2004 (Romeo et al. 2007)
5-Methyl furancarboxaldehyde 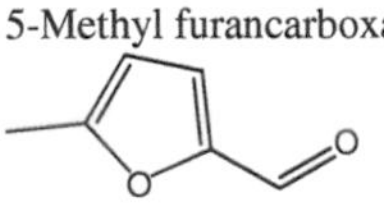	55 PPM (mg/kg) by HS-SPME/GC–MS in lactic-fermented *C. spinosa* buds collected on the island of Salina (Eolian Archipelago, Sicily) in August 2004 (Romeo et al. 2007)
(E,E)-2,4-Nonadienal 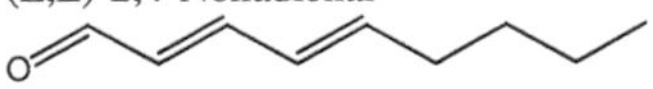	77 PPM (mg/kg) by HS-SPME/GC–MS in lactic-fermented *C. spinosa* buds collected on the island of Salina (Eolian Archipelago, Sicily) in August 2004 (Romeo et al. 2007)

(*Continued*)

TABLE 10.1 (*Continued*)
Volatile Oils of *Capparis* spp.

Compound	Notes
(E,Z)-2,6 Nonadienal 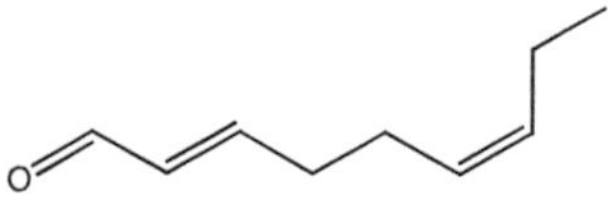	58 PPM (mg/kg) by HS-SPME/GC–MS in lactic-fermented *C. spinosa* buds collected on the island of Salina (Eolian Archipelago, Sicily) in August 2004 (Romeo et al. 2007)
Nonanal 	32 PPM (mg/kg) by HS-SPME/GC–MS in lactic-fermented *C. spinosa* buds collected on the island of Salina (Eolian Archipelago, Sicily) in August 2004 (Romeo et al. 2007)
(E)-2-Nonenal 	50 PPM (mg/kg) by HS-SPME/GC–MS in lactic-fermented *C. spinosa* buds collected on the island of Salina (Eolian Archipelago, Sicily) in August 2004 (Romeo et al. 2007)
Octanal 	25 PPM (mg/kg) by HS-SPME/GC–MS in lactic-fermented *C. spinosa* buds collected on the island of Salina (Eolian Archipelago, Sicily) in August 2004 (Romeo et al. 2007)
Pentanal 	4 PPM (mg/kg) by HS-SPME/GC–MS in lactic-fermented *C. spinosa* buds collected on the island of Salina (Eolian Archipelago, Sicily) in August 2004 (Romeo et al. 2007)
(E)-2-Pentenal 	9 PPM (mg/kg) by HS-SPME/GC–MS in lactic-fermented *C. spinosa* buds collected on the island of Salina (Eolian Archipelago, Sicily) in August 2004 (Romeo et al. 2007)
Phenylacetaldehyde 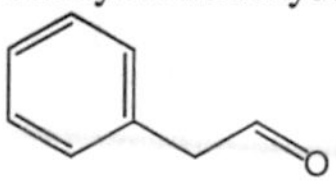	0.8% of volatile oil extd. and purified from *C. spinosa* seeds by steam dist., GC-MS (Meiliwan et al. 2009)
Volatile Ketones of *Capparis*	
Acetoin 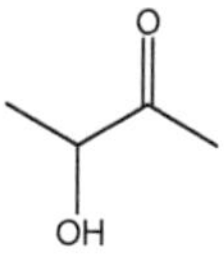	24 PPM (mg/kg) by HS-SPME/GC–MS in lactic-fermented *C. spinosa* buds collected on the island of Salina (Eolian Archipelago, Sicily) in August 2004 (Romeo et al. 2007)
Acetophenone 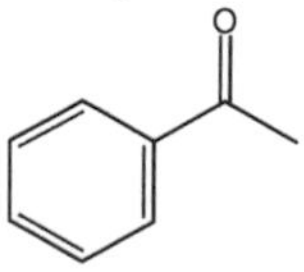	69 PPM (mg/kg) by HS-SPME/GC–MS in lactic-fermented *C. spinosa* buds collected on the island of Salina (Eolian Archipelago, Sicily) in August 2004 (Romeo et al. 2007)
4-Methoxy acetophenone 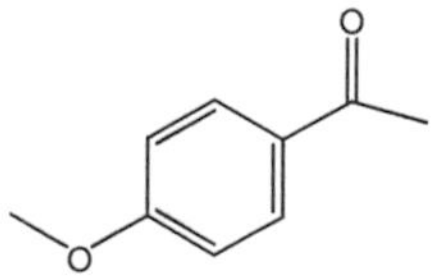	121 PPM (mg/kg) by HS-SPME/GC–MS in lactic-fermented *C. spinosa* buds collected on the island of Salina (Eolian Archipelago, Sicily) in August 2004 (Romeo et al. 2007)

TABLE 10.1 (*Continued*)
Volatile Oils of *Capparis* spp.

Compound	Notes
6-Methyl 5-hepten-2-one	29 PPM (mg/kg) by HS-SPME/GC–MS in lactic-fermented *C. spinosa* buds collected on the island of Salina (Eolian Archipelago, Sicily) in August 2004 (Romeo et al. 2007); 3.5% of volatile oil extd. and purified from *C. spinosa* seeds by steam dist., GC-MS (Meiliwan et al. 2009)
(E,E) 3,5-Octadiene-2-one 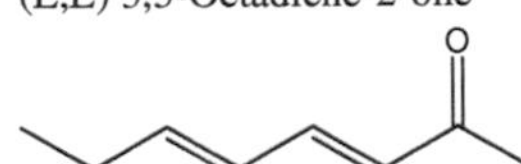	47 PPM (mg/kg) by HS-SPME/GC–MS in lactic-fermented *C. spinosa* buds collected on the island of Salina (Eolian Archipelago, Sicily) in August 2004 (Romeo et al. 2007)
Volatile Alcohols of *Capparis*	
Aliphatic alcohols in general	45 mg/kg of total extd. lipids in *C. spinosa* seeds (Tlili et al. 2011)
p-Allylanisole	92 PPM (mg/kg) by HS-SPME/GC–MS in lactic-fermented *C. spinosa* buds collected on the island of Salina (Eolian Archipelago, Sicily) in August 2004 (Romeo et al. 2007)
Benzyl alcohol	20.4% of *C. ovata* bud oil, steam dist., solvent ext., GC-MS (El-Ghorab et al. 2007)
1-Butanol	10 PPM (mg/kg) by HS-SPME/GC–MS in lactic-fermented *C. spinosa* buds collected on the island of Salina (Eolian Archipelago, Sicily) in August 2004 (Romeo et al. 2007)
4-Decanol	56 PPM (mg/kg) by HS-SPME/GC–MS in lactic-fermented *C. spinosa* buds collected on the island of Salina (Eolian Archipelago, Sicily) in August 2004 (Romeo et al. 2007)
2-Ethyl 1-hexanol	44 PPM (mg/kg) by HS-SPME/GC–MS in lactic-fermented *C. spinosa* buds collected on the island of Salina (Eolian Archipelago, Sicily) in August 2004 (Romeo et al. 2007)
Hexadecanol	Identified in *C. spinosa* seeds (Tlili et al. 2011)
2-Methyl 1-butanol	15 PPM (mg/kg) by HS-SPME/GC–MS in lactic-fermented *C. spinosa* buds collected on the island of Salina (Eolian Archipelago, Sicily) in August 2004 (Romeo et al. 2007)

(*Continued*)

TABLE 10.1 (*Continued*)
Volatile Oils of *Capparis* spp.

Compound	Notes
4-(1-Methyl propyl) phenol	110 PPM (mg/kg) by HS-SPME/GC–MS in lactic-fermented *C. spinosa* buds collected on the island of Salina (Eolian Archipelago, Sicily) in August 2004 (Romeo et al. 2007)
Octadecanol	28 PPM of total extd. lipids in *C. spinosa* seeds (Tlili et al. 2011)
1-Octanol	52 PPM (mg/kg) by HS-SPME/GC–MS in lactic-fermented *C. spinosa* buds collected on the island of Salina (Eolian Archipelago, Sicily) in August 2004 (Romeo et al. 2007)
3-Octanol	30 PPM (mg/kg) by HS-SPME/GC–MS in lactic-fermented *C. spinosa* buds collected on the island of Salina (Eolian Archipelago, Sicily) in August 2004 (Romeo et al. 2007)
1-Octen-3-ol	38 PPM (mg/kg) by HS-SPME/GC–MS in lactic-fermented *C. spinosa* buds collected on the island of Salina (Eolian Archipelago, Sicily) in August 2004 (Romeo et al. 2007)
Phenyl ethyl alcohol	100 PPM (mg/kg) by HS-SPME/GC–MS in lactic-fermented *C. spinosa* buds collected on the island of Salina (Eolian Archipelago, Sicily) in August 2004 (Romeo et al. 2007)
Volatile Sulfur-Containing Compounds (Other than Isothiocyanates) of *Capparis*	
Benzothiazole 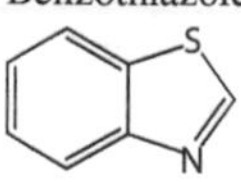	106 PPM (mg/kg) by HS-SPME/GC–MS in lactic-fermented *C. spinosa* buds collected on the island of Salina (Eolian Archipelago, Sicily) in August 2004 (Romeo et al. 2007)
Carbon disulfide	1 PPM (mg/kg) by HS-SPME/GC–MS in lactic-fermented *C. spinosa* buds collected on the island of Salina (Eolian Archipelago, Sicily) in August 2004 (Romeo et al. 2007)
Dimethyl sulfide	2 PPM (mg/kg) by HS-SPME/GC–MS in lactic-fermented *C. spinosa* buds collected on the island of Salina (Eolian Archipelago, Sicily) in August 2004 (Romeo et al. 2007)
2-Ethyl thiophene	2 PPM (mg/kg) by HS-SPME/GC–MS in lactic-fermented *C. spinosa* buds collected on the island of Salina (Eolian Archipelago, Sicily) in August 2004 (Romeo et al. 2007)
S-containing volatile principle	From seeds of *C. decidua*, active against *Vibrio cholera* (Gaind et al. 1972)

TABLE 10.1 (*Continued*)
Volatile Oils of *Capparis* spp.

Compound	Notes
	Volatile Hydrocarbons of *Capparis*
Cyclopentadecane 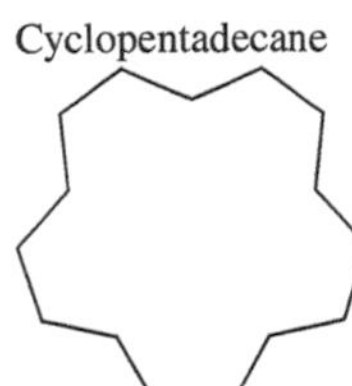	51 PPM (mg/kg) by HS-SPME/GC–MS in lactic-fermented *C. spinosa* buds collected on the island of Salina (Eolian Archipelago, Sicily) in August 2004 (Romeo et al. 2007)
Cyclotetradecane 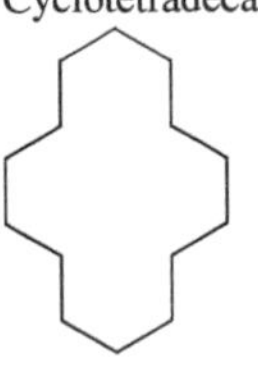	36 PPM (mg/kg) by HS-SPME/GC–MS in lactic-fermented *C. spinosa* buds collected on the island of Salina (Eolian Archipelago, Sicily) in August 2004 (Romeo et al. 2007)
Docosane	125 PPM (mg/kg) by HS-SPME/GC–MS in lactic-fermented *C. spinosa* buds collected on the island of Salina (Eolian Archipelago, Sicily) in August 2004 (Romeo et al. 2007)
Eicosane	109 PPM (mg/kg) by HS-SPME/GC–MS in lactic-fermented *C. spinosa* buds collected on the island of Salina (Eolian Archipelago, Sicily) in August 2004 (Romeo et al. 2007)
Heneicosane	116 PPM (mg/kg) by HS-SPME/GC–MS in lactic-fermented *C. spinosa* buds collected on the island of Salina (Eolian Archipelago, Sicily) in August 2004 (Romeo et al. 2007)
Heptadecane	76 PPM (mg/kg) by HS-SPME/GC–MS in lactic-fermented *C. spinosa* buds collected on the island of Salina (Eolian Archipelago, Sicily) in August 2004 (Romeo et al. 2007)
Hexadecane	60 PPM (mg/kg) by HS-SPME/GC–MS in lactic-fermented *C. spinosa* buds collected on the island of Salina (Eolian Archipelago, Sicily) in August 2004 (Romeo et al. 2007)
4-Methyl heneicosane	123 PPM (mg/kg) by HS-SPME/GC–MS in lactic-fermented *C. spinosa* buds collected on the island of Salina (Eolian Archipelago, Sicily) in August 2004 (Romeo et al. 2007)
4-Methyl heptadecane	85 PPM (mg/kg) by HS-SPME/GC–MS in lactic-fermented *C. spinosa* buds collected on the island of Salina (Eolian Archipelago, Sicily) in August 2004 (Romeo et al. 2007)

(*Continued*)

TABLE 10.1 (*Continued*)
Volatile Oils of *Capparis* spp.

Compound	Notes
4-Methyl hexacosane 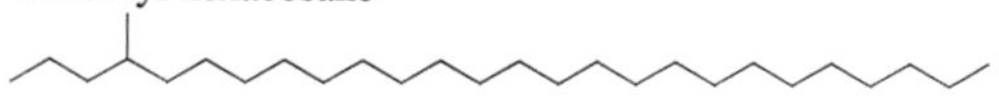	145 PPM (mg/kg) by HS-SPME/GC–MS in lactic-fermented *C. spinosa* buds collected on the island of Salina (Eolian Archipelago, Sicily) in August 2004 (Romeo et al. 2007)
4-Methyl hexadecane 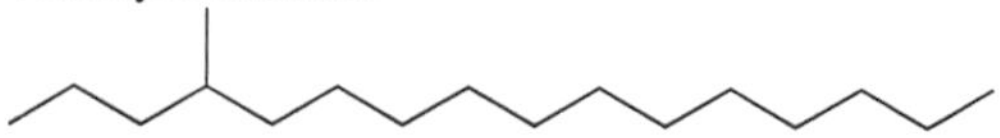	68 PPM (mg/kg) by HS-SPME/GC–MS in lactic-fermented *C. spinosa* buds collected on the island of Salina (Eolian Archipelago, Sicily) in August 2004 (Romeo et al. 2007)
4-Methyl nonadecane 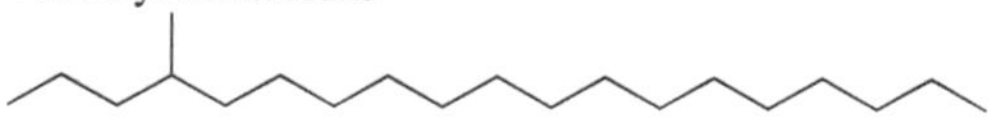	105 PPM (mg/kg) by HS-SPME/GC–MS in lactic-fermented *C. spinosa* buds collected on the island of Salina (Eolian Archipelago, Sicily) in August 2004 (Romeo et al. 2007)
4-Methyl octadecane 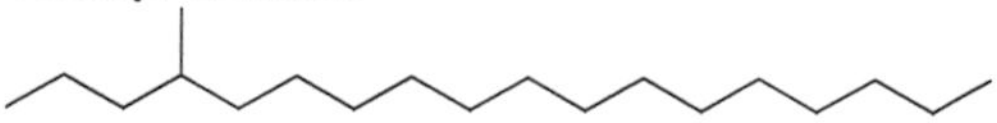	94 PPM (mg/kg) by HS-SPME/GC–MS in lactic-fermented *C. spinosa* buds collected on the island of Salina (Eolian Archipelago, Sicily) in August 2004 (Romeo et al. 2007)
4-Methyl pentadecane 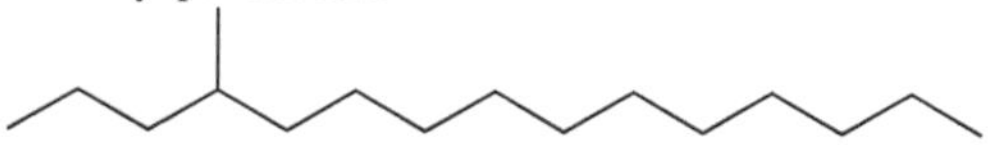	53 PPM (mg/kg) by HS-SPME/GC–MS in lactic-fermented *C. spinosa* buds collected on the island of Salina (Eolian Archipelago, Sicily) in August 2004 (Romeo et al. 2007)
4-Methyl tetracosane 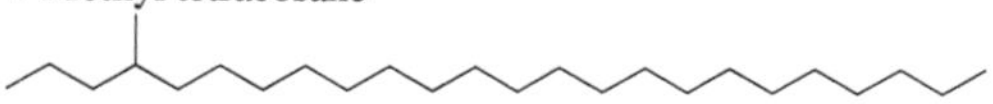	137 PPM (mg/kg) by HS-SPME/GC–MS in lactic-fermented *C. spinosa* buds collected on the island of Salina (Eolian Archipelago, Sicily) in August 2004 (Romeo et al. 2007)
4-Methyl tetradecane 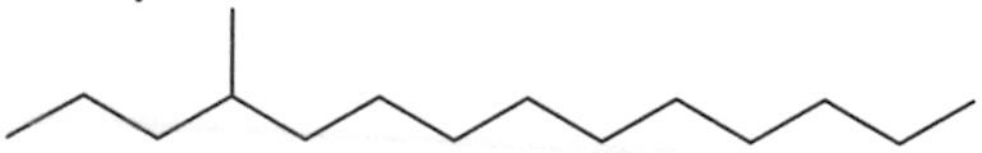	40 PPM (mg/kg) by HS-SPME/GC–MS in lactic-fermented *C. spinosa* buds collected on the island of Salina (Eolian Archipelago, Sicily) in August 2004 (Romeo et al. 2007)
4-Methyl tricosane 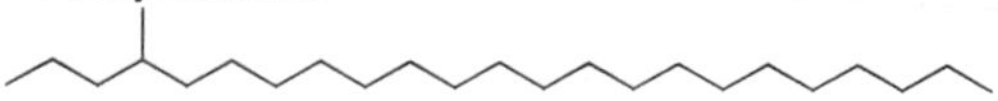	133 PPM (mg/kg) by HS-SPME/GC–MS in lactic-fermented *C. spinosa* buds collected on the island of Salina (Eolian Archipelago, Sicily) in August 2004 (Romeo et al. 2007)
Nonadecane 	99 PPM (mg/kg) by HS-SPME/GC–MS in lactic-fermented *C. spinosa* buds collected on the island of Salina (Eolian Archipelago, Sicily) in August 2004 (Romeo et al. 2007)
Octadecane 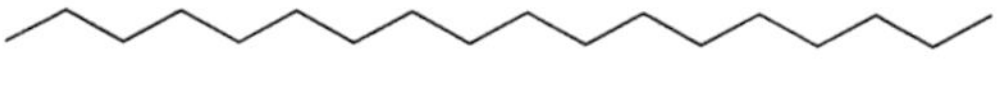	89 PPM (mg/kg) by HS-SPME/GC–MS in lactic-fermented *C. spinosa* buds collected on the island of Salina (Eolian Archipelago, Sicily) in August 2004 (Romeo et al. 2007)
Pentacosane 	140 PPM (mg/kg) by HS-SPME/GC–MS in lactic-fermented *C. spinosa* buds collected on the island of Salina (Eolian Archipelago, Sicily) in August 2004 (Romeo et al. 2007)
Pentadecane 	46 PPM (mg/kg) by HS-SPME/GC–MS in lactic-fermented *C. spinosa* buds collected on the island of Salina (Eolian Archipelago, Sicily) in August 2004 (Romeo et al. 2007)

TABLE 10.1 (*Continued*)
Volatile Oils of *Capparis* spp.

Compound	Notes
Styrene	21 PPM (mg/kg) by HS-SPME/GC–MS in lactic-fermented *C. spinosa* buds collected on the island of Salina (Eolian Archipelago, Sicily) in August 2004 (Romeo et al. 2007)
Tetradecane	33 PPM (mg/kg) by HS-SPME/GC–MS in lactic-fermented *C. spinosa* buds collected on the island of Salina (Eolian Archipelago, Sicily) in August 2004 (Romeo et al. 2007)
Tricosane	132 PPM (mg/kg) by HS-SPME/GC–MS in lactic-fermented *C. spinosa* buds collected on the island of Salina (Eolian Archipelago, Sicily) in August 2004 (Romeo et al. 2007)
Tridecane	27 PPM (mg/kg) by HS-SPME/GC–MS in lactic-fermented *C. spinosa* buds collected on the island of Salina (Eolian Archipelago, Sicily) in August 2004 (Romeo et al. 2007)
Other Volatile Compounds of *Capparis*	
Diphenyl ether 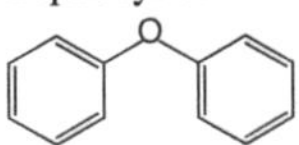	111 PPM (mg/kg) by HS-SPME/GC–MS in lactic-fermented *C. spinosa* buds collected on the island of Salina (Eolian Archipelago, Sicily) in August 2004 (Romeo et al. 2007)
1-Methoxy-4-(1-propenyl) benzene 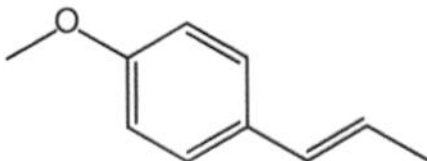	12.5% of volatile oil extd. and purified from *C. spinosa* seeds by steam dist., GC-MS (Meiliwan et al. 2009)
p-Methoxystyrene 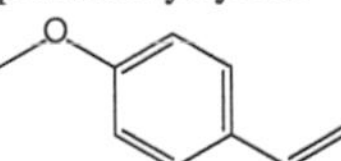	73 PPM (mg/kg) by HS-SPME/GC–MS in lactic-fermented *C. spinosa* buds collected on the island of Salina (Eolian Archipelago, Sicily) in August 2004 (Romeo et al. 2007)
4-Methyl-4-pentenenitrile 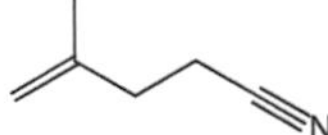	From volatile oil of *C. flexuosa*, by IR, MS, NMR, and chem. derivatization (Gramosa et al. 1997)
Naphthalene	83 PPM (mg/kg) by HS-SPME/GC–MS in lactic-fermented *C. spinosa* buds collected on the island of Salina (Eolian Archipelago, Sicily) in August 2004 (Romeo et al. 2007)
2-Naphthalenon	95 PPM (mg/kg) by HS-SPME/GC–MS in lactic-fermented *C. spinosa* buds collected on the island of Salina (Eolian Archipelago, Sicily) in August 2004 (Romeo et al. 2007)

REFERENCES

Abdel-Mogib, M., S.T. Ezmirly, and S.A. Basaif. 2000. Phytochemistry of *Dipterygium glaucum* and *Capparis decidua*. *Journal of Saudi Chemical Society* 4(1): 103–108.

Anon. 2001. Final report on the safety assessment of *Juniperus communis* extract, *Juniperus oxycedrus* extract, *Juniperus oxycedrus* tar, *Juniperus phoenicea* extract, and *Juniperus virginiana* extract. *International Journal of Toxicology* 20(Suppl. 2): 41–56.

Bayliss, D.A. and P.Q. Barrett. 2008. Emerging roles for two-pore-domain potassium channels and their potential therapeutic impact. *Trends in Pharmacological Science* 29(11): 566–575.

Beauchamp, J. 2011. Inhaled today, not gone tomorrow: pharmacokinetics and environmental exposure of volatiles in exhaled breath. *Journal of Breath Research* 5(3): 037103.

Boutin-Forzano, S., Y. Hammou, M. Gouitaa, and D. Charpin. 2005. Air pollution and atopy. *European Annals of Allergy and Clinical Immunology* 37(1): 11–16.

Calway, T., G.J. Du, C.Z. Wang, W.H. Huang, J. Zhao, S.P. Li, and C.S. Yuan. 2012. Chemical and pharmacological studies of *Oplopanax horridus*, a North American botanical. *Journal of Natural Medicine* 66(2): 249–256.

Chaudhury, N.A. and D. Ghosh. 1970. Insecticidal plants: chemical examination of the leaves of *Capparis sepiaria*. *Journal of the Indian Chemical Society* 47(8): 751–754.

Crinnion, W.J. 2012. Do environmental toxicants contribute to allergy and asthma? *Alternative Medical Review* 17(1): 6–18.

de Sousa, D.P. 2011. Analgesic-like activity of essential oils constituents. *Molecules* 16(3): 2233–2252.

Dhawan, K., S. Dhawan, and A. Sharma. 2004. Passiflora: a review update. *Journal of Ethnopharmacology* 94(1): 1–23.

Dutt, H.C., S. Singh, B. Avula, I.A. Khan, and Y.S. Bedi. 2012. Pharmacological review of Caralluma R.Br. with special reference to appetite suppression and anti-obesity. *Journal of Medicinal Food* 15(2): 108–119.

Efferth, T. and E. Koch. 2011. Complex interactions between phytochemicals: the multi-target therapeutic concept of phytotherapy. *Current Drug Targets* 12(1): 122–132.

El-Ghorab, A., T. Shibamoto, and M. Özcan. 2007. Chemical composition and antioxidant activities of buds and leaves of capers (*Capparis ovata* Desf. var. *canescens*) cultivated in Turkey. *Journal of Essential Oil Research* 19(1): 72–77.

Gaind, K.N., T.R. Juneja, and P.N. Bhandarkar. 1972. Volatile principle from seeds of *Capparis decidua*: kinetics of in vitro antibacterial activity against *Vibrio cholerae* Ogava, Inaba, and Eltor. *Indian Journal of Pharmacy* 34(4): 86–88.

Gali-Muhtasib, H., A. Roessner, and R. Schneider-Stock. 2006. Thymoquinone: a promising anti-cancer drug from natural sources. *International Journal of Biochemistry and Cell Biology* 38(8): 1249–1253.

Goldstraw, P., D. Ball, J.R. Jett, T. Le Chevalier, E. Lim, A.G. Nicholson, and F.A. Shepherd. 2011. Non-small-cell lung cancer. *Lancet* 378(9804): 1727–1740.

Gramosa, N.V., T.L.G. Lemos, and R. Braz-Filho. 1997. Volatile constituents isolated from *Capparis flexuosa* of Brazil. *Journal of Essential Oil Research* 9(6): 709–712.

Herrmann, A. 2007. Controlled release of volatiles under mild reaction conditions: from nature to everyday products. *Angew Chemie International Edition in English* 46(31): 5836–5863.

Horváth, I., Z. Lázár, N. Gyulai, M. Kollai, and G. Losonczy. 2009. Exhaled biomarkers in lung cancer. *European Respiratory Journal* 34(1): 261–275.

Kalemba, D. and A. Kunicka. 2003. Antibacterial and antifungal properties of essential oils. *Current Medical Chemistry* 10(10): 813–829.

Landoni, G., S. Turi, E. Bignami, and A. Zangrillo. 2009. Organ protection by volatile anesthetics in non-coronary artery bypass grafting surgery. *Future Cardiology* 5(6): 589–603.

Lertsutthiwong, P. and P. Rojsitthisak. 2011. Chitosan-alginate nanocapsules for encapsulation of turmeric oil. *Pharmazie* 66(12): 911–915.

Li, G.Q., J. Li, Z.G. Han, and S. Ababakri. 2009. Chemical constituents of volatile oil in the leaves of *Capparis spinosa* L. from Xinjiang. *Shengwu Jishu* 19(1): 46–48.

Li, Y.M., L. Zhu, J.G. Jiang, L. Yang, and D.Y. Wang. 2009. Bioactive components and pharmacological action of *Wikstroemia indica* (L.) C. A. Mey and its clinical application. *Current Pharmaceutical Biotechnology* 10(8): 743–752.

Li Volti, G., F. Basile, P. Murabito, F. Galvano, C. Di Giacomo, D. Gazzolo, S. Vadalà, R. Azzolina, N. D'Orazio, H. Mufeed, L. Vanella, A. Nicolosi, G. Basile, and A. Biondi. 2008. Antioxidant properties of anesthetics: the biochemist, the surgeon and the anesthetist. *Clinica Terapeutica* 159(6): 463–469.

Liu, S.J., A.P. Tu, and J.X. Dong. 2009. Studies on chemical components of extraction of oil from *Capparis spinosa* by supercritical CO_2. *Jiefangjun Yaoxue Xuebao* 25(4): 321–323.

Loesche, W.J. and C. Kazor. 2002. Microbiology and treatment of halitosis. *Periodontol 2000* 28: 256–79.

McLuhan, M. 1964. *Understanding Media; The Extensions of Man*. New York: McGraw-Hill.

Meiliwan, A., A. Ainiwaer, Y. Wang, and Hajinisha. 2009. Extraction of volatile oil from *Capparis spinosa* L. seeds and analysis by GC-MS. *Huaxi Yaoxue Zazhi* 24(1): 5–6.

Mennella, J.A. and G.K. Beauchamp. 2008. Optimizing oral medications for children. *Clinical Therapeutics* 30(11): 2120–2132.

Mimica-Dukic, N. and B. Bozin. 2008. *Mentha* L. species (Lamiaceae) as promising sources of bioactive secondary metabolites. *Current Pharmaceutical Design* 14(29): 3141–3150.

Nomicos, E.Y. 2007. Myrrh: medical marvel or myth of the Magi? *Holistic Nursing Practice* 21(6): 308–323.

Pauli, A. 2006. Anticandidal low molecular compounds from higher plants with special reference to compounds from essential oils. *Medical Research Review* 26(2): 223–268.

Pauluhn, J. 2004. Acute inhalation studies with irritant aerosols: technical issues and relevance for risk characterization. *Archives of Toxicology* 78(5): 243–251.

Pisseri, F., A. Bertoli, and L. Pistelli. 2008. Essential oils in medicine: principles of therapy. *Parassitologia* 50(1–2): 89–91.

Porter, S.R. 2011. Diet and halitosis. *Current Opinion on Clinical Nutrition and Metabolic Care* 14(5): 463–468.

Quirynen, M., H. Zhao, and D. van Steenberghe. 2002. Review of the treatment strategies for oral malodour. *Clinical Oral Investigations* 6(1): 1–10.

Ren, Y., J. Xu, J. Zhao, F. Xu, W. Yang, and Y. Liu. 2009. Chemical components of volatile oil and fatty acid by supercritical carbon dioxide fluid extracts from fruit of *Capparis spinosa* L. *Xinjiang Yike Daxue Xuebao* 32(12): 1659–1660, 1663.

Romeo, V., M. Ziino, D. Giuffrida, C. Condurso, and A. Verzera. 2007. Flavour profile of capers (*Capparis spinosa* L.) from the Eolian Archipelago by HS-SPME/GC–MS. *Food Chemistry* 101: 1272–1278.

Sakurada, T., H. Kuwahata, S. Katsuyama, T. Komatsu, L.A. Morrone, M.T. Corasaniti, G. Bagetta, and S. Sakurada. 2009. Intraplantar injection of bergamot essential oil into the mouse hindpaw: effects on capsaicin-induced nociceptive behaviors. *International Review of Neurobiology* 85: 237–248.

Sorg, B.A., M.L. Tschirgi, S. Swindell, L. Chen, and J. Fang. 2001. Repeated formaldehyde effects in an animal model for multiple chemical sensitivity. *Annals of the New York Academy of Sciences* 933: 57–67.

Soukup, J., K. Schärff, K. Kubosch, C. Pohl, M. Bomplitz, and J. Kompardt. 2009. State of the art: sedation concepts with volatile anesthetics in critically ill patients. *Journal of Critical Care* 24(4): 535–544.

Swiderski, F., M. Dabrowska, A. Rusaczonek, and B. Waszkiewicz-Robak. 2007. Bioactive substances of garlic and their role in dietoprophylaxis and dietotherapy. *Rocz Panstw Zakl Hig* 58(1): 41–46.

Tedesco, P. and J. Cicchetti. 2001. Like cures like: homeopathy. *American Journal of Nursing* 101(9): 43–49; quiz 49–50.

Tlili, N., T. El Guizani, N. Nasri, A. Khaldi, and S. Triki. 2011. Protein, lipid, aliphatic and triterpenic alcohol content of caper seeds "Capparis spinosa." *Journal of the American Oil Chemists' Society* 88(2): 265–270.

van Vuuren, S.F. 2008. Antimicrobial activity of South African medicinal plants. *Journal of Ethnopharmacology* 119(3): 462–472.

Vijverberg, S.J., L. Koenderman, E.S. Koster, C.K. van der Ent, J.A. Raaijmakers, A.H. Maitland-van der Zee. 2011. Biomarkers of therapy responsiveness in asthma: pitfalls and promises. *Clinical & Experimental Allergy* 41(5): 615–629.

Willc, S.M. and W.E. Lambert. 2004. Volatile substance abuse—post-mortem diagnosis. *Forensic Science International* 142(2–3): 135–156.

Wilson, A.D. and M. Baietto. 2011. Advances in electronic-nose technologies developed for biomedical applications. *Sensors (Basel)* 11(1): 1105–1176.

Woronuk, G., Z. Demissie, M. Rheault, and S. Mahmoud. 2011. Biosynthesis and therapeutic properties of *Lavandula* essential oil constituents. *Planta Medica* 77(1): 7–15.

Yang, T., H. Liu, X. Cheng, F. Yu, G. Chou, C. Wang, and Z. Wang. 2011. The chemical constituents from stems and leaves of *Capparis spinosa* L. *Xibei Yaoxue Zazhi* 26(1): 16–18.

11 Vitamins

Capparis fruits and berries (Figure 11.1) contain both water-soluble vitamins such as vitamins C and B and fat-soluble vitamins E (tocopherols) and vitamin A precursors (carotenoids). All of these are presented in Table 11.1.

First, we briefly review functions of these vitamins in plants, and second touch on pharmacological uses and implications of these vitamins in human and veterinary medicine. Let us begin with water-soluble vitamin C.

Vitamin C often occurs in plant salts as ascorbates, resulting from combinations of ascorbic acid with electrolytes such as sodium, potassium, calcium, and the like. In plants, ascorbates act as antioxidants, protecting the plant in cooperation with other phytochemicals from oxidative damage sustained during aerobic metabolism, photosynthesis, and pollutants, partially through the mechanism of the ascorbate glutathione cycle. In addition, vitamins function as coenzymes, providing material support to hydroxylase enzymes (e.g., prolyl hydroxylase) and violaxanthin deepoxidase, the latter enzyme linking ascorbate to the photoprotective xanthophyll cycle. Ascorbate plays a role in regulating electron transport, in concert with cell wall-localized ascorbate oxidase, modulating cell extensibility and growth and cell cycle progression from G-1 to S (Smirnoff 1996).

Vitamin E refers to a family of compounds known as tocochromanols, or tocols, which include alpha, beta, gamma, and delta tocopherols; alpha, beta, gamma, and delta tocotrienols; and tocomonoenols, the latter having been characterized primarily in palm oil and seafoods, including salmon and krill and their eggs (Shen et al. 2011). In higher plants, tocols provide a range of protective functions. In the model grass *Arabidopsis thaliana*, for example, vitamin E in chloroplasts protects two major sites of singlet oxygen production in conjunction with the xanothophyll cycle: against photoinactivation of photosystem II (PSII) and against photooxidation of membrane lipids, both associated with environmental conditions of extremely high light, especially when concurrent with depressed ambient temperature (Havaux et al. 2005). Transgenic explorations in altering tocochromanol biosynthesis in tobacco, potato, and maize highlighted the role of tocols in germination, export of photoassimilates, growth, leaf senescence, and plant responses to abiotic stresses, all beyond antioxidant actions (Falk and Munné-Bosch 2010). In *A. thaliana*, enhanced expression of tocopherols in response to extreme cold occurs (Maeda et al. 2006). *Tocotrienol* function in tobacco plants includes antioxidant and lipid protection (Matringe et al. 2008), and other roles may exist. To date, among the tocochromanols, only tocopherols have been identified in capers, most prominently in the seed oil of various species.

To round out the fat-soluble vitamins of *Capparis*, a discussion of the large class of vitamin A precursors, the carotenoids, is needed. Over 400 different carotenoids occur in higher plants, with the carotenoid β-carotenene one of the most important and also the best known. Carotenoids guard against oxidative damage in plants and microorganisms by a triplet energy transfer mechanism to protect against potential formation of singlet oxygen, with a singlet energy transfer mechanism, from excited chlorophyll to zeaxanthin preventing oxidative disruption of photosynthesis (Denunig-Adams et al. 1996). Deepoxidation products, such as zeaxanthin, help to dissipate surplus light energy not required for photosynthesis. An analogous mechanism may occur in the macular area of primate retina, in which carotenoids zeaxanthin and the related lutein occur (see comments of contributing editor Norman I. Krinsky in Denunig-Adams et al. 1996). So, β-carotene, lutein, and zeaxanthin rule ophthalmologic nutraceuticals (Bartlett and Eperjesi 2004). Dermatologically

FIGURE 11.1 Flowers and fruit of *Capparis decidua* contain considerable amounts of vitamins. (24.3.2010, Rajasthan, India: by Ryan Brookes.)

TABLE 11.1
Vitamins of *Capparis* spp.

Compound	Notes
Ascorbic acid	40 and 66 mg/g dry wt. of *C. decidua* flowers and immature fruits, resp. (Dahot 1993); 1.2 mg/g dry wt. of dried immature fruit of *C. decidua* (Chauhan et al. 1986)
β-Carotene (pro-vitamin A)	375 μg/100 g in *C. spinosa* seed oil (Tlili et al. 2009); in petroleum ether ext. of *C. zeylanica* leaves (Laddha and Jolly 1985); 5.40 mg/100 g in unripe fruit (*ker*) of *C. decidua* (Chauhan et al. 1986)
Carotenoids in general	457 μg/100 g in *C. spinosa* seed oil (Tlili et al. 2009); ident. in *C. decidua*, commonly known as *Karer* (Hindi), by paper chromatography and TLC (Dhar et al. 1972)
Folic acid	300 and 900 μg/g dry wt. of *C. decidua* flowers and immature fruits, resp. (Dahot 1993)

TABLE 11.1 (*Continued*)
Vitamins of *Capparis* spp.

Compound	Notes
Riboflavin 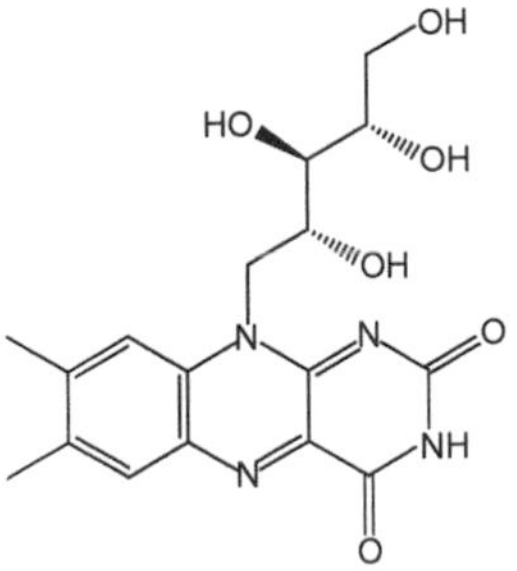	160 and 270 μg/g dry wt. of *C. decidua* flowers and immature fruits, resp. (Dahot 1993)
Thiamin 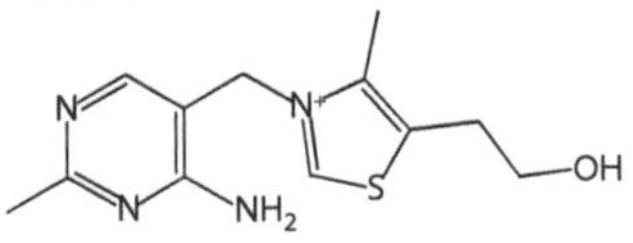	490 and 720 μg/g dry wt. of *C. decidua* flowers and immature fruits, resp. (Dahot 1993)
α-Tocopherol 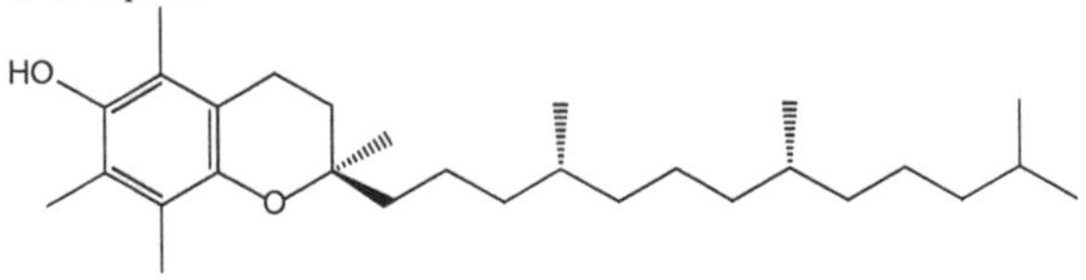	In *C. spinosa* seed oil tocopherol fraction (4%) (Tlili et al. 2009); in *C. spinosa* and *C. ovata* seed oils 0.6–13.8 mg/100 g (Matthäus and Özcan 2005)
γ-Tocopherol 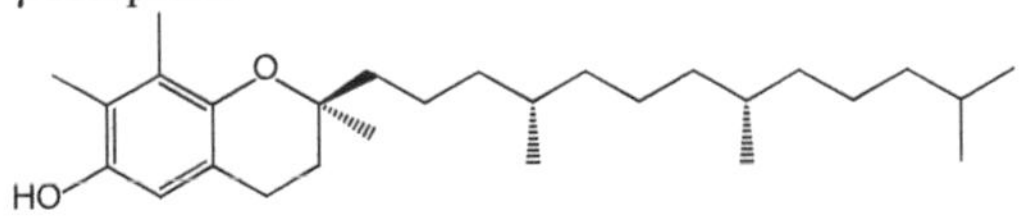	Major tocol in *C. spinosa* seed oil tocopherol fraction (92%) (Tlili et al. 2009); in *C. spinosa* and *C. ovata* seed oils, 124.3–1,944.9 mg/100 g (Matthäus and Özcan 2005); α-tocopherol, 0.6–13.8 mg/100 g
δ-Tocopherol 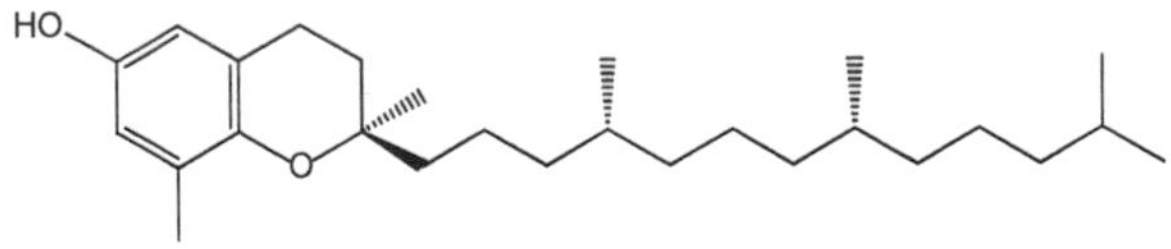	In *C. spinosa* seed oil tocopherol fraction (2%) (Tlili et al. 2009); in *C. spinosa* and *C. ovata* seed oils, 2.7–269.5 mg/100 g (Matthäus and Özcan 2005)
Vitamin E in general	Detection, isolation, and capillary GC-GC/MS quant. evaluation of vitamin E in *C. orientalis* and *C. sicula* ssp. *sicula* (Conforti et al. 2011a); *C. spinosa* seed oil → tocopherols (628 mg/100 g) (Tlili et al. 2009)

applied carotenoids may also attenuate the carcinogenic influence of ultraviolet (UV) radiation on skin through antioxidative mechanisms (Black and Lambert 2001).

Overall, vitamins A, C, E, and B9 (folate) have antioxidant activity linked to their protective and putatively therapeutic activity against numerous diseases and pathogens or their products, including aflatoxin (Alpsoy and Yalvac 2011), breast cancer (Gerber 2001), carcinogenesis in general (Guz et al. 2007), herpes simplex (Gaby 2006), and diseases affecting the cardiovascular system (Núñez-Córdoba and Martínez-González 2011).

Although synthetically produced vitamins figure prominently in modern conceptions and practices of antioxidant therapy (Firuzi et al. 2011), medicinal value of vitamins within their complex fruit and vegetable milieux has been emphasized (Esfahani et al. 2011, Hunter et al. 2011, Sroka et al. 2005). Detailed pharmacologic mechanisms of the antioxidant activity of vitamins C and E have been described (Chaudière and Ferrari-Iliou 1999, Traber and Stevens 2011), and the specific regulatory hurdles to be negotiated for having specific vitamins approved as medications have been elaborated (Moyad 2002).

The vitamins of *Capparis*, indicated in Table 11.1, although not unique to *Capparis*, undoubtedly contribute to its multiple pharmacologic effects, considered elsewhere in Chapter 13. The value of the vitamins of *Capparis* specifically for cosmetology has also been described (Lemmi Cena and Rovesti 1979).

REFERENCES

Alpsoy, L. and M.E. Yalvac. 2011. Key roles of vitamins A, C, and E in aflatoxin B1-induced oxidative stress. *Vitamins & Hormones* 86: 287–305.

Bartlett, H. and F. Eperjesi. 2004. An ideal ocular nutritional supplement? *Ophthalmic and Physiological Optics* 24(4): 339–349.

Black, H.S. and C.R. Lambert. 2001. Radical reactions of carotenoids and potential influence on UV carcinogenesis. *Current Problems in Dermatology* 29: 140–156.

Chaudière, J. and R. Ferrari-Iliou. 1999. Intracellular antioxidants: from chemical to biochemical mechanisms. *Food and Chemical Toxicology* 37(9–10): 949–962.

Chauhan, E.M., A. Duhan, and C.M. Bhat. 1986. Nutritional value of ker (*Capparis decidua*) fruit. *Journal of Food Science and Technology* 23(2): 106–108.

Conforti, F., S. Modesto, F. Menichini, G.A. Statti, D. Uzunov, U. Solimene, P. Duez, and F. Menichini. 2011a. Correlation between environmental factors, chemical composition, and antioxidative properties of caper species growing wild in Calabria (South Italy). *Chemistry & Biodiversity* 8(3): 518–531.

Dahot, M.U. 1993. Chemical evaluation of the nutritive value of flowers and fruits of *Capparis decidua*. *Journal of the Chemical Society of Pakistan* 15(1): 78–81.

Denunig-Adams, B., A.M. Gilmore, and W.W. Adams III, 1996. In vivo functions of carotenoids in higher plants. *FASEB J* 10: 403–412.

Dhar, D.N., R.P. Tewari, R.D. Tripathi, and A.P. Ahuja. 1972. Chemical examination of *C. decidua*. *Proceedings of the National Academy of Science, India: Section A* 42(1): 24–27.

Esfahani, A., J.M. Wong, J. Truan, C.R. Villa, A. Mirrahimi, K. Srichaikul, and C.W. Kendall. 2011. Health effects of mixed fruit and vegetable concentrates: a systematic review of the clinical interventions. *Journal of the American College of Nutrition* 30(5): 285–294.

Falk, J. and S. Munné-Bosch. 2010. Tocochromanol functions in plants: antioxidation and beyond. *Journal of Experimental Botany* 61(6): 1549–1566.

Firuzi, O., R. Miri, M. Tavakkoli, and L. Saso. 2011. Antioxidant therapy: current status and future prospects. *Current Medical Chemistry* 18(25): 3871–3888.

Gaby, A.R. 2006. Natural remedies for Herpes simplex. *Alternative Medicine Review* 11(2): 93–101.

Gerber, M. 2001. [Protective vegetal micronutrients and microcomponents for breast cancer]. *Bull Cancer* 88(10): 943–53.

Guz, J., T. Dziaman, and A. Szpila. 2007. [Do antioxidant vitamins influence carcinogenesis?] [in Polish]. *Postepy Hig Med Dosw (Online)* 61: 185–198.

Havaux, M., F. Eymery, S. Porfirova, P. Rey, and P. Dormann. 2005. Vitamin E protects against photoinhibition and photooxidative stress in *Arabidopsis thaliana*. *Plant Cell* 17: 3451–3469.

Hunter, D.C., J. Greenwood, J. Zhang, and M.A. Skinner. 2011. Antioxidant and "natural protective" properties of kiwifruit. *Current Topics in Medicinal Chemistry* 11(14): 1811–1820.

Laddha, K.S. and C.I. Jolly. 1985. Preliminary phytochemical studies on the leaves of *Capparis zeylanica* Linn. *Indian Drugs* 22(9): 499.

Lemmi Cena, T. and P. Rovesti. 1979. Experimental studies on the cosmetological uses of *Capparis*. *Rivista Italiana Essenze, Profumi, Piante Officinali, Aromatizzanti, Syndets, Saponi, Cosmetici, Aerosols* 61(1): 2–9.

Maeda, H., W. Song, T.L. Sage, and D. DellaPenna. 2006. Tocopherols play a crucial role in low-temperature adaptation and phloem loading in *Arabidopsis*. *Plant Cell* 18: 2710–2732.

Matringe, M., B. Ksas, P. Rey, and M. Havaux. 2008. Tocotrienols, the unsaturated forms of vitamin E, can function as antioxidants and lipid protectors in tobacco leaves. *Plant Physiology* 147: 764–778.

Matthäus, B. and M. Özcan. 2005. Glucosinolates and fatty acid, sterol, and tocopherol composition of seed oils from *Capparis spinosa* var. *spinosa* and *Capparis ovata* Desf. var. *canescens* (Coss.) Heywood. *Journal of Agricultural and Food Chemistry* 53(18): 7136–7141.

Moyad, M.A. 2002. The placebo effect and randomized trials: analysis of alternative medicine. *Urologic Clinics of North America* 29(1): 135–55, x.

Núñez-Córdoba, J.M. and M.A. Martínez-González. 2011. Antioxidant vitamins and cardiovascular disease. *Current Topics in Medicinal Chemistry* 11(14): 1861–1869.

Shen, Y., K. Lebold, E.P. Lansky, M.G. Traber, and E. Nevo. 2011. "Tocol-omic" diversity in wild barley, short communication. *Chemistry & Biodiversity* 8(12): 2322–2330.

Smirnoff, N. 1996. The function and metabolism of ascorbic acid in plants. *Annals of Botany* 78: 661–669.

Sroka, Z., A. Gamian, and W. Cisowski. 2005. [Low-molecular antioxidant compounds of natural origin] [in Polish]. *Postepy Hig Med Dosw (Online)* 59: 34–41.

Tlili, N., S. Munne-Bosch, N. Nasri, E. Saadaoui, A. Khaldi, and S. Triki. 2009. Fatty acids, tocopherols and carotenoids from seeds of Tunisian caper "Capparis spinosa." *Journal of Food Lipids* 16(4): 452–464.

Traber, M.G. and J.F. Stevens. 2011. Vitamins C and E: beneficial effects from a mechanistic perspective. *Free Radical Biology & Medicine* 51(5): 1000–1013.

12 Other Compounds

"Entourage compounds" refer to molecules in a chemical milieu that may augment or otherwise help "balance" the "active compounds" in the same mix (Ben Shabat et al. 1998). From the previous chapters, we have some idea of the tremendous amount and diversity of chemical compounds with known or putative biological activity in *Capparis* in a number of major classes. However, due to our focus on selected classes such as flavonoids, complementary and related compounds such as tannins and phenolic acids have been left out and hence are presented in this chapter. Based on earlier work with pomegranate (*Punica granatum*), phenolic acids were shown to enhance pharmacological potency against invasion of prostate cancer cells across an artificial membrane, an *in vitro* model of metastasis (Lansky et al. 2005). Also of interest are the common sucrose, glucosides, and various polysaccharides. The last are well known to include compounds with potent immunopotentive activity and, from *C. spinosa*, to possess anti-inflammatory and analgesic virtues (Zhang et al. 2011). This information is also included in Table 12.1. Some of the compounds are also novel, limited to *Capparis* or a particular *Capparis* species.

TABLE 12.1
Other Compounds of *Capparis* spp.

Compound	Notes
Adenosine 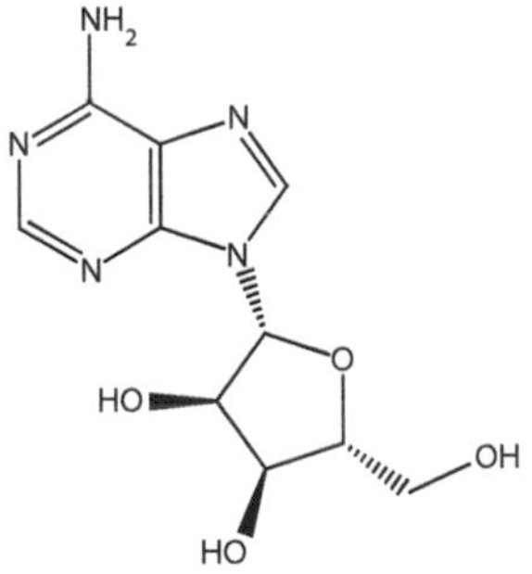 NH_2 N N N N HO O OH HO	From aerial parts of *C. flavicans* via NMR and library (Luecha et al. 2009); from *C. spinosa* fruits (Jiang et al. 2005)
Alangilignoside B 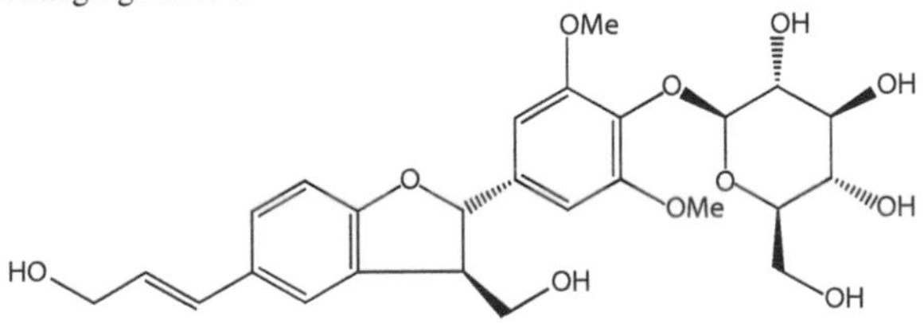 OMe OH O OH O OMe OH HO O OH OH	From aerial parts of *C. flavicans* via NMR and library (Luecha et al. 2009)

(*Continued*)

TABLE 12.1 (*Continued*)
Other Compounds of *Capparis* spp.

Compound	Notes
Alangilignoside C 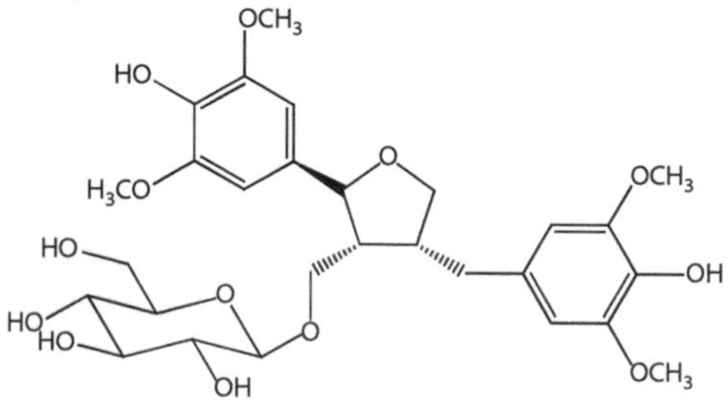	From aerial parts of *C. flavicans* via NMR and library (Luecha et al. 2009)
Alangilignoside D 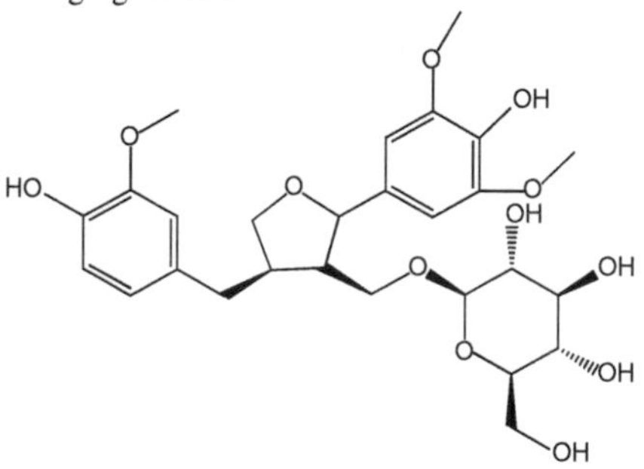	From aerial parts of *C. flavicans* via NMR and library (Luecha et al. 2009)
Alangionoside C 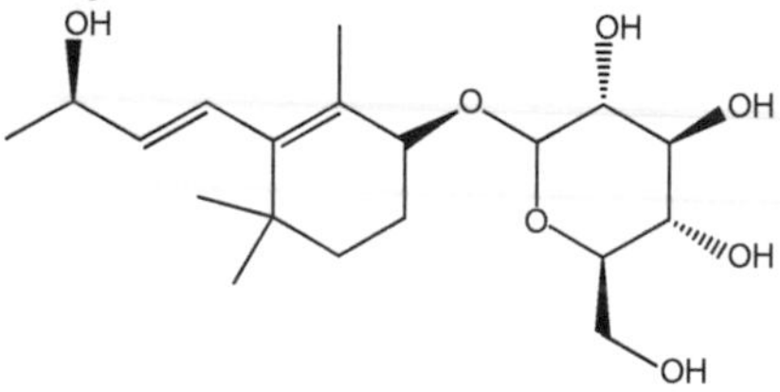	From aerial parts of *C. flavicans* via NMR and library (Luecha et al. 2009)
Americanol A 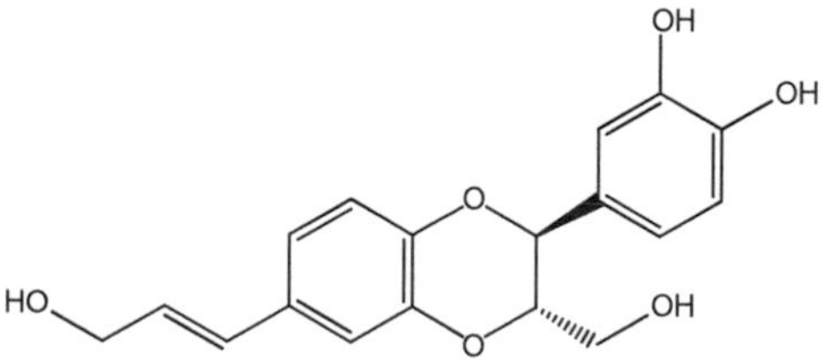	From aerial parts of *C. flavicans* via NMR and library (Luecha et al. 2009)
Arabinose 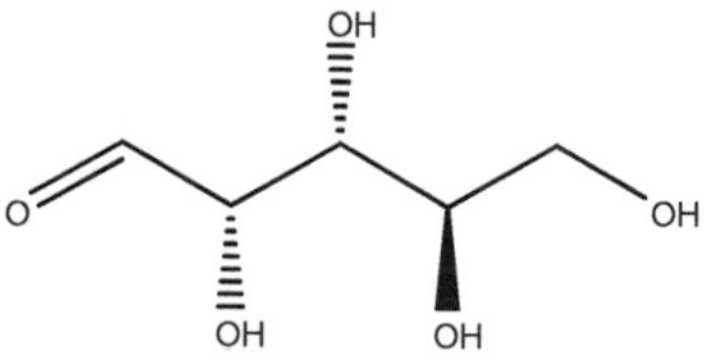	By paper and TLC from *C. decidua* (Dhar et al. 1972)

TABLE 12.1 (*Continued*)
Other Compounds of *Capparis* spp.

Compound	Notes
Betulin 28-acetate 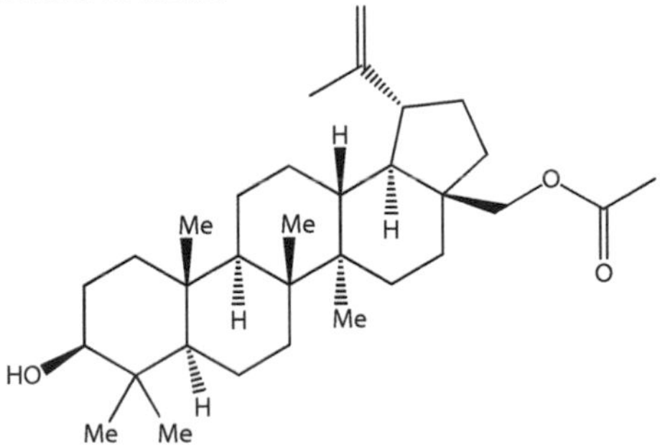	Determined by ^{1}H- and ^{13}C-NMR spectra from *C. sepiaria* (Saraswathy et al. 1991)
(±)-3′,3″-Bisdemethylpinoresinol	From aerial parts of *C. flavicans* via NMR and library (Luecha et al. 2009)
bis(2-Ethylhexyl) phthalate 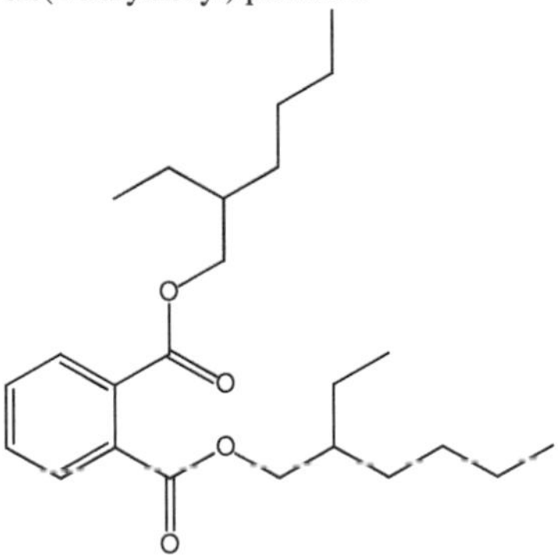	From *C. spinosa* fruit pericarp (Xiao et al. 2008)
Butanedioic acid 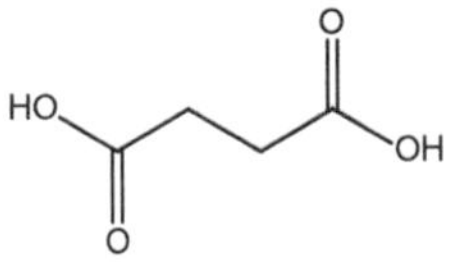	From *C. spinosa* fruit, chromatography, spectroscopy (Yu et al. 2006)
Butyl-3-oxoeicosanoate 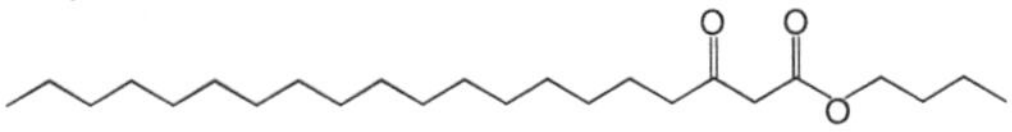	Aliphatic compound in *C. decidua* root bark alc. ext. (Gupta and Ali 1998)
C-29 hydrocarbons	From *C. decidua* young fruits, flowers, buds, seeds (Rai 1987a)
C-31 hydrocarbons	From *C. decidua* young fruits, flowers, buds, seeds (Rai 1987a)
Caffeic acid ethyl ester 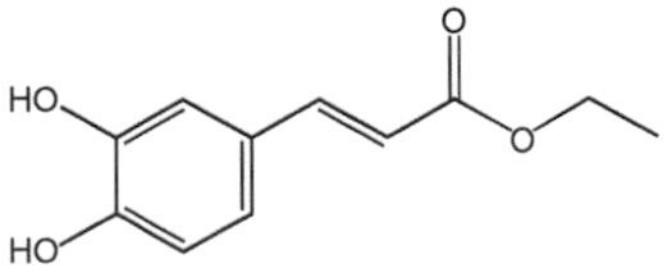	From *C. formosana* stems ether extract after removal of N-hexane fraction (Liu et al. 1977)

(*Continued*)

TABLE 12.1 (*Continued*)
Other Compounds of *Capparis* spp.

Compound	Notes
Cappamensin A (2H-1,4-Benzoxazin-3(4H)-one, 6-methoxy-2-methyl-4-carbaldehyde) 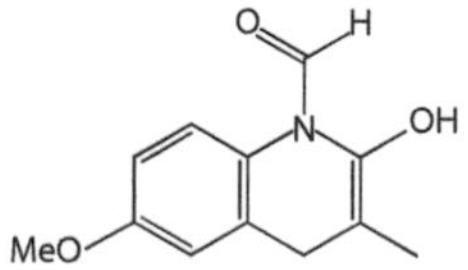	From roots of *C. sikkimensis* var. *formosana*, bioactivity-guided fractionation, 2D NMR, *in vitro* anticancer angst. ovarian (1A9), lung (A549), ileocecal (HCT-8), breast (MCF-7), nasopharyngeal (KB), vincristine resistant (KB-VIN) human tumor cell lines, ED50 ≤ 4 µg/mL (mean GI50 15.1 µ*M*) (Wu et al. 2003)
Cappaprenol 12	Isol. by prep. HPLC from *C. spinosa* alc. exts., 12 isoprene units (Al-Said et al. 1988)
Cappaprenol 13	Isol. by prep. HPLC from *C. spinosa* alc. exts., inhibited carrageenan-induced rat paw edema 44% vs. 67% for oxyphenbutazone, 13 isoprene units (Al-Said et al. 1988)
Cappaprenol 14	Isol. by prep. HPLC from *C. spinosa* alc. exts., 14 isoprene units (Al-Said et al. 1988)
Cappariloside A (1H-Indole-3-acetonitrile 4-O-β-glucopyranoside) 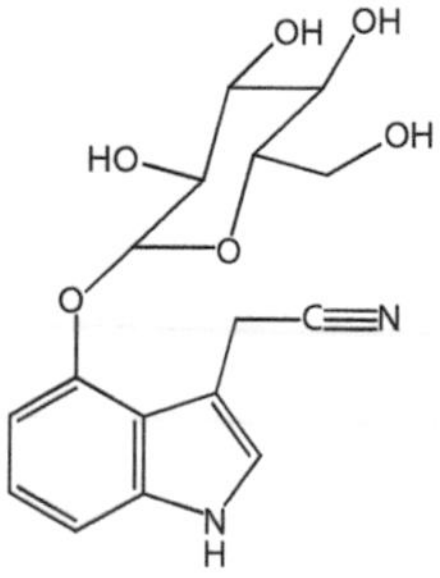	From mature *C. spinosa* fruits (Calis et al. 1999); from *C. spinosa* fruit (Fu et al. 2007); from *C. spinosa* fruits (Jiang et al. 2005)
Cappariloside B (1H-Indole-3-acetonitrile 4-O-β-(6′-O-β-glucopyranosyl)-glucopyranoside) 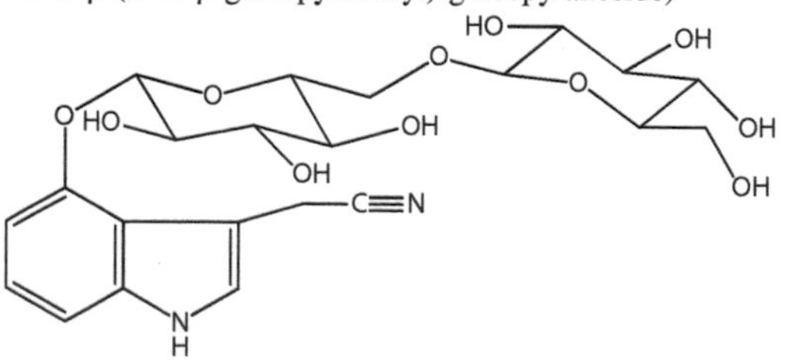	From mature *C. spinosa* fruits (Calis et al. 1999)
Capparisesterpenolide (3-Carboxy-6,17-dihydroxy-7,11,15,19-tetramethyleicos-13-ene-δ-lactone)	Oxygenated heterocyclic in alc. ext. of *C. decidua* root bark (Gupta and Ali 1997)

TABLE 12.1 (*Continued*)
Other Compounds of *Capparis* spp.

Compound	Notes
Capparoside A	From aerial parts of *C. flavicans* via NMR and library, putative inhibitory effect on excess estrogen in women who experience insufficient milk for breastfeeding (Luecha et al. 2009)
2-Carboxy-1,1-dimethylpyrrolidine	Upadhyay et al. 2006
Catalpol	Iridoid with anti-inflammatory activity found in stem bark of *C. ovata* (Villasenor 2007)
trans-Cinnamic acid	From *C. spinosa* leaves and stems, column chrom., spectroscop. (Yang et al. 2011)
Corchoionoside C ((6S,9S)-roseoside) (+)-(6S,9S)-9-O-β-d-Glucopyranosyloxy-6-hydroxy -3-oxo-α-ionol 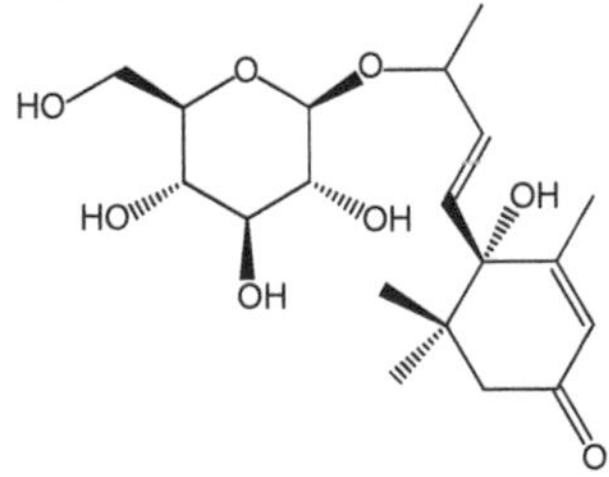	From *C. spinosa* mature fruits, (+)-(S)-abscisic acid deriv., ^{13}C-resonance of C-9 for abs. configuration at that center (Calis et al. 2002)
Deciduaterpenolide A	Oxygenated heterocyclic δ-lactone deriv. of 1,3,3-trimethyl-1,4-cyclohexadien-6-one from alc. ext. of *C. decidua* root bark (Gupta and Ali 1997)

(*Continued*)

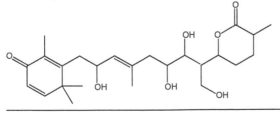

TABLE 12.1 (*Continued*)
Other Compounds of *Capparis* spp.

Compound	Notes
Deciduaterpenolide B 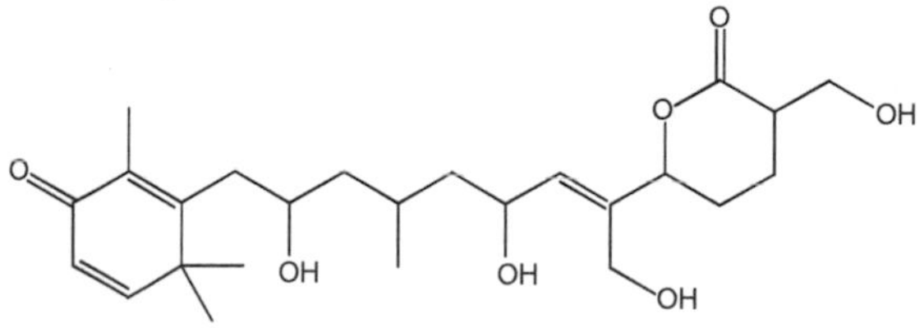	Oxygenated heterocyclic δ-lactone deriv. of 1,3,3-trimethyl-1,4-cyclohexadien-6-one from alc. ext. of *C. decidua* root bark (Gupta and Ali 1997)
Deciduaterpenolide C 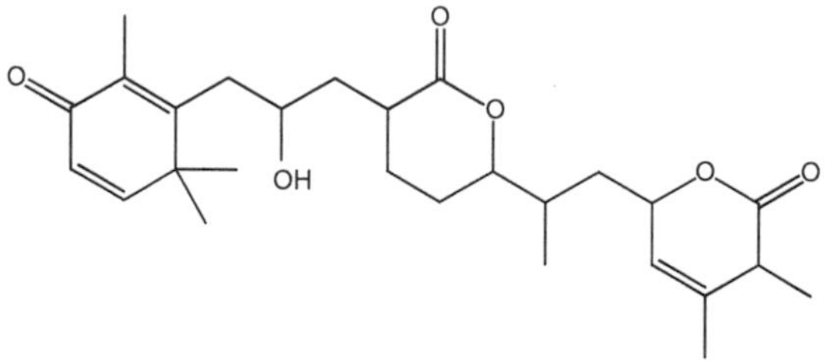	Oxygenated heterocyclic δ-lactone deriv. of 1,3,3-trimethyl-1,4-cyclohexadien-6-one from alc. ext. of *C. decidua* root bark (Gupta and Ali 1997)
Deciduaterpenolide D 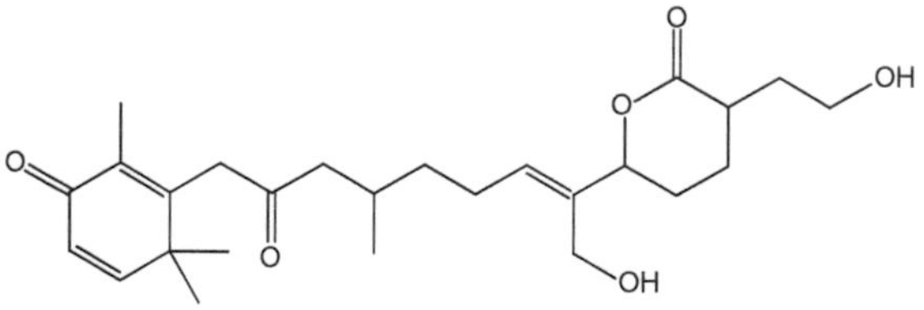	Oxygenated heterocyclic δ-lactone deriv. of 1,3,3-trimethyl-1,4-cyclohexadien-6-one from alc. ext. of *C. decidua* root bark (Gupta and Ali 1997)
Deciduaterpenolide E 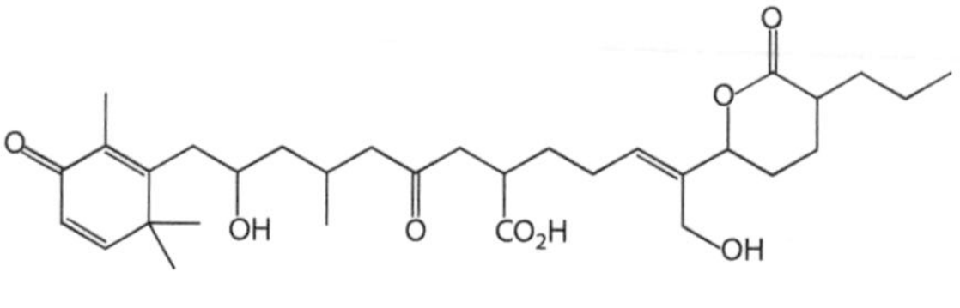	Oxygenated heterocyclic δ-lactone deriv. of 1,3,3-trimethyl-1,4-cyclohexadien-6-one from alc. ext. of *C. decidua* root bark (Gupta and Ali 1997)
(-)-Dehydrodiconiferyl alcohol 4-O-β-D-glucopyranoside 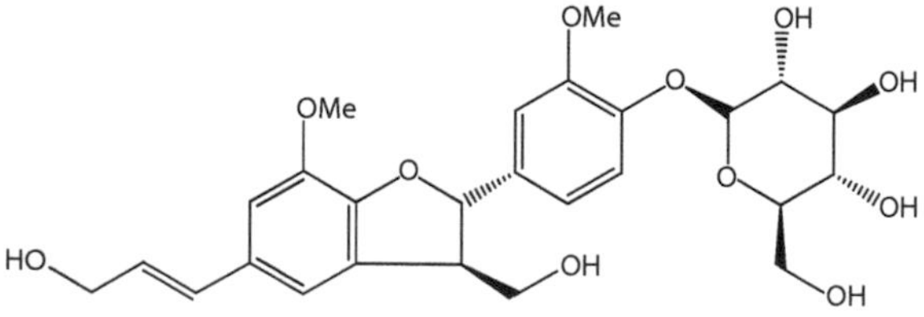	From aerial parts of *C. flavicans* via NMR and library (Luecha et al. 2009)
2, 4-Dihydroxybenzoic acid 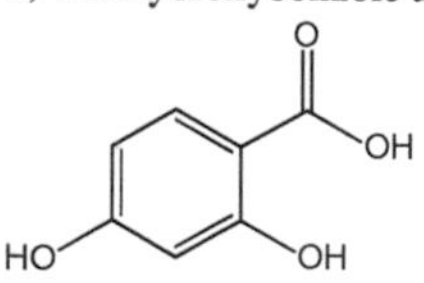	From *C. spinosa* fruit pericarp (Xiao et al. 2008)
(-)-5,5′-Dimethoxy-4-O-(β-D-glucopyranosyl) lariciresinol	From acrial parts of *C. flavicans* via NMR and library (Luecha et al. 2009)

TABLE 12.1 (*Continued*)
Other Compounds of *Capparis* spp.

Compound	Notes
(+/-)-8,8′-Dimethoxy-1-O-(β-D glucopyranosyl) secoisolariciresinol 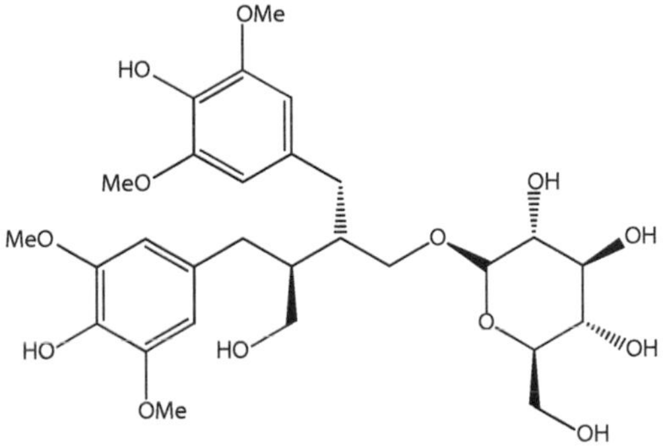	From aerial parts of *C. flavicans* via NMR and library (Luecha et al. 2009)
Dolichol-like polyprenols w/hydrogenated OH terminal isoprene unit	In *C. coriacea* leaves (Jankowski and Chojnacki 1991)
(-)-8′-Epilyoniresin-4-yl β-glucopyranoside 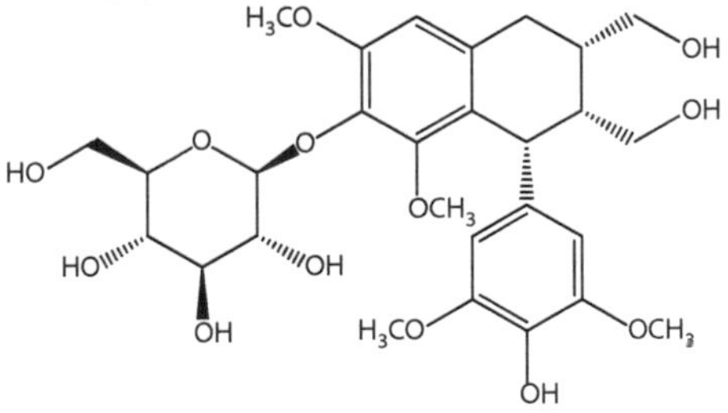	From roots of *C. tenera* (Su et al. 2007)
Ethyl hydroxybenzoate 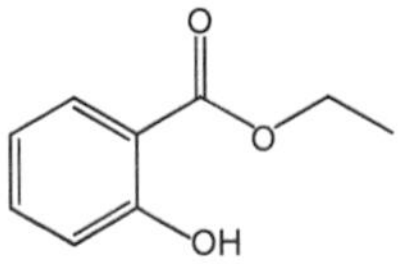	From *C. spinosa* leaves and stems, column chrom., spectroscop. (Yang et al. 2011)
Flower surface wax in general	14% in *C. decidua* (Rai 1987a)
Galactose 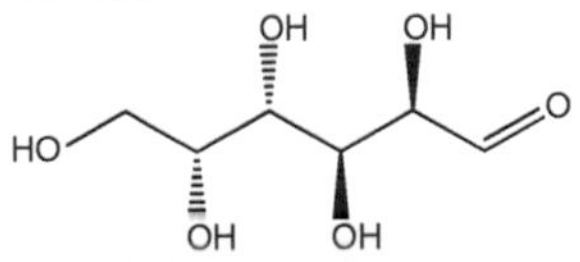	By paper and TLC from *C. decidua* (Dhar et al. 1972)
Gallotannin A 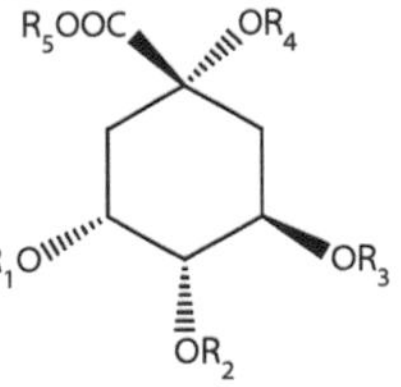	Fruits of *C. moonii* hydro-alc. ext. → 2 new chebulinic acid derivs. w/insulinomimetic actions (Kanaujia et al. 2010)

(*Continued*)

TABLE 12.1 (*Continued*)
Other Compounds of *Capparis* spp.

Compound	Notes
Gallotannin B 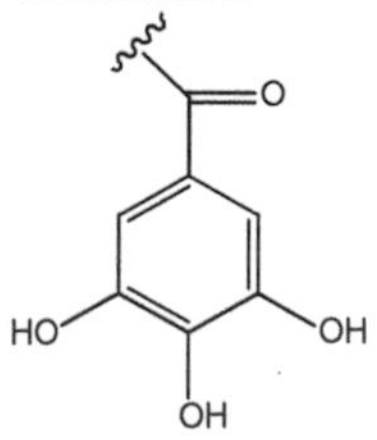	
3R-O-β-D-Glucopyranosyl lyoniresinol	From aerial parts of *C. flavicans* via NMR and library (Luecha et al. 2009)
4-(β-Glucopyranosyloxy)-1H-indole-3-acetamide 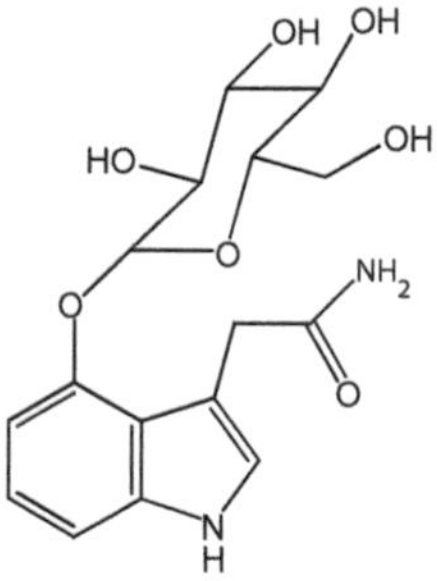	From roots of *C. tenera* (Su et al. 2007)
4-(β-Glucopyranosyloxy)-1H-indole-3-carbaldehyde 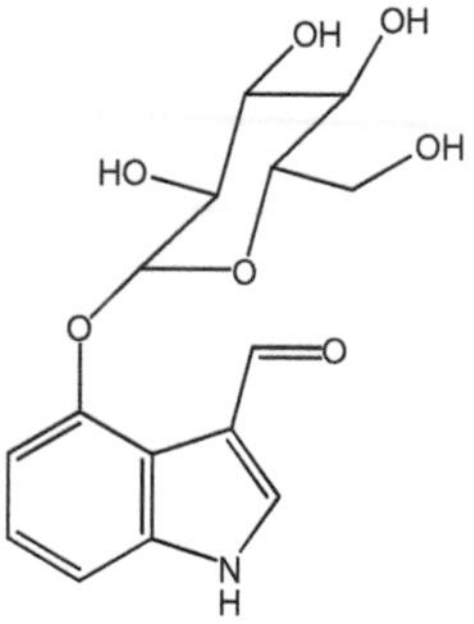	From roots of *C. tenera* (Su et al. 2007)
(-)-1-O-β-d-Glucopyranosyloxy-3-methyl-2-buten-1-ol 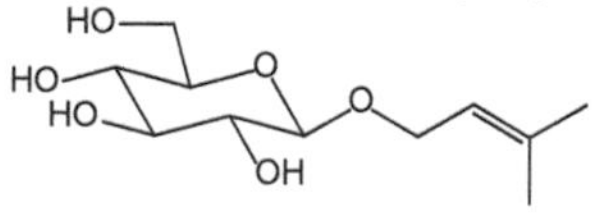	From *C. spinosa* mature fruits (Calis et al. 2002)
n-Hentriacontane 	From *C. decidua* flowers (Rai 1987a)
Histamine 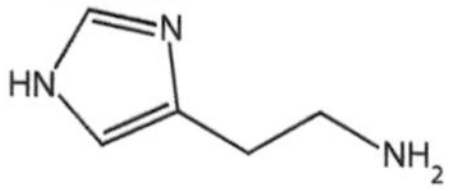	38 mg/kg in *C. spinosa* fruit (Garcia-Garcia et al. 2001)

TABLE 12.1 (*Continued*)
Other Compounds of *Capparis* spp.

Compound	Notes
4-Hydroxybenzoic acid (p-Hydroxybenzoic acid) 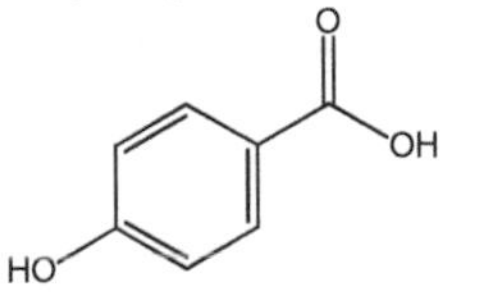	From *C. spinosa* fruit pericarp (Xiao et al. 2008), *C. spinosa* fruit, chromatography, spectroscopy (Yu et al. 2006), 33% of MeOH sol. fraction of aq. ext. of *C. spinosa*, protected liver after paracetamol or CCl_4 or rat hepatocytes after thioacetamide or galactosamine exposure in vitro (Gadgoli and Mishra 1999)
6-Hydroxy-12-carboxy-blumenol A-β-D-glucopyranoside	From *C. spinosa* fruits (Jiang et al. 2005)
4-Hydroxy-2,6-dimethoxyphenol 1-O-β-D-glucopyranoside, leonuriside 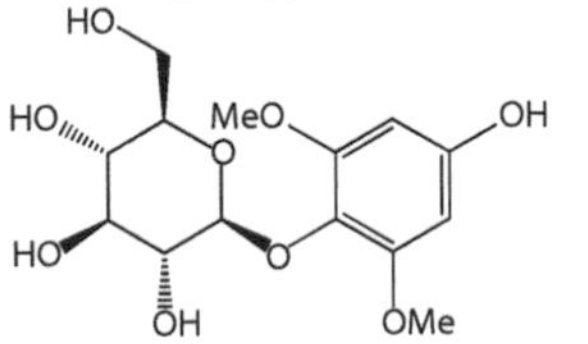	From aerial parts of *C. flavicans* via NMR and library (Luecha et al. 2009)
s4-({6-O-[(4-Hydroxy-3,5-dimethoxyphenyl)carbonyl]-β-glucopyranosyl}oxy)-3,5-dimethoxybenzoic acid 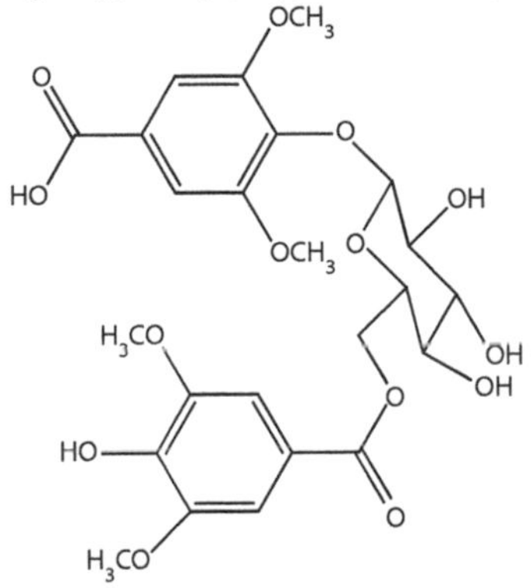	From roots of *C. tenera* (Su et al. 2007)
4-Hydroxy-2,6-dimethoxyphenyl β-glucopyranoside 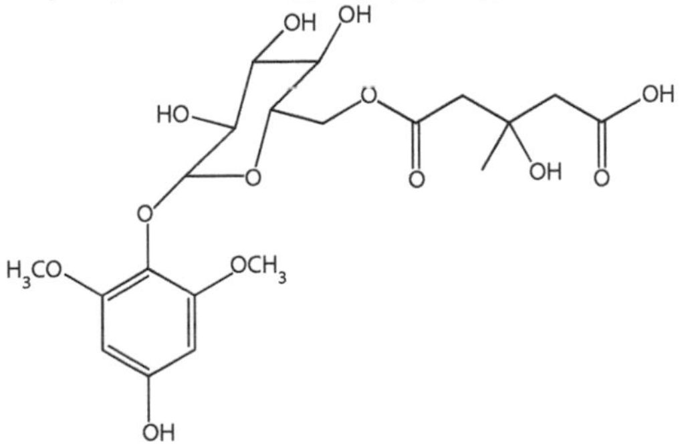	From roots of *C. tenera* (Su et al. 2007)
4-Hydroxy-5-(hydroxymethyl) dihydrofuran-2(3H)-one 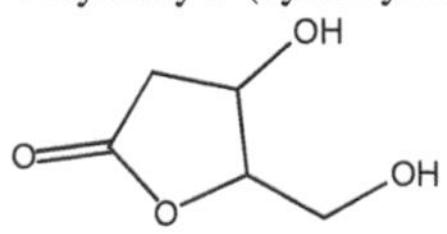	From *C. spinosa* fruits ethanol ext., column chrom., spectroscop. (Yang et al. 2010a)

(*Continued*)

TABLE 12.1 (*Continued*)
Other Compounds of *Capparis* spp.

Compound	Notes
(6S,9S)-6-Hydroxyinamoside ((-)-(6S,9S)-9-O-β-D-glucopyranosyloxy-6,13-dihydroxy-3-oxo-α-ionol) (Spionoside A) 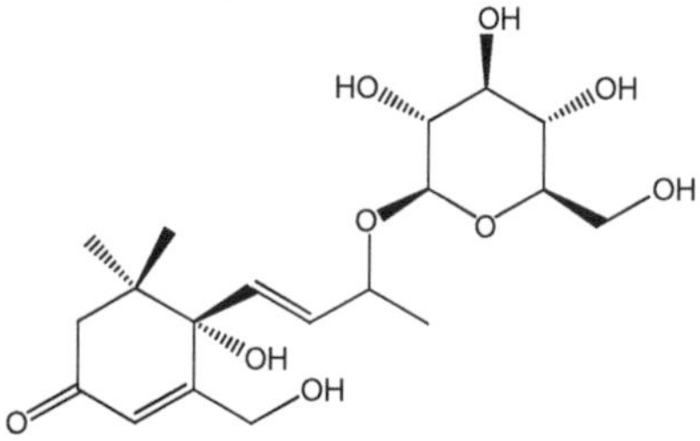	From *C. spinosa* mature fruits, (+)-(S)-abscisic acid deriv., ^{13}C-resonance of C-9 for abs. configuration at that center (Calis et al. 2002)
4-Hydroxy-3-methoxyphenyl 1-O-β-D-xylopyranosyl-(1f6)-β-D-glucopyranoside	From aerial parts of *C. flavicans* via NMR and library (Luecha et al. 2009)
4-Hydroxy-5-methylfuran-3-carboxylic acid 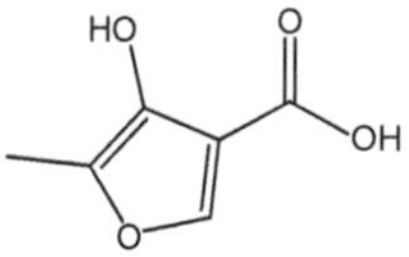	From EtOAc ext. of *C. spinosa* fruits by column chromatog., NMR (Yang et al. 2010b)
6-(1-Hydroxy-non-3-enyl)tetrahydropyran-2-one 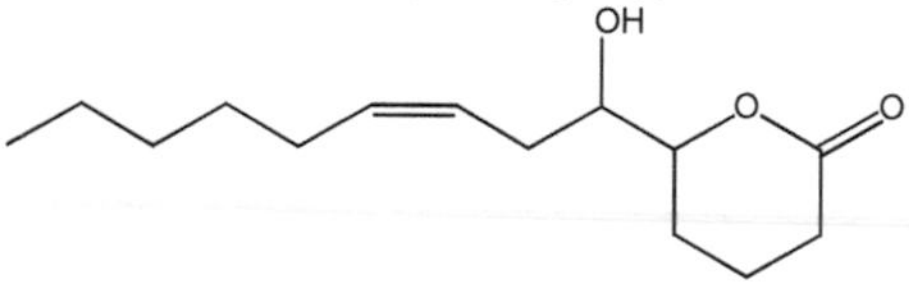	Upadhyay et al. 2006
7-Hydroxy-2-oxoindol-3-yl acetic acid 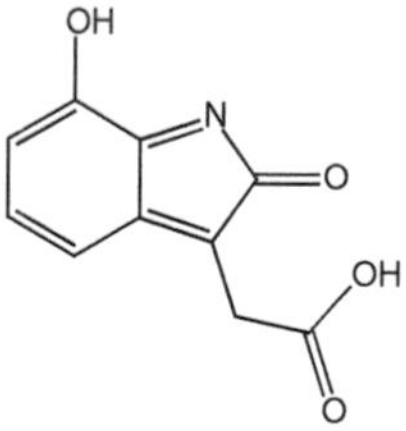	From *C. spinosa* fruits ethanol ext., column chrom., spectroscop. (Yang et al. 2010a)
(6S)-Hydroxy-3-oxo-α-ionol 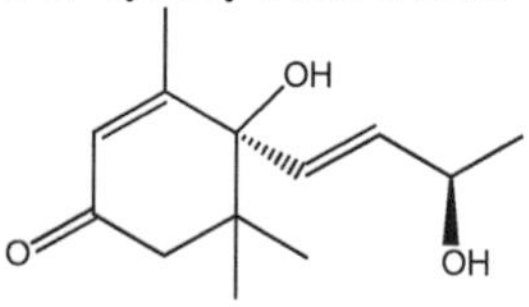	From *C. spinosa* fruits (Jiang et al. 2005); from *C. spinosa* fruits ethanol ext., column chrom., spectroscop. (Yang et al. 2010a)
Hypoxanthine 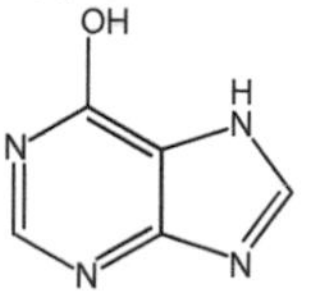	From *C. spinosa* fruit (Fu et al. 2007)

TABLE 12.1 (*Continued*)
Other Compounds of *Capparis* spp.

Compound	Notes
Icariside B1 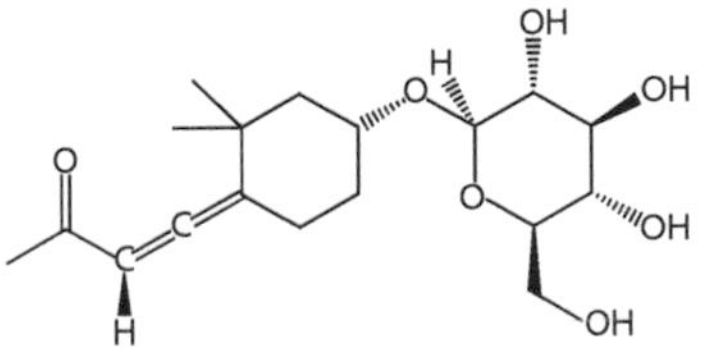	From aerial parts of *C. flavicans* via NMR and library (Luecha et al. 2009)
3-Indolcarboxaldehyde 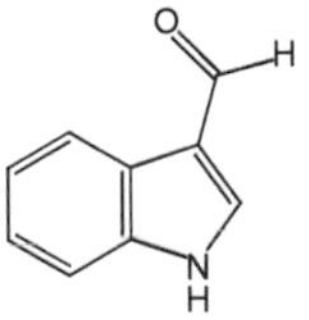	From *C. spinosa* fruits (Jiang et al. 2005)
1H-Indole-3-carbaldehyde 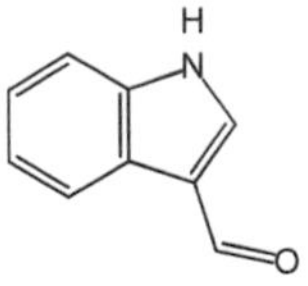	From *C. spinosa* fruits ethanol ext, column chrom., spectroscop. (Yang et al. 2010a)
Inosine 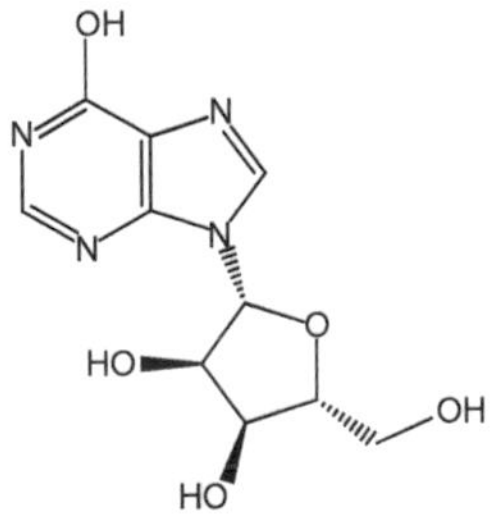	From *C. spinosa* fruits (Jiang et al. 2005)
Isoamericanol A 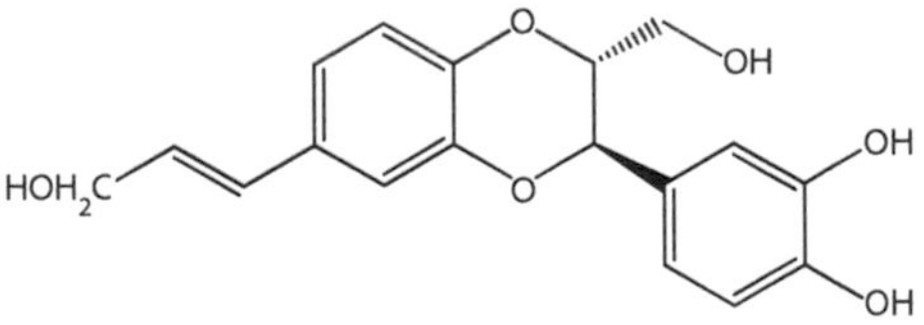	From aerial parts of *C. flavicans* via NMR and library (Luecha et al. 2009)
Isoprincepin 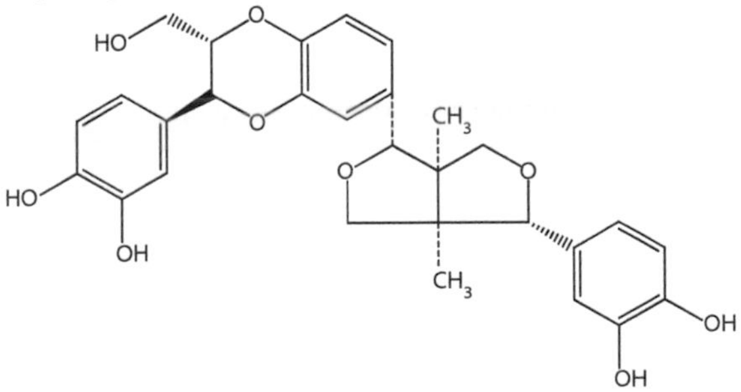	From aerial parts of *C. flavicans* via NMR and library (Luecha et al. 2009)

(*Continued*)

TABLE 12.1 (*Continued*)
Other Compounds of *Capparis* spp.

Compound	Notes
Isopsoralen 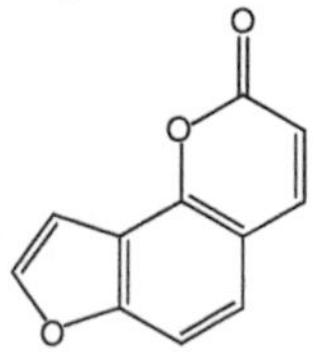	From *C. spinosa* leaves and stems, column chrom., spectroscop. (Yang et al. 2011)
Isotachioside 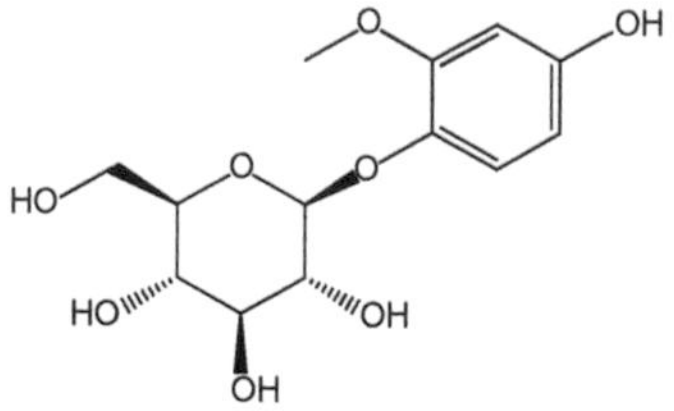	From aerial parts of *C. flavicans* via NMR and library (Luecha et al. 2009)
Kalanchoe glycosides	From EtOAc ext. of *C. spinosa* fruits (Yu et al. 2011)
Koaburaside 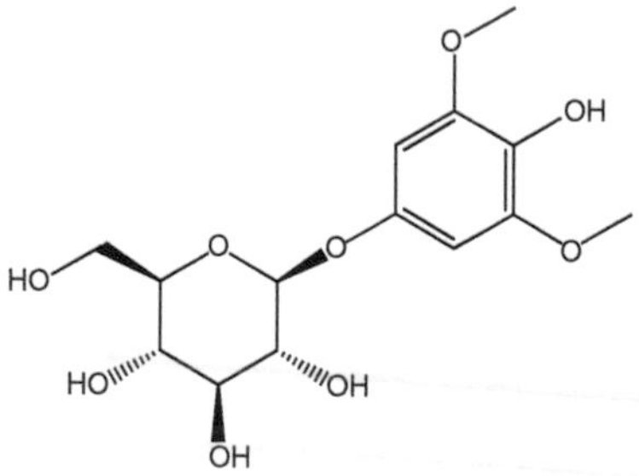	From aerial parts of *C. flavicans* via NMR and library (Luecha et al. 2009)
(-)-Lariciresinol 4-O-β-D-glucopyranoside 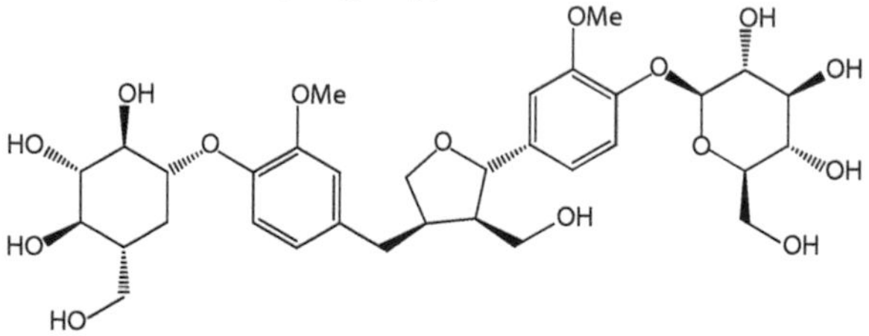	From aerial parts of *C. flavicans* via NMR and library (Luecha et al. 2009)
Long-chain polyisoprenoid alcohols in general	Up to 0.3% of dried leaves in 12 *Capparis* species, usually w/12, 13, 14, or 15 isoprenoid units, in some species also minor components with 19, 20, or 21 isoprenoid units (Jankowski and Chojnacki 1991)

TABLE 12.1 (*Continued*)
Other Compounds of *Capparis* spp.

Compound	Notes
(+)-Lyoniresin-4′-yl β-glucopyranoside 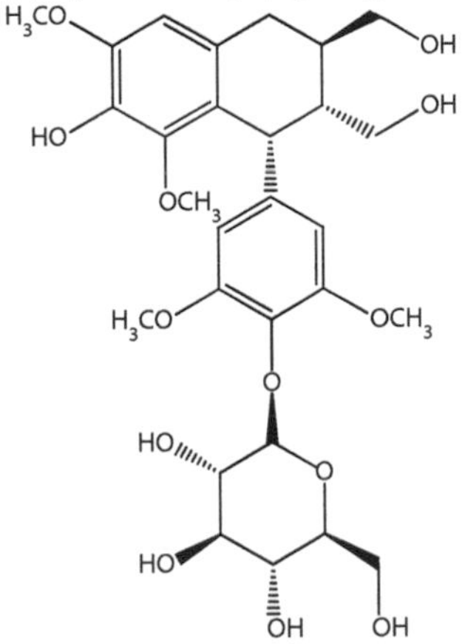	From roots of *C. tenera* (Su et al. 2007)
3-Methyl-2-butenyl-β-glucoside 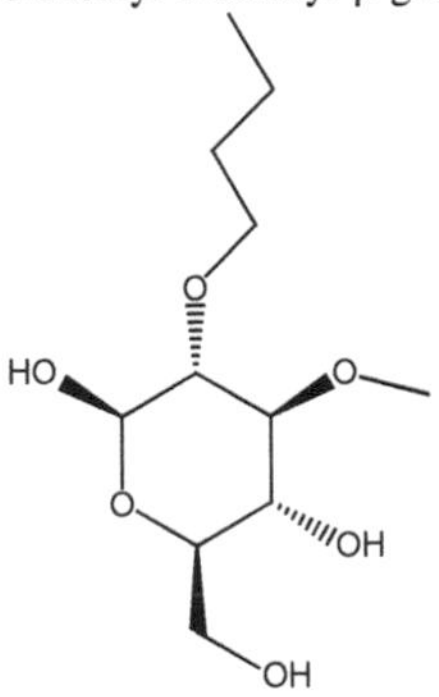	In *C. spinosa* (Khanfar et al 2003)
3-Methyl-7-hydroxymethylene-10-(12,16,16-trimethylcyclohex-11-ene-yl)-dec-9-ene-5-one-8-ol	Diterpene alc. in *C. decidua* root bark alc. ext. (Gupta and Ali 1998)
Methylparaben 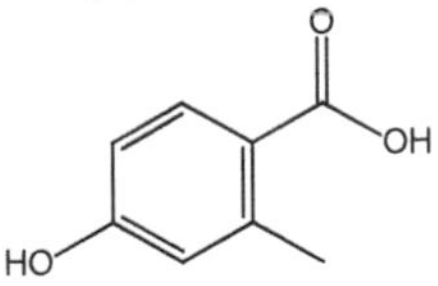	From *C. spinosa* leaves and stems, column chrom., spectroscop. (Yang et al. 2011)
myo-Inositol 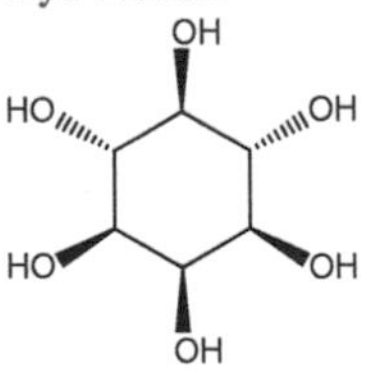	From *C. spinosa* fruits ethanol ext., column chrom., spectroscop. (Yang et al. 2010a)
New polysaccharide	From ripe fruit of *C. spinosa* by HPCE (Ji et al. 2006)

(*Continued*)

TABLE 12.1 (*Continued*)
Other Compounds of *Capparis* spp.

Compound	Notes
n-Nonacosane 	From *C. decidua* flowers (Rai 1987a)
Oxazolidine-2-thione 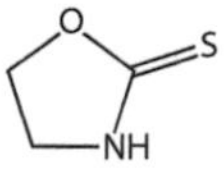	Bitter-tasting component from *C. masaikai* seeds aq. ext., mol. configuration and bond angles and distances presented (Hu et al. 1987)
25-Oxo-octacosan-1,20-diol 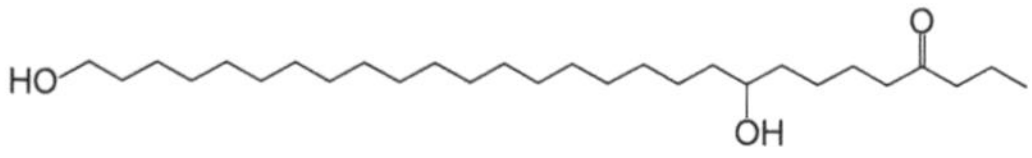	Aliphatic compound in *C. decidua* root bark alc. ext. (Gupta and Ali 1998)
Phytic acid 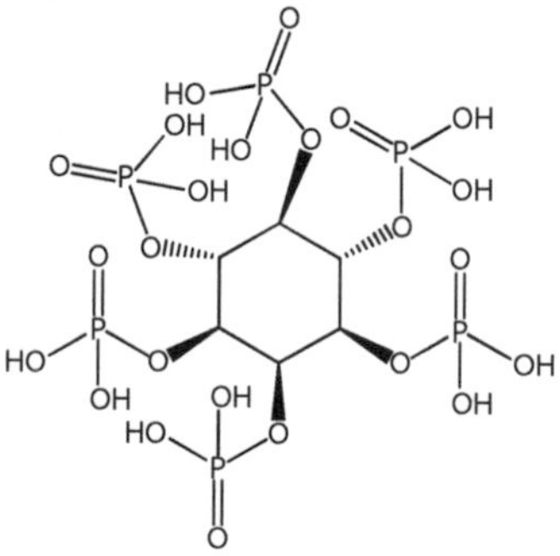	304 mg/100 g in unripe fruit of *C. decidua* (Chauhan et al. 1986)
Princepin 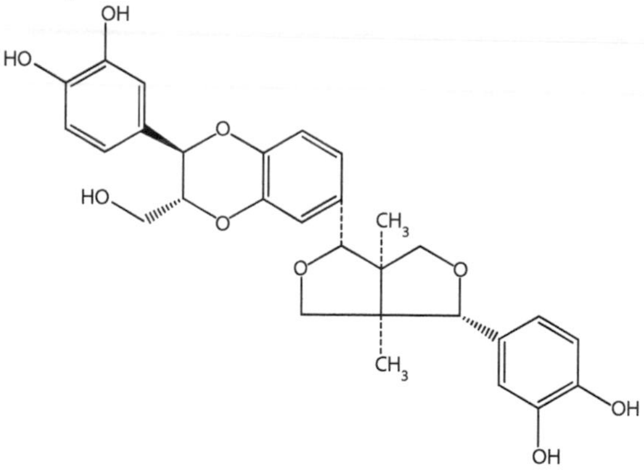	From aerial parts of *C. flavicans* via NMR and library (Luecha et al. 2009)
Protocatechuic acid 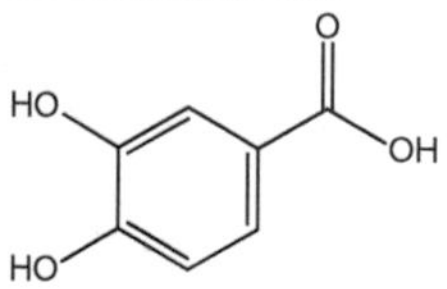	From *C. spinosa* fruit pericarp (Xiao et al. 2008); *C. spinosa* fruit, chromatography, spectroscopy (Yu et al. 2006); from *C. spinosa* leaves and stems, column chrom., spectroscop. (Yang et al. 2011)

TABLE 12.1 (*Continued*)
Other Compounds of *Capparis* spp.

Compound	Notes
Psoralen (Furocoumarin) 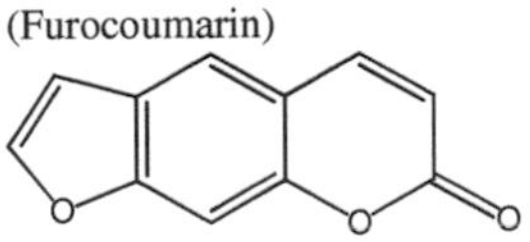	From *C. spinosa* leaves and stems, column chrom., spectroscop. (Yang et al. 2011)
Putrescine 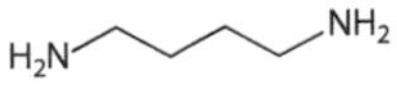	In *C. spinosa* fruit (Garcia-Garcia et al. 2001)
4-L-Rhamnose-benzoic acid 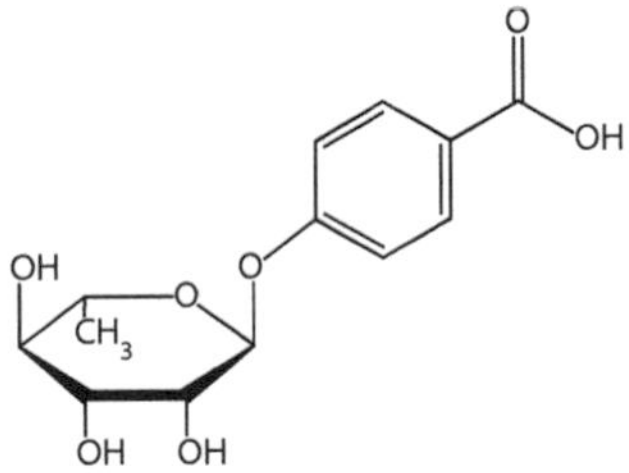	From *C. spinosa* fruits (Jiang et al. 2005)
Roseoside 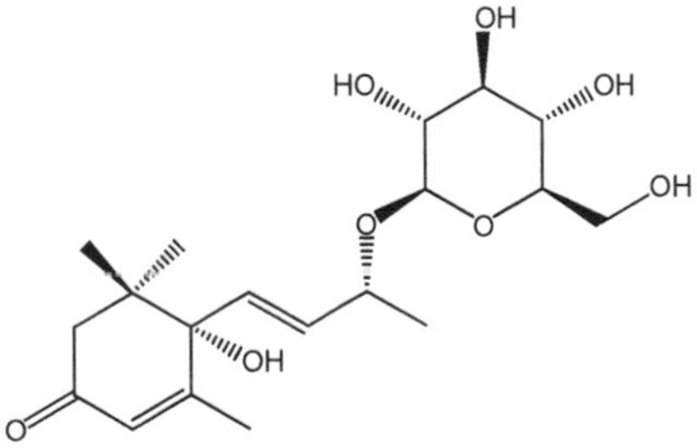	From aerial parts of *C. flavicans* via NMR and library (Luecha et al. 2009)
Scopoletin 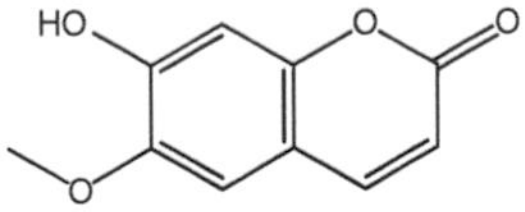	From *C. formosana* stems ether extract after removal of N-hexane fraction (Liu et al. 1977)
Scopolin 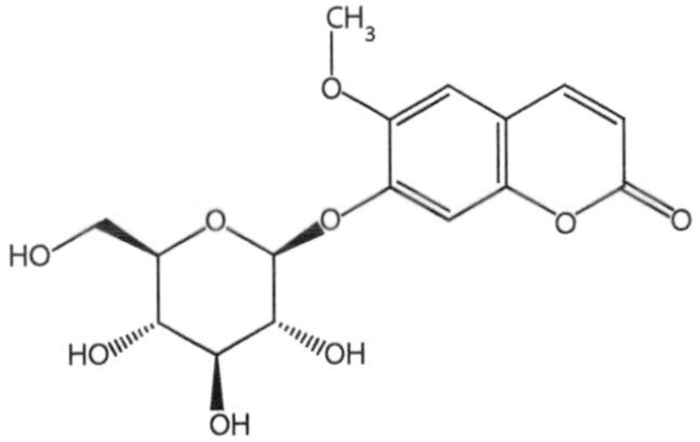	Glucoside from *C. formosana* stems ether extract after removal of N-hexane fraction (Liu et al. 1977)
Seed fat in general	5.7–50% in *C. spinosa* from Western Rajasthan, India, arid areas (Rai 1987b); 18% in *C. decidua* (Rai 1987a)

(*Continued*)

TABLE 12.1 (*Continued*)
Other Compounds of *Capparis* spp.

Compound	Notes
Spionoside B 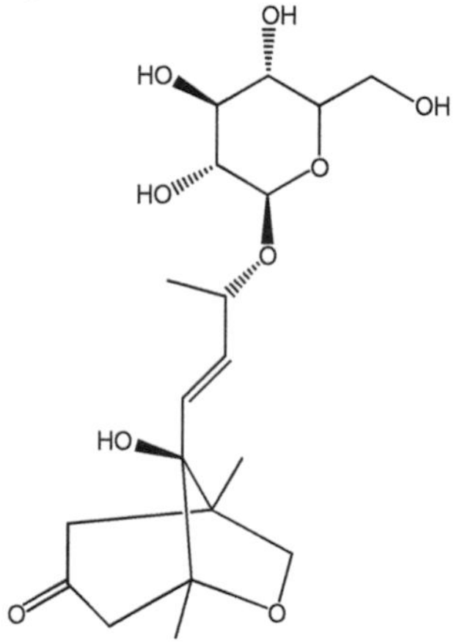	From aerial parts of *C. flavicans* via NMR and library (Luecha et al. 2009); from *C. spinosa* fruits ethanol ext., column chrom., spectroscop. (Yang et al. 2010a)
Succinic acid 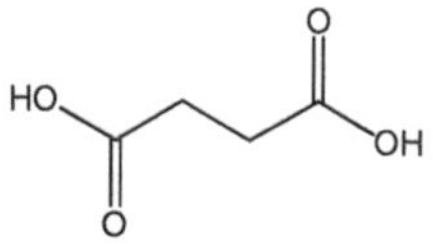	From *C. spinosa* fruit pericarp (Xiao et al. 2008)
Sucrose 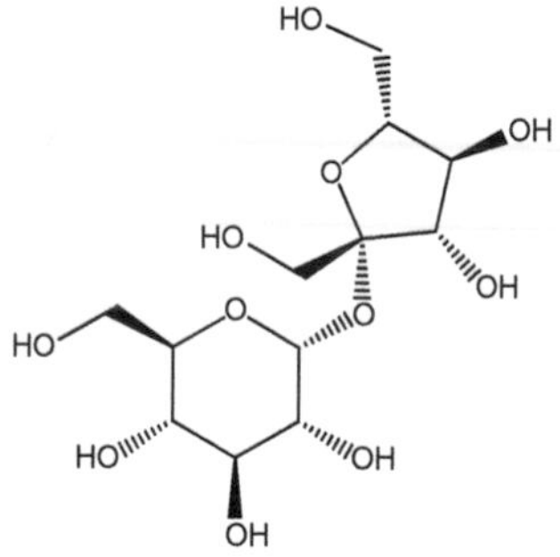	From *C. spinosa* fruits (Jiang et al. 2005)
(+/-)-Syringaresinol 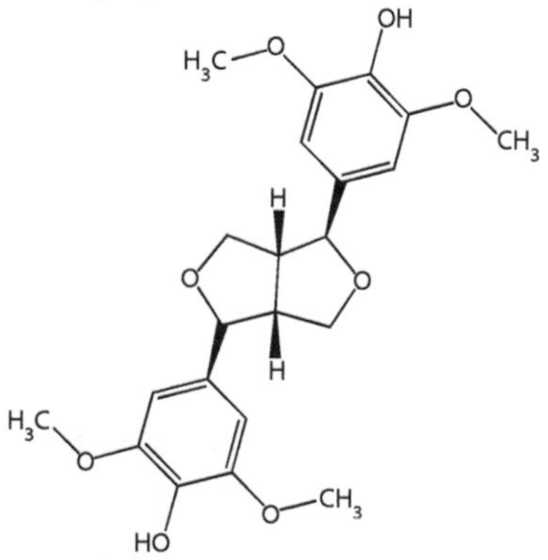	From aerial parts of *C. flavicans* via NMR and library (Luecha et al. 2009)

TABLE 12.1 (*Continued*)
Other Compounds of *Capparis* spp.

Compound	Notes
Tachioside 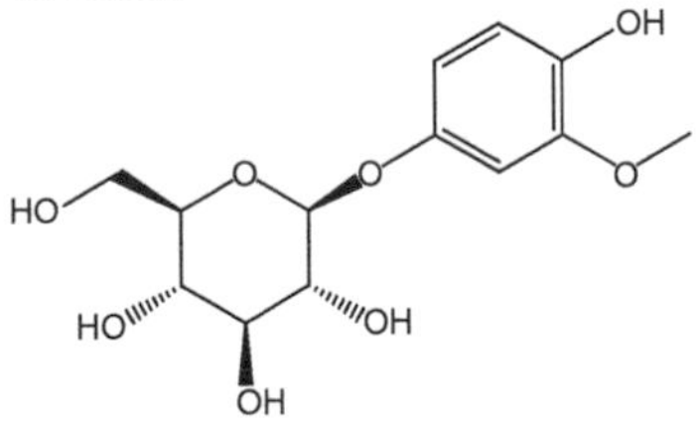	From aerial parts of *C. flavicans* via NMR and library (Luecha et al. 2009)
Tamarixetin 3-O-β-D-galactoside	From aerial parts of *C. flavicans* via NMR and library (Luecha et al. 2009)
(7*R*,8*R*)-threo-Guaiacylglycerol 8-O-4′-sinapyl ether 7-O-β-D-glucopyranoside 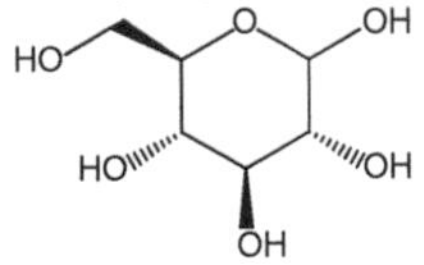	From aerial parts of *C. flavicans* via NMR and library (Luecha et al. 2009)
Triacontanol 	Upadhyay et al. 2006
n-Tricosa-17-on-1-oic acid (Capparitricosanoic acid) 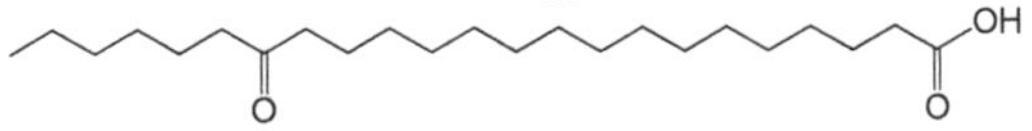	From fruits of *C. moonii* (Ramachandram et al. 2004)
1,3,6-Tri-O-galloyl-2-chebuloyl-β-D-glucopyranoside 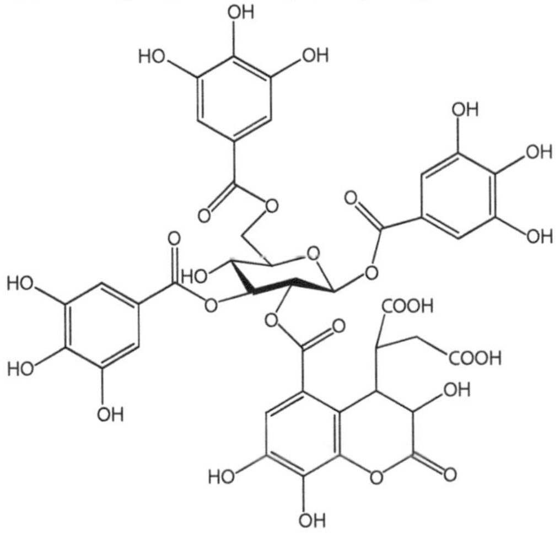	From hydro-alc. ext. of *C. moonii* fruits, ↑ glucose uptake 223% and 219% over the control at 10 ng/mL and 100 ng/mL concn. in L6 cells, resp., w/signif. IR and IRS-1 phosphorylation, GLUT4 and PI3-kinase mRNA expression (Kanaujia et al. 2010)
1,3,6-Tri-O-galloyl-2-chebuloyl ester-β-D-glucopyranoside	From hydro-alc. ext. of *C. moonii* fruits, ↑ glucose uptake 223% and 219% over the control at 10 ng/mL and 100 ng/mL concn. in L6 cells, resp., w/signif. IR and IRS-1 phosphorylation, GLUT4 and PI3-kinase mRNA expression (Kanaujia et al. 2010)

(*Continued*)

TABLE 12.1 (*Continued*)
Other Compounds of *Capparis* spp.

Compound	Notes
9-(11,15,15-Trimethylcyclohex-11-ene-13-one-yl)-one-6-hydroxymethylene-7-one-yl,4′-Me heptanoate	Diterpenic ester in *C. decidua* root bark alc. ext. (Gupta and Ali 1998)
Uracil 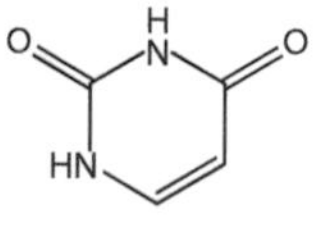	From *C. spinosa* fruit, chromatography, spectroscopy (Yu et al. 2006, Fu et al. 2007; in the ethanolic extract, Yang et al. 2010a)
Uridine 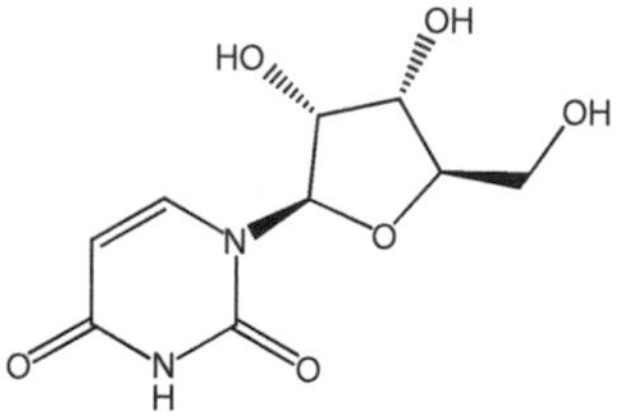	From *C. spinosa* fruits ethanol ext., column chrom., spectroscop. (Yang et al. 2010a); from *C. spinosa* fruit, chromatography, spectroscopy (Yu et al. 2006)
Vanillic acid 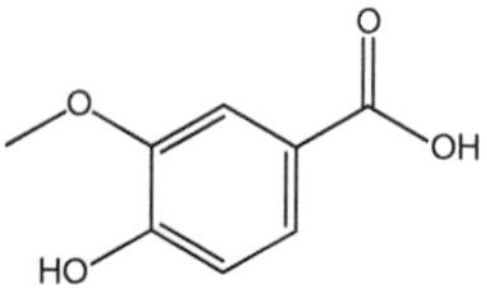	From *C. spinosa* fruit pericarp (Xiao et al. 2008); *C. spinosa* fruit, chromatography, spectroscopy (Yu et al. 2006)
Vanilloloside 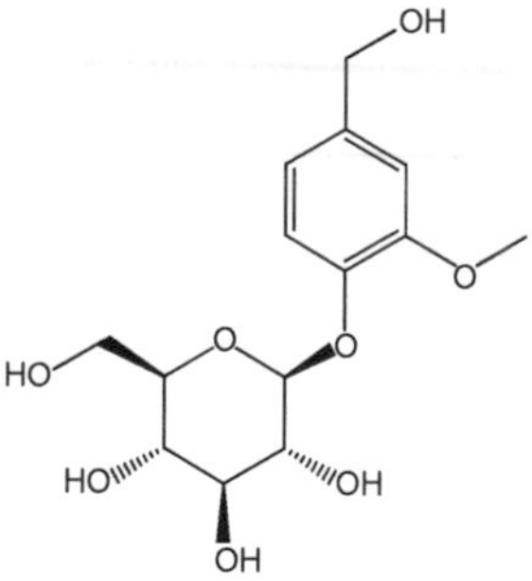	From aerial parts of *C. flavicans* via NMR and library (Luecha et al. 2009)

REFERENCES

Al-Said, M.S., E.A. Abdelsattar, S.I. Khalifa, and F.S. El-Feraly. 1988. Isolation and identification of an anti-inflammatory principle from *Capparis spinosa*. *Pharmazie* 43(9): 640–641.

Ben-Shabat, S., E. Fride, T. Sheskin, T. Tamiri, M.H. Rhee, Z. Vogel, T. Bisogno, L. De Petrocellis, V. Di Marzo, and R. Mechoulam. 1998. An entourage effect: inactive endogenous fatty acid glycerol esters enhance 2-arachidonoyl-glycerol cannabinoid activity. *European Journal of Pharmacology* 353(1): 23–31.

Calis, I., A. Kuruuzum-Uz, P.A. Lorenzetto, and P. Ruedi. 2002. (6S)-Hydroxy-3-oxo-α-ionol glucosides from *Capparis spinosa* fruits. *Phytochemistry* 59(4): 451–457.

Calis, I., A. Kuruuzum, and P. Ruedi. 1999. 1H-indole-3-acetonitrile glycosides from *Capparis spinosa* fruits. *Phytochemistry* 50(7): 1205–1208.

Chauhan, E.M., A. Duhan, and C.M. Bhat. 1986. Nutritional value of ker (*Capparis decidua*) fruit. *Journal of Food Science and Technology* 23(2): 106–108.

Dhar, D.N., R.P. Tewari, R.D. Tripathi, and A.P. Ahuja. 1972. Chemical examination of *Capparis decidua*. *Proceedings of the National Academy of Science, India: Section A* 42(1): 24–27.

Fu, X.P., H.A. Aisa, M. Abdurahim, A. Yili, S.F. Aripova, and B. Tashkhodzhaev. 2007. Chemical composition of *Capparis spinosa* fruit. *Chemistry of Natural Compounds* 43(2): 181–183.

Gadgoli, C. and S.H. Mishra. 1999. Antihepatotoxic activity of p-methoxybenzoic acid from *Capparis spinosa*. *Journal of Ethnopharmacology* 66(2): 187–192.

Garcia-Garcia, P., M. Brenes-Balbuena, C. Romero-Barranco, and A. Garrido-Fernandez. 2001. Biogenic amines in packed table olives and pickles. *Journal of Food Protection* 64(3): 374–378.

Gupta, J. and M. Ali. 1997. Oxygenated heterocyclic constituents from *Capparis decidua* root-barks. *Indian Journal of Heterocyclic Chemistry* 6(4): 295–302.

Gupta, J. and M. Ali. 1998. Phytoconstituents of *Capparis decidua* root barks. *Journal of Medicinal and Aromatic Plant Sciences* 20(3): 683–689.

Hu, Z., L. Liang, and M. He. 1987. Isolation and identification of oxazolidine-2-thione in the seeds of *Capparis masaikai*. *Yunnan Zhiwu Yanjiu* 9(1): 113–115.

Jankowski, W.J. and T. Chojnacki. 1991. Long-chain polyisoprenoid alcohols in leaves of *Capparis* species. *Acta Biochimica Polonica* 38(2): 265–276.

Ji, Y.B., S.D. Guo, and C.F. Ji. 2006. Determination and analysis of polysaccharide from *Capparis spinosa* ripe fruits by HPCE. *Zhongguo Yaoxue Zazhi (Beijing, China)* 41(15): 1186–1189.

Jiang, W., W.H. Lin, and S.D. Guo. 2005. Study on chemical constituents of *Capparis spinosa*. *Harbin Shangye Daxue Xuebao, Ziran Kexueban* 21(6): 684–686, 689.

Kanaujia, A., R. Duggar, S.T. Pannakal, S.S. Yadav, C.K. Katiyar, V. Bansal, S. Anand, S. Sujatha, and B.S. Lakshmi. 2010. Insulinomimetic activity of two new gallotannins from the fruits of *Capparis moonii*. *Bioorganic & Medicinal Chemistry* 18(11): 3940–3945.

Khanfar, M.A., S.S. Sabri, M.H. Abu Zarga, and K.P. Zeller. 2003. The chemical constituents of *Capparis spinosa* of Jordanian origin. *Natural Prodroduct Research* 17(1): 9–14.

Lansky, E.P., G. Harrison, P. Froom, and W.G. Jiang. 2005. Pomegranate (*Punica granatum*) pure chemicals show possible synergistic inhibition of human PC-3 prostate cancer cell invasion across Matrigel. *Investigational New Drugs* 23(2): 121–122.

Liu, K.C., C.J. Chou, and W.C. Pan. 1977. Studies on the constituents of the stems of *Capparis formosana* Hemsl. *Taiwan Yaoxue Zazhi* 28(1–2): 62–65.

Luecha, P., K. Umehara, T. Miyase, and H. Noguchi. 2009. Antiestrogenic constituents of the Thai medicinal plants *Capparis flavicans* and *Vitex glabrata*. *Journal of Natural Products* 72(11): 1954–1959.

Rai, S. 1987a. Chemical examination of edible plants of Rajasthan desert with special reference to Capparidaceae. *Current Agriculture* 11(1–2): 15–23.

Rai, S. 1987b. Oils and fats in arid plants with particular reference to *Capparis decidua* L. *Transactions of Indian Society of Desert Technology* 12(2): 99–105.

Ramachandram, R., M. Ali, and S.R. Mir. 2004. Phytoconstituents from *Capparis moonii* fruits. *Indian Journal of Natural Products* 20(1): 40–42.

Saraswathy, A., E. Sasikala, A. Patra, and A.B. Kundu. 1991. Betulin 28-acetate from *Capparis sepiaria* L. *Journal of the Indian Chemical Society* 68 (11): 633–634.

Su, D.M., Y.H. Wang, S.S. Yu, D.Q. Yu, Y.C. Hu, W.Z. Tang, G.T. Liu, and W.J. Wang. 2007. Glucosides from the roots of *Capparis tenera*. *Chemistry & Biodiversity* 4(12): 2852–2862.

Upadhyay, R.K., L. Rohatgi, M.K. Chaubey, and S.C. Jain. 2006. Ovipositional responses of the pulse beetle, *Bruchus chinensis* (*Coleoptera: Bruchidae*) to extracts and compounds of *Capparis decidua*. *Journal of Agricultural and Food Chemistry* 54(26): 9747–9751.

Villasenor, I.M. 2007. Bioactivities of iridoids. *Anti-Inflammatory & Anti-Allergy Agents in Medicinal Chemistry* 6(4): 307–314.

Wu, J.H., F.R. Chang, K.I. Hayashi, H. Shiraki, C.C. Liaw, Y. Nakanishi, K.F. Bastow, D. Yu, I.S. Chen, and K.H. Lee. 2003. Antitumor agents. Part 218: Cappamensin A, a new in vitro anticancer principle, from *Capparis sikkimensis*. *Bioorganic & Medicinal Chemistry Letters* 13(13): 2223–2225.

Xiao, W., N. Li, and X. Li. 2008. Isolation and identification of organic acids from pericarp of *Capparis spinosa* L. *Shenyang Yaoke Daxue Xuebao* 25(10): 790–792.

Yang, T., X. Cheng, F. Yu, G. Chou, C. Wang, and Z. Wang. 2010a. The chemical constituents of *Capparis spinosa* L. fruits. *Xibei Yaoxue Zazhi* 25(4): 260–263.

Yang, T., H. Liu, X. Cheng, F. Yu, G. Chou, C. Wang, and Z. Wang. 2011. The chemical constituents from stems and leaves of *Capparis spinosa* L. *Xibei Yaoxue Zazhi* 26(1): 16–18.

Yang, T., C. Wang, H. Liu, G. Chou, X. Cheng, and Z. Wang. 2010b. A new antioxidant compound from *Capparis spinosa*. *Pharmaceutical Biology* 48(5):589–94.
Yu, L., C.P. He, X.M. Zhang, N. Yu, and L.Q. Xie. 2011. Extraction and antioxidant activity of chemical compositions in *Capparis spinosa* L. *Harbin Shangye Daxue Xuebao, Ziran Kexueban* 27(4): 524–7.
Yu, Y., H. Gao, Z. Tang, X. Song, and L. Wu. 2006. Several phenolic acids from the fruit of *Capparis spinosa*. *Asian Journal of Traditional Medicines* 1(2): 101–4.
Zhang, Y., H. Zhang, B. Han, and W. Chen. 2011. Extraction of polysaccharides in *Capparis spinosa* L. and anti-inflammatory and analgesic effects. *Shihezi Daxue Xuebao, Ziran Kexueban* 29(2): 205–209.

Section II

Medical Uses

13 Oxidative Tension

Oxidative tension or, more commonly, *oxidative stress* is nothing less than the exhaust of living, a side effect of operations ranging from photosynthesis in plants to humans being "cut down" at work by an insensitive boss. In plants, antioxidant defense mechanisms are triggered by many stressors, for example, the abiotic stress of ultraviolet radiation from the sun (Vyšniauskiene and Ranceliene 2008). Oxidative tension is also at the root of nearly every human disease and malfunction, after and during physical or mental trauma, cancer, schizophrenia, Alzheimer's disease, cardiovascular disease, hepatic disease, lung disease, digestive disorder, and on and on. Oxidative tension is also the most important trigger for inflammation, which is considered subsequently. The ubiquity of oxidative tension underlying physical and mental maladies of all kinds underlies the public's fascination with antioxidants and physicians gradually (and sometimes grudgingly) coming around to realize their importance.

Oxidation involves, as is obvious to chemists, an acceptance of electrons by electron donor compounds, known as oxidizing agents or free radicals. The new compound is now in a lower redox (reduction-oxidation) state, or more oxidized state, than the compound was before the acceptance of electrons occurred. These extra electrons, however, are not particularly stable in their position, and thus the new oxidized compound seeks out other compounds on which to donate or "dump" its electrons. These newer compounds consequently seek out even more compounds on which to relieve themselves of *their* electrons, and the process becomes a kind of runaway positive-feedback process toward greater oxidation of compounds. In a biological system, such as a human brain, liver, or kidney, the process may be nothing less than catastrophic. Fortunately, though, all these systems (but unfortunately also systems defined as "cancers" or "neoplasms") possess antioxidant defense mechanisms, which we now know are mediated by genes that are turned on or activated by a specialized gene-protein cascade known as Nrf2. This system can reverse the effects of oxidation and "reduce" compounds that have already been oxidized, relieving them of their extra electrons, and in turn can donate protons, which are positively charged hydrogen atoms (lacking an electron). The movement back and forth of electrons and protons between compounds is known as redox cycle and determines the overall redox potential. Redox modulation is a key chemical mechanism that further regulates many physiological processes at all levels in both health and disease. Although we stress the importance of antioxidants for maintaining health and fighting (or healing) disease, sometimes oxidation is a desirable process both for overcoming pathology (as occurs by free radicals interfering with cancer cell growth) or for triggering normal and desired physiological processes, as occurs in the regulation of ovulation and the succeeding events in the female reproductive cycle. In short, a certain amount of strategic oxidation is sometimes required in medicine, and the wise physician works with the wisdom of the body to achieve this goal.

However, in many, if not the majority, of cases, antioxidant influence on the body is indicated both for the maintenance of vibrant health and the resolution of illness. Antioxidants specifically benefit the health of all the vital organs, reduce atherosclerosis and inflammation; keep the brain and its function, the mind, nimble; and allow the immune system to hone its surveillance to a supercompetent level of operation. Thus, plant foods in general, and *Capparis* extracts in particular, can play an important global role in preventing oxidation and spoilage in living tissue. The effect may resonate deeply in the animal kingdom, and possibly, through both direct and indirect effects, in the plant kingdom as well (Figure 13.1).

The examples in Table 13.1 provide a brief overview of selected research that has succeeded to demonstrate antioxidant effects in chemical, *in vitro, in vivo,* and clinical settings. This short body of work provides a clear and modern basis for the numerous anti-inflammatory and general health promoting effects that follow.

FIGURE 13.1 Fruits and leaves of *Capparis spinosa* are antioxidant, as seen both in its traditional use and in modern research. (25.7.2010, Ein Karem, Jerusalem, Israel: by Helena Paavilainen.)

TABLE 13.1
Antioxidant Properties of *Capparis* spp.

Capparis sp.	Plant Part	Pharmacological Action	Compounds Responsible (If Known)	Reference(s)
aphylla	Stem methanol extract, 30 mg/kg single p.o. dose	In streptozocin-induced diabetic rat, ↓ glutathione level, superoxide dismutase (SOD), catalase (CAT), glutathione peroxidase activity in liver, heart, kidney; rapid lipid peroxidation prevention, secondary free radical scavenging		Dangi and Mishra 2011a
decidua	Powdered fruit	↓ rat erythrocyte, liver, heart alloxan-induced lipid peroxidation and erythrocyte SOD act., ↑ CAT activity in erythrocyte, liver, kidney and heart, possibly helping neutralize H_2O_2		Yadav et al. 1997
decidua	Stems, MeOH and aq. exts.	200, 400 mg kg^{-1} body wt for 10 days methanol and aq. exts. masked fatty liver changes secondary CCl_4 in paraffin oil (1:9 vol./vol.) 0.2 ml/kg for 10 days comparable to silymarin, w/↓ serum aspartate amino transferase, alanine amino transferase, alkaline phosphatase (ALP), and bilirubin	Alkaloids, flavonoids, tannins, sterols, saponins, cyanogenic glycosides, and cumarins	Ali et al. 2009

TABLE 13.1 (*Continued*)
Antioxidant Properties of *Capparis* spp.

Capparis sp.	Plant Part	Pharmacological Action	Compounds Responsible (If Known)	Reference(s)
spinosa	Aq. alc. ext.	Antiox. in 1,1-diphenyl-2-picrylhydrazyl (DPPH) assay, prepg. 90, 60, 30, and 15 μ*M* of DPPH·dissolved in solns. that measured spectrophotometrically at 1,515 nm		Akay and Yildiz 2001
spinosa	Buds	Antioxidant activity *in vitro*, reduced post-UV skin erythema in human volunteers	Flavonols (kaempferol and quercetin derivatives) and hydroxycinnamic acids (caffeic acid, ferulic acid, p-cumaric acid, and cinnamic acid)	Bonina et al. 2002
spinosa	Buds, MeOH extract	*In vitro* ↓ oxidative bleaching of stable DPPH radical (DPPH test), peroxidation by water-soluble radical initiator 2,2'-azobis (2-amidinopropane) hydrochloride from mixed dipalmitoylphosphatidylcholine and linoleic acid unilamellar vesicles (LUVs) (LP-LUV test), UV-induced peroxidation of phosphatidylcholine multilamellar vesicles (UV-IP test), ↓ UVB-induced skin erythema in healthy human volunteers	Kaempferol and quercetin derivatives, hydroxycinnamic acids (caffeic acid, ferulic acid, p-cumaric acid, cinnamic acid) by chromatography, spectroscopy	Bonina et al. 2002
spinosa	MeOH ext. of buds	Strong antioxidant action *in vitro* by ascorbate/Fe^{2+}-mediated lipid peroxidn. of rat liver microsomes, bleaching of 1,1-diphenyl-2-picrylhydrazyl radical, autoxidation of Fe^{2+} ion exposed to bathophenanthroline disulfonate	Phenols but not glucosinolates	Germano et al. 2002
spinosa	Lactic fermented buds (in brine)	Total antioxidant potential 25.8 μmol Trolox equiv. w/[2,2'-azinobis(3-ethylbenzothiazoline-6-sulfonic acid)] diammonium salt [ABTS, 2,2'-azino-bis(3-ethylbenzothiazoline-6-sulfonic acid] cation radical decolorization assay, at 3.5 and 7.0 μ*M* gallic acid equivalent (GAE) dose-dependent peroxyl radical scavenging activity, at 70–280 μ*M* GAE → dose-dependent lipid autoxidn. inhib. in heated red meat, incubated with simulated gastric fluid for 180 min, rutin at same amt. as concn. ext. was ineffective, α-tocopherol at 25 μ*M* poorly effective, 70 μ*M* GAE ext. prevented consumption of coincubated α-tocopherol, but lipid oxidn. inhibited for full exptl. time, suggesting cooperation between ext. components and vitamin	In 8.6 g capers, 13.76 mg rutin, 42.14 μmol isothiocyanates, 4.19 mg total phenols as GAE	Tesoriere et al. 2007

(*Continued*)

TABLE 13.1 (*Continued*)
Antioxidant Properties of *Capparis* spp.

Capparis sp.	Plant Part	Pharmacological Action	Compounds Responsible (If Known)	Reference(s)
spinosa	Fruit	Fractional extraction of meal, by DPPH, Et acetate > water > n-butanol > chloroform > petroleum ether		Yu et al. 2011
spinosa	Fruit, aqueous and ethyl acetate extracts	Antioxidant, DPPH radical scavenging *in vitro*	Cappariside (4-hydroxy-5-methylfuran-3-carboxylic acid) by NMR	Yang et al. 2010
spinosa	Fruit, ethanolic extract	Decreases oxidative tension in dermal fibroblasts *in vitro*		Cao et al. 2010
spinosa	Fruits	Aq. and EtOH extracts showed potent *in vitro*/in silico effects of 95–98% action of BHA or BHT, compared to 88% for α-tocopherol by ABTS, DPPH, ferric thiocyanate method, superoxide anion radical scavenging, and metal chelating		Nadaroglu et al. 2009
spinosa	Fruits	EtOAc and aq. exts. greater DPPH scavenging act. than pet. ether fractions		Yang et al. 2010
spinosa	Leaf	Antioxidant effect variable from different trans-Himalaya sites (*Skuru* >> *Tirchey*), by ABTS, DPPH, and FRAP	ABTS and FRAP well correlated with total flavonoids (as quercetin equivalent), DPPH poorly correlated	Bhoyar et al. 2011
spinosa	MeOH fraction of aqueous extract	Antihepatotoxic activity after CCl_4 or paracetamol exposure in rats *in vivo* or thioacetamide or galactosamine exposure of rat hepatocytes *in vitro*		Gadgoli and Mishra 1999
spinosa	Stem, total flavonoids extract	Antioxidant by DPPH in rape seed oil > BHT	Flavonoids	Chen et al. 2010
spinosa		By DPPH, Et acetate fraction (cited as rich in mono- and diglycosides) stronger antioxidant than butanolic fraction (cited rich in polyglycosides)	For example, kalanchoe glycosides	El-Haci et al. 2010
zeylanica	Stem	In Wistar albino rats, lowered indices of hepatic damage, serum glutamic oxaloacetic transaminase (SGOT), serum glutamic pyruvate transaminase (SGPT), alkaline phosphatase (ALP), total bilirubin (TB), and total protein compared with toxic control following 7 days of oral dose 100 mg/kg of ext., and CCl_4 on the 7th day		Satyanarayana et al. 2010

The following overview of the modern ethnomedical uses of various *Capparis* species is based on the wide scientific and descriptive literature on the subject and in particular on the extremely useful databases NAPRALERT, Dr. Duke's Phytochemical and Ethnobotanical Databases, and U.K. Cropnet's EthnobotDB. As mentioned, most of the medical indications in the literature, both modern and ancient, for *Capparis* are connected with oxidative tension (Figures 13.2 and 13.3). Therefore, Table 13.2 includes only material that does not appear in the more specific tables in other chapters. The reader is referred to them, especially to those discussing inflammation.

FIGURE 13.2 Root, root bark, and bark of *Capparis spinosa* are used in ethnomedicine for gynecological problems. (20.5.2010, Jerusalem, Israel: by Helena Paavilainen.)

FIGURE 13.3 *Capparis fascicularis* DC. var. *fascicularis* branches and fruit. The root of *C. fascicularis* is used in Kenya against gastrointestinal problems. (13.11.2008: by Linda Loffler via Swaziland's Flora Database, Swaziland National Trust Commission, http://www.sntc.org.sz/flora/pho to.asp?phid=2729.)

TABLE 13.2
Contemporary Ethnomedical Uses of *Capparis* spp.

Indication	Geography	*Capparis* sp.	Part Used	References
(A) Modern Ethnomedical Uses of *Capparis* spp., General				
Alterative		*sepiaria*		*Wealth of India* 1985–1992
Alterative	Turkey	*spinosa*		Steinmetz 1957
Antiperspirant		*horrida*		Uphof 1968
Antiperspirant		*zeylanica*		*Wealth of India* 1985–1992
Astringent	Saudi Arabia	*decidua*	Dried aerial parts	Al-Yahya 1986
Astringent	Turkey	*spinosa*		Steinmetz 1957
Bitter taste	Saudi Arabia	*spinosa*	Dried bark	Agha 1984
Caustic	Israel	*spinosa*	Fresh fruit + root	Dafni et al. 1984
Diaphoretic	India	*decidua*	Dried root bark	Gaind and Juneja 1969
Stimulant	Morocco	*spinosa*	Fruit	Bellakhdar et al. 1991
Stimulant	Turkey	*spinosa*		Steinmetz 1957
Tonic: as a tonic in menopause complaints	Iran	*spinosa*	Dried root	Miraldi et al. 2001
Tonic		*sepiaria*		*Wealth of India* 1985–1992
Tonic	India	*spinosa*	Dried root bark	Shirwaikar et al. 1996
Tonic	Turkey	*spinosa*		Steinmetz 1957
Vesicant (= counterirritant)	Venezuela	*flexuosa*		Pittier 1926
(B) Modern Ethnomedical Uses of *Capparis* spp., Skin Problems				
Skin diseases	Mexico	*baducca*		Standley 1920–1926
Skin diseases	Saudi Arabia	*decidua*	Dried aerial parts	Al-Yahya 1986
Skin problems	Sudan	*decidua*		Broun and Massey 1929
Skin problems	Mexico	*flexuosa*		Martínez 1933, Standley 1920–1926
Skin diseases	India	*sepiaria*	Dried leaf	Sebastian and Bhandari 1984
Against sores	India	*zeylanica*		Jain and Tarafder 1970
(C) Modern Ethnomedical Uses of *Capparis* spp., Neurological Problems				
Paralysis		*spinosa*		Duke 1993
Hemiplegia	India	*zeylanica*		Jain and Tarafder 1970
Neuritis	Peru	*cordata*		Ramirez et al. 1988
Neuritis	Peru	*ovalifolia*	Fruit	Ramirez et al. 1988
Neuralgia	India	*zeylanica*		Jain and Tarafder 1970
Neurotonic (nervine)	South America	*amygdalina*	Fruit	Dragendorff 1898
Neurotonic (nervine)	West Indies	*amygdalina*	Root	Dragendorff 1898
Against vertigo	Java	*micracantha*		Burkill et al. 1966
Against vertigo	Java	*pyrifolia*		Duke 1993, EthnobotDB 1996
Against giddiness	Java	*acuminata*		Burkill et al. 1966
Sedative	Mexico	*baducca*, *flexuosa*		Standley 1920–1926
Sedative		*horrida*		Uphof 1968
Sedative	India	*zeylanica*		*Wealth of India* 1985–1992

TABLE 13.2 (*Continued*)
Contemporary Ethnomedical Uses of *Capparis* spp.

(D) Modern Ethnomedical Uses of *Capparis* spp., Gastrointestinal

Appetite stimulant	Europe	*spinosa*	Dried fruit	De Baïracli-Levy 1974
Appetite stimulant	Dominican Republic	*cynophallophora*		Liogier 1974
Digestive	Italy	*spinosa*	Buds	De Feo et al. 1992, De Feo and Senatore 1993
Digestive stimulant	Europe	*spinosa*	Dried fruit	De Baïracli- Levy 1974
Emetic	South Africa	*critifolia*		Simon and Lamla 1991
Emetic, said to be	Egypt	*decidua*	Fresh seeds	Goodman and Hobbs 1988
Gastrointestinal problems	Kenya	*fascicularis*	Root	Johns et al. 1995
Gastrointestinal problems		*horrida*		Uphof 1968
Stomach pains	Kenya	*fascicularis*	Dried root	Johns et al. 1990
Stomachache	Java	*acuminata*		Burkill et al. 1966
Stomachache	Java	*pyrifolia*		Duke 1993, EthnobotDB 1996
Abdominal pain	Haiti	*flexuosa*	Dried wood	Weniger et al. 1986
Colic	India	*zeylanica*		Jain and Tarafder 1970
Laxative	Saudi Arabia	*decidua*	Dried aerial parts	Al-Yahya 1986
Laxative		*heyneana*		*Wealth of India* 1985–1992
Purgative	Uganda	*corymbosa*	Dried root bark	Sawadogo et al. 1981
Purgative	Spain	*spinosa*	Dried leaf + root	Rios et al. 1987
Purgative	India	*spinosa*	Dried root bark	Shirwaikar et al. 1996
Purgative	Rwanda	*tomentosa*	Dried entire plant	Chagnon 1984
Aperient	Italy	*spinosa*	Buds	De Feo and Senatore 1993
Stomachic	Saudi Arabia	*decidua*	Dried aerial parts	Al-Yahya 1986
Stomachic		*horrida*		Uphof 1968
Stomachic		*zeylanica*		*Wealth of India* 1985–1992

(E) Modern Ethnomedical Uses of *Capparis* spp., Liver and Spleen

Cholagogue	Pakistan	*zeylanica*	Root bark	Baquar and Tasnif 1967
Jaundice	India	*spinosa*	Dried root	Jain and Puri 1984
Jaundice	South Africa	*tomentosa*	Fresh root bark	Watt and Breyer-Brandwijk 1962
Liver afflictions	India	*spinosa*	Dried root bark	Shirwaikar et al. 1996
Hepatitis	Rwanda	*tomentosa*	Dried leaf	Vlietinck et al. 1995
Liver sclerosis		*jamaicensis*		Hartwell 1967–1971
Gall sickness (human)	South Africa	*critifolia*		Simon and Lamla 1991
Biliousness	Java	*acuminata*		Burkill et al. 1966
Biliousness		*decidua*		Duke 1993
Biliousness	Java	*pyrifolia*		Duke 1993, EthnobotDB 1996
Splenitis		*spinosa*		Duke 1993
Spleen sclerosis	Egypt	*spinosa*		Hartwell 1967–1971

(*Continued*)

TABLE 13.2 (*Continued*)
Contemporary Ethnomedical Uses of *Capparis* spp.

(F) Modern Ethnomedical Uses of *Capparis* spp., Kidney				
Diuretic	Mexico	*baducca*		Standley 1920–1926
Diuretic	Uganda	*corymbosa*	Dried root bark	Sawadogo et al. 1981
Diuretic	Saudi Arabia	*decidua*	Dried aerial parts	Al-Yahya 1986, Tanira et al. 1989
Diuretic	Mexico, Venezuela	*flexuosa*		Martínez 1933, Pittier 1926, Standley 1920–1926
Diuretic	Indonesia	*micracantha*	Root	Burkill et al. 1966
As a diuretic in menopause complaints	Iran	*spinosa*	Dried root	Miraldi et al. 2001
Diuretic	Jordan	*spinosa*	Root bark	Al-Khalil 1995
Diuretic	India	*spinosa*	Dried root bark	Shirwaikar et al. 1996
Diuretic	Spain, Turkey	*spinosa*		Font Quer 1979, Steinmetz 1957
Diuretic	Rwanda	*tomentosa*	Dried entire plant	Chagnon 1984
(G) Modern Ethnomedical Uses of *Capparis* spp., Gynecological/Urogenital Problems				
Abortifacient	India	*decidua*		Casey 1960
Abortifacient	Rwanda	*tomentosa*	Dried entire plant	Chagnon 1984
Antifertility agent (female)	India	*decidua*		Casey 1960
Aphrodisiac (male)	South Africa (Zulu)	*corymbifera*	Root	Bryant 1966
Emmenagogue	Mexico	*baducca*		Standley 1920–1926
Emmenagogue	Cuba	*cynophallophora*	Root	Roig y Mesa 1945
Emmenagogue	Haiti	*cynophallophora*		Liogier 1974
Emmenagogue	India	*decidua*	Entire plant	Saha et al. 1961
Emmenagogue	Mexico	*flexuosa*	Bark	Martínez 1933
Emmenagogue	Peru	*flexuosa*		Moreno 1975
Emmenagogue	Venezuela	*flexuosa*		Gonzalez and Silva 1987
Emmenagogue	Venezuela, West Indies	*flexuosa*	Root	Roig y Mesa 1945
Emmenagogue	Pakistan	*galeata*	Leaf	Baquar and Tasnif 1967
Emmenagogue	Pakistan	*spinosa*	Bark	Ahmad 1957
Emmenagogue	Jordan	*spinosa*	Dried entire plant	Khanfar et al. 2003
Emmenagogue	India	*spinosa*	Root bark	Saha et al. 1961
Emmenagogue	Jordan	*spinosa*	Root bark	Al-Khalil 1995
Emmenagogue	India	*spinosa*	Dried root bark	Shirwaikar et al. 1996
To facilitate menstruation	Vietnam	*bariensis*	Root	Petelot 1952–1954
To induce menses	Europe	*spinosa*		Jöchle 1974
Uterotonic	Philippines	*micrantha*		Uphof 1968
To cure impotency (male)	South Africa (Zulu)	*tomentosa*	Dried root	Watt and Breyer-Brandwijk 1962
Against impotence (male)	South Africa	*tomentosa*	Dried root	Dekker et al. 1987
For male erection	Israel	*spinosa*	Dried flowers	Dafni et al. 1984
To provoke milk secretion	Congo	*erythrocarpos*	Root	Vasileva 1969
Galactogogue	Zaire	*erythrocarpos*		Ayensu 1978
As a diuretic in menopause complaints	Iran	*spinosa*	Dried root	Miraldi et al. 2001

TABLE 13.2 (*Continued*)
Contemporary Ethnomedical Uses of *Capparis* spp.

As a tonic in menopause complaints	Iran	*spinosa*	Dried root	Miraldi et al. 2001
Menstrual troubles/complaints	South America	*amygdalina*	Fruit	Dragendorff 1898
Menstrual troubles/complaints	West Indies	*amygdalina*	Root	Dragendorff 1898
Painful menstruation	Morocco	*spinosa*	Fruit	Bellakhdar et al. 1991
Menstrual troubles/complaints	Europe (Southern), North Africa, Saudi Arabia	*spinosa*	Root bark	Dragendorff 1898
Dysmenorrhea	India	*spinosa*	Fruit	Uniyal 1990
Against sterility (female)	Egypt, India, Yemen	*decidua*	Fruit	Dragendorff 1898
Against infertility (female)	Venda	*tomentosa*	Dried root	Arnold and Gulumian 1984
Against barrenness (female)	South Africa	*tomentosa*	Dried root	Dekker et al. 1987
For reproduction enhancement (female)	Israel	*spinosa*		Friedman et al. 1986
Against male and female sterility	Israel	*spinosa*	Dried root	Dafni et al. 1984
Against gonorrhea	Somalia	*tomentosa*	Fresh root	Samuelsson et al. 1991
Antisyphilitic	Argentina	*salicifolia*	Root	Filipoy 1994
Against illness due to miscarriage	Zimbabwe	*tomentosa*	Dried root	Nyazema 1984
(H) Modern Ethnomedical Uses of *Capparis* spp., Poison Related				
Antidote	Saudi Arabia	*decidua*	Dried aerial parts	Al-Yahya 1986
Antidote to poisons	Saudi Arabia	*decidua*	Dried aerial parts	Shah et al. 1989
Antivenin	India	*cartilaginea*	Leaf	Selvanayahgam et al. 1994
Antivenin	India	*sepiaria*	Seed	Selvanayahgam et al. 1994
Antivenin	India	*tomentosa*	Leaf	Selvanayahgam et al. 1994
Against snakebite	India	*sepiaria*	Shade dried root	Nyman et al. 1998
Against snakebite	South Africa	*tomentosa*	Fresh leaf	Watt and Breyer-Brandwijk 1962
Antidote for snakebite (prophylactic)	India	*zeylanica*	Fresh pickled fruit	Siddiqui and Husain 1990
Poisonous	Guatemala, Mexico	*baducca*	Fruit	Standley 1920–1926, Standley and Steyermark 1952
Poisonous	Africa	*fascicularis*		Lewis and Elvin-Lewis 1977
Poisonous	Guatemala	*incana*		Standley 1920–1926, Standley and Steyermark 1952
Poisonous	Africa, Sudan	*tomentosa*		Broun and Massey 1929, Lewis and Elvin-Lewis 1977

(*Continued*)

TABLE 13.2 (*Continued*)
Contemporary Ethnomedical Uses of *Capparis* spp.

Used as a poison (for humans)	South Africa	*hereroensis*	Dried root	Watt and Breyer-Brandwijk 1962
Used in arrow poison	Peru	*sola*		Duke and Vasquez 1994
Toxic to livestock	South Africa	*hereroensis*	Fresh aerial parts	Watt and Breyer-Brandwijk 1962
		(I) Modern Ethnomedical Uses of *Capparis* spp., Veterinary		
Tonic to treat redness of eye of cattle	India	*decidua*	Root	Sikarwar et al. 1994
Edible to livestock	South Africa	*oleoides*	Fresh aerial parts	Watt and Breyer-Brandwijk 1962
		(J) Modern Ethnomedical Uses of *Capparis* spp., Other		
For healing fractures	Hawaii	*sandwichiana*		Nagata 1971
Against gland problems		*spinosa*		Duke 1993
Against scurvy	Saudi Arabia	*deserti*	Leaf	Al-Said 1993
Against scurvy	Turkey	*spinosa*		Steinmetz 1957
To cause embedded thorns to come out of skin	Paraguay	*speciosa*	Dried bark	Arenas 1987
Food	Saudi Arabia	*cartilaginea*	Fresh fruit	Agha 1984
Food	Egypt	*decidua, galeata*	Fresh fruit	Goodman and Hobbs 1988
Food	India	*grandiflora*	Fresh fruit	Ramachandran and Nair 1981
Food	Thailand	*micrantha*	Fresh stem	Murakami et al. 1993
Food	Paraguay	*retusa, salicifolia, tweediana*	Boiled fruit (unripe)	Schmeda-Hirschmann 1994
Food	India	*spinosa*	Fruit	Uniyal 1990
Eaten by humans	South Africa	*oleoides*	Fresh flower buds	Watt and Breyer-Brandwijk 1962
Eaten in times of famine	South Africa	*oleoides*	Fresh root	Watt and Breyer-Brandwijk 1962
As a condiment	Saudi Arabia	*decidua*	Dried aerial parts	Al-Yahya 1986
Food dressing	Italy	*spinosa*	Buds	De Feo et al. 1992
Used as a flavoring in cooking	Saudi Arabia	*spinosa*	Fresh (pickled) buds	Agha 1984
Spice	Iraq	*spinosa*		Al-Rawi and Chakravarty 1964
Sweet taste	China	*masakai*	Dried seed	Hu and He 1983, Liu et al. 1993
As a medicine	Saudi Arabia	*cartilaginea*	Fresh fruit	Agha 1984
As a medicine	Thailand	*micrantha*	Fresh stem	Murakami et al. 1993
As a herbal mixture (purpose not specified in NAPRALERT)	India	*spinosa* (in a herbal mixture)	Part not specified in NAPRALERT	Upadhya et al. 1988

REFERENCES

Agha, Z.M. 1984. Medicinal uses of plants in Saudi Arabia. Personal communication with the author. Via NAPRALERT[SM]. Natural Products Alert Database. 1975. [Online database.] Program for Collaborative Research in the Pharmaceutical Sciences (PCRPS), College of Pharmacy, University of Illinois at Chicago. Chicago, IL. http://www.napralert.org (accessed 24 June 2012).

Ahmad, Y.S. 1957. *A Note on the Plants of Medicinal Value Found in Pakistan.* Karachi, Pakistan: Government of Pakistan Press.

Akay, F. and F. Yildiz. 2001. Estimation of total antioxidant capacity of pomegranate, apricot, caper, eggplant and oils. *Royal Society of Chemistry* 269: 368–370.

Ali, S.A., T.H. Al-Amin, A.H. Mohamed, and A.A. Gameel. 2009. Hepatoprotective activity of aqueous and methanolic extracts of *Capparis decidua* stems against carbon tetrachloride induced liver damage in rats. *Journal of Pharmacology and Toxicology* 4(4): 167–172.

Al-Khalil, S. 1995. A survey of plants used in Jordanian traditional medicine. *International Journal of Pharmacognosy* 33(4): 317–323.

Al-Rawi, A. and H. Chakravarty. 1964. *Medicinal Plants of Iraq.* Baghdad, Iraq: Government Press.

Al-Said, M.S. 1993. Traditional medicinal plants of Saudi Arabia. *American Journal of Chinese Medicine* 21(3–4): 291–298.

Al-Yahya, M.A. 1986. Phytochemical studies of the plants used in traditional medicine of Saudi Arabia. *Fitoterapia* 57(3): 179–182.

Arenas, P. 1987. Medicine and magic among the Maka Indians of the Paraguayan Chaco. *Journal of Ethnopharmacology* 21 (3): 279–295.

Arnold, H.J. and M. Gulumian. 1984. Pharmacopoeia of traditional medicine in Venda. *Journal of Ethnopharmacology* 12(1): 35–74.

Ayensu, E.S. 1978. *Medicinal Plants of West Africa.* Algonac, MI: Reference Publications.

Baquar, S.R. and M. Tasnif. 1967. *Medicinal Plants of Southern West Pakistan.* Pak P C S I R Bull Monogr 3. Islamabad, Pakistan: Pakistan Council of Scientific and Industrial Research.

Bellakhdar, J., R. Claisse, J. Fleurentin, and C. Younos. 1991. Repertory of standard herbal drugs in the Moroccan pharmacopoea. *Journal of Ethnopharmacology* 35(2): 123–143.

Bhoyar, M.S., G.P. Mishra, P.K. Naik, and R.B. Srivastava. 2011. Estimation of antioxidant activity and total phenolics among natural populations of caper (*Capparis spinosa*) leaves collected from cold arid desert of trans-Himalayas. *Australian Journal of Crop Science* 5(7): 912–919.

Bonina, F., C. Puglia, D. Ventura, R. Aquino, S. Tortora, et al. 2002. In vitro antioxidant and in vivo photoprotective effects of a lyophilized extract of *Capparis spinosa* L. buds. *Journal of Cosmetic Science* 53(6): 321–335.

Broun, A.F. and R.E. Massey. 1929. *Flora of the Sudan: With a Conspectus of Groups of Plants and Artificial Key to Families.* London: [s.n.].

Bryant, A.T. 1966. *Zulu Medicine and Medicine-Men.* Cape Town, South Africa: Struik.

Burkill, I.H., W. Birtwistle, F.W. Foxworthy, J.B. Scrivenor, and J.G. Watson. 1966. *A Dictionary of the Economic Products of the Malay Peninsula.* Kuala Lumpur, Malaysia: Published on behalf of the governments of Malaysia and Singapore by the Ministry of Agriculture and cooperatives.

Cao, Y.L., X. Li, and M. Zheng. 2010. *Capparis spinosa* protects against oxidative stress in systemic sclerosis dermal fibroblasts. *Archives of Dermatological Research* 302(5): 349–355.

Casey, R.C.D. 1960. 298 alleged antifertility plants of India. *Indian Journal of Medical Sciences* 14: 590–601.

Chagnon, M. 1984. General pharmacologic inventory of medicinal plants of Rwanda. *Journal of Ethnopharmacology* 12(3): 239–251.

Chen, Y., G. Ma, and H. Bai. 2010. Antioxidative activities in vitro of total flavonoids extracted solution from *Capparis spinosa* L. stem. *Xinjiang Nongye Kexue* 47(12): 2489–2495.

Dafni, A., Z. Yaniv, and D. Palevitch. 1984. Ethnobotanical survey of medicinal plants in northern Israel. *Journal of Ethnopharmacology* 10(3): 295–310.

Dangi, K.S. and S.N. Mishra. 2011a. Antioxidative and β cell regeneration effect of *Capparis aphylla* stem extract in streptozotocin induced diabetic rat. *Biology and Medicine (Aligarh)* 3(3): 82–91.

De Baïracli-Levy, J. 1974. *Common Herbs for Natural Health.* New York: Schocken Books.

De Feo, V., R. Aquino, A. Menghini, E. Ramundo, and F. Senatore. 1992. Traditional phytotherapy in the Peninsula Sorrentina, Campania, Southern Italy. *Journal of Ethnopharmacology* 36(2): 113–125.

De Feo, V. and F. Senatore. 1993. Medicinal plants and phytotherapy in the Amalfitan Coast, Salerno Province, Campania, Southern Italy. *Journal of Ethnopharmacology* 39(1): 39–51.

Dekker, T.G., T.G. Fourie, E. Matthee, and F.O. Snyckers. 1987. An oxindole from the roots of *Capparis tomentosa. Phytochemistry* 26(6): 1845–1846.

Dragendorff, G. 1898. *Die Heilpflanzen der verschiedenen Völker und Zeiten, ihre Anwendung, wesentlichen Bestandtheile und Geschichte; ein Handbuch für Ärzte, Apotheker, Botaniker und Droguisten*. Stuttgart, Germany: Enke.

Duke, J.A. 1993. Dr. Duke's Phytochemical and Ethnobotanical Databases. [Online database.] USDA–ARS–NGRL, Beltsville Agricultural Research Center, Beltsville, MD. http://www.ars-grin.gov/duke/ethnobot.html (accessed 12 May 2011).

Duke, J.A. and R. Vasquez. 1994. *Amazonian Ethnobotanical Dictionary*. Boca Raton, FL: CRC Press.

El-Haci, I.A., A. Didi, F.A. Bekkara, and M.A. Didi. 2010. Total phenolic contents and antioxidant activity of organic fractions from *Capparis spinosa* and *Limoniastrum feei*. *Natural Products: An Indian Journal* 6(3): 118–124.

EthnobotDB. EthnobotDB—Worldwide Plant Uses. 1996. [Online database.] National Agricultural Library, Agricultural Genome Information System, Washington, DC. http://ukcrop.net/perl/ace/search/EthnobotDB (accessed 24 July 2009).

Filipoy, A. 1994. Medicinal plants of the Pilaga of Central Chaco. *Journal of Ethnopharmacology* 44(3): 181–193.

Font Quer, P. 1979. *Plantas medicinales: el Dioscórides renovado*. Barcelona: Editorial Labor.

Friedman, J., Z. Yaniv, A. Dafni, and D. Palevitch. 1986. A preliminary classification of the healing potential of medicinal plants, based on a rational analysis of an ethnopharmacological field survey among Bedouins in the Negev Desert, Israel. *Journal of Ethnopharmacology* 16(2–3): 275–287.

Gadgoli, C., and S.H. Mishra. 1999. Antihepatotoxic activity of p-methoxybenzoic acid from *Capparis spinosa*. *J Ethnopharmacol* 66(2): 187–92.

Gaind, K.N. and T.R. Juneja. 1969. Investigations on *Capparis decidua*. *Planta Med* 17: 95–98.

Germano, M.P., R. De Pasquale, V. D'Angelo, S. Catania, V. Silvari, and C. Costa. 2002. Evaluation of extracts and isolated fraction from *Capparis spinosa* L. buds as an antioxidant source. *Journal of Agricultural and Food Chemistry* 50(5): 1168–1171.

Gonzales, F. and M. Silva. 1987. A survey of plants with antifertility properties described in the South American folk medicine. In *Proceedings of the Princess Congress on Natural Products*, 20. 10–13 December, Bangkok.

Goodman, S.M. and J.I. Hobbs. 1988. The ethnobotany of the Egyptian Eastern desert: a comparison of common plant usage between two culturally distinct Bedouin groups. *Journal of Ethnopharmacology* 23 (1): 73–89.

Hartwell, J.L. 1967–1971. Plants used against cancer. A survey. *Lloydia* 30–34.

Hu, Z. and M. He. 1983. Studies on mabinlin, a sweet protein from the seeds of *Capparis masaikai* Levl. I. Extraction, purification and certain characteristics. *Yunnan Zhiwu Yanjiu* 5(2): 207–212.

Jain, S.K. and C.R. Tarafder. 1970. Medicinal plant-lore of the Santals. *Economic Botany* 24(3): 241–278.

Jain, S.P. and H.S. Puri. 1984. Ethnomedicinal plants of Jaunsar-Bawar Hills, Uttar Pradesh, India. *Journal of Ethnopharmacology* 12(2): 213–222.

Jöchle, W. 1974. Menses-inducing drugs: their role in antique, medieval and Renaissance gynecology and birth control. *Contraception* 10: 425–439.

Johns, T., G.M. Faubert, J.O. Kokwaro, R.L.A. Mahunnah, and E.K. Kimanani. 1995. Anti-giardial activity of gastrointestinal remedies of the Luo of East Africa. *Journal of Ethnopharmacology* 46(1): 17–23.

Johns, T., J.O. Kokwaro, and E.K. Kimanani. 1990. Herbal remedies of the Luo of Siaya District, Kenya: establishing quantitative criteria for consensus. *Economic Botany* 44(3): 369–381.

Khanfar, M.A., S.S. Sabri, M.H. Abu Zarga, and K.P. Zeller. 2003. The chemical constituents of *Capparis spinosa* of Jordanian origin. *Natural Products Research* 17(1): 9–14.

Lewis, W.H. and M.P.F. Elvin-Lewis. 1977. *Medical Botany: Plants Affecting Man's Health*. New York: Wiley.

Liogier, A.H. 1974. *Diccionario botanico de nombres vulgares de la española*. Santo Domingo, Dominican Republic: Impresora UNPHU.

Liu, X., S. Maeda, Z. Hu, T. Aiuchi, K. Nakaya, and Y. Kurihara. 1993. Purification, complete amino acid sequence and structural characterization of the heat-stable sweet protein, mabinlin II. *European Journal of Biochemistry* 211(1–2): 281–287.

Martínez, M. 1933. *Las plantas medicinales de México*. Mexico City: Ediciones Botas.

Miraldi, E., S. Ferri, and V. Mostaghimi. 2001. Botanical drugs and preparations in the traditional medicine of West Azerbaijan (Iran). *Journal of Ethnopharmacology* 75(2/3): 77–87.

Moreno, A.R. 1975. Two hundred sixty-eight medicinal plants used to regulate fertility in some countries of South America. Unpublished (stenciled) review in Spanish. Via NAPRALERT[SM]. Natural Products Alert Database. 1975. [Online database.] Program for Collaborative Research in the Pharmaceutical Sciences (PCRPS), College of Pharmacy, University of Illinois at Chicago. Chicago, IL. http://www.napralert.org (accessed 24 June 2012).

Murakami, A., A. Kondo, Y. Nakamura, H. Ohigashi, and K. Koshimizu. 1993. Possible anti-tumor promoting properties of edible plants from Thailand, and identification of an active constituent, cardamonin, of *Boesenbergia pandurata. Bioscience, Biotechnology, and Biochemistry* 57(11): 1971–1973.

Nadaroglu, H., N. Demir, and Y. Demir. 2009. Antioxidant and radical scavenging activities of capsules of caper (*Capparis spinosa*). *Asian Journal of Chemistry* 21(7): 5123–5134.

Nagata, K.M. 1971. Hawaiian medicinal plants. *Economic Botany* 25(3): 245–254.

NAPRALERT[SM]. Natural Products Alert Database. 1975. [Online database.] Program for Collaborative Research in the Pharmaceutical Sciences (PCRPS), College of Pharmacy, University of Illinois at Chicago. Chicago, IL. http://www.napralert.org (accessed 24 June 2012).

Nyazema, N.Z. 1984. Poisoning due to traditional remedies. *Cent Afr J Med* 30(5): 80–3.

Nyman, U., P. Joshi, L.B. Madsen, T.B. Pedersen, M. Pinstup, S. Rajasekharan, V. George, and P. Pushpangadan. 1998. Ethnomedical information and in vitro screening for angiotensin-coverting enzyme inhibition of plants utilized as traditional medicines in Gujarat, Rajasthan and Kerala (India). *Journal of Ethnopharmacology* 60(3): 247–263.

Petelot, A. 1952–1954. *Les plantes médicinales du Cambodge, du Laos et du Viêtnam*. Saigon: Centre de recherches scientifiques et techniques.

Pittier, H. 1926. *Manual de las plantas usuales de Venezuela*. Caracas, Venezuela: Litografia del comercio.

Ramachandran, V.S. and N.C. Nair. 1981. Ethnobotanical observations on Irulars of Tamil Nadu (India). *Journal of Economic and Taxonomic Botany* 2: 183–190.

Ramirez, V.R., L.J. Mostacero, A.E. Garcia, C.F. Mejia, P.F. Pelaez, C.D. Medina, and C.H. Miranda. 1988. *Vegetales empleados en medicina tradicional norperuana*. Trujillo, Peru: Banco Agrario del Peru and Universidad Nacional de Trujillo.

Rios, J.L., M.C. Recio, and A. Villar. 1987. Antimicrobial activity of selected plants employed in the Spanish Mediterranean area. *Journal of Ethnopharmacology* 21(2): 139–152.

Roig y Mesa, J.T. 1945. *Plantas medicinales, aromáticas o venenosas de Cuba*. Havana: [Editorial Guerrero Casamayor y cía].

Saha, J.C., E.C. Savini, and S. Kasinathan. 1961. Ecbolic properties of Indian medicinal plants. Part 1. *Indian Journal of Medical Research* 49: 130–151.

Samuelsson, G., M.H. Farah, P. Claeson, M. Hagos, M. Thulin, O. Hedberg, A.M. Warfa, A.O. Hassan, A.H. Elmi, A.D. Abdurahman, A.S. Elmi, Y.A. Abdi, and M.H. Alin. 1991. Inventory of plants used in traditional medicine in Somalia. I. Plants of the families Acanthaceae-Chenopodiaceae. *Journal of Ethnopharmacology* 35(1): 25–63.

Satyanarayana, T., A.A. Mathews, and E.M. Chinna. 2010. Prevention of carbon tetrachloride induced hepatotoxicity in rats by alcohol extract of *Capparis zeylanica* stem. *J Pharmaceutical Chemistry* 4(2): 37–39.

Sawadogo, M., A.M. Tessier, and P. Delaveau. 1981. Studies on the roots of *Capparis corymbosa* Lam. *Plantes Medicinales et Phytotherapie* 15: 234–239.

Schmeda-Hirschmann, G. 1994. Plant resources used by the Ayoreo of the Paraguayan Chaco. *Economic Botany* 48(3): 252–258.

Sebastian, M.K. and M.M. Bhandari. 1984. Medico-ethno botany of Mount Abu, Rajasthan, India. *Journal of Ethnopharmacology* 12(2): 223–2230.

Selvanayahgam, Z.E., S.G. Gnanevendhan, K. Balakrishna, and R.B. Rao. 1994. Antisnake venom botanicals from ethnomedicine. *Journal of Herbs, Spices, and Medicinal Plants* 2(4): 45–100.

Shah, A.H., M. Tariq, A.M. Ageel, and S. Qureshi. 1989. Cytological studies on some plants used in traditional Arab medicine. *Fitoterapia* 60(2): 171–173.

Shirwaikar, A., K.K. Sreenivasan, B.R. Krishnanand, and A.V. Kumar. 1996. Chemical investigation and antihepatotoxic activity of the root bark of *Capparis spinosa*. *Fitoterapia* 67(3): 200–204.

Siddiqui, M.B. and W. Husain. 1990. Traditional antidotes of snake poison. *Fitoterapia* 61(1): 41–44.

Sikarwar, R.L.S., A.K. Bajpai, and R.M. Painuli. 1994. Plants used as veterinary medicines by aboriginals of Madhya Pradesh, India. *International Journal of Pharmacognosy* 32(3): 251–255.

Simon, C. and M. Lamla. 1991. Merging pharmacopoeia: understanding the historical origins of incorporative pharmacopoeial processes among Xhosa healers in Southern Africa. *Journal of Ethnopharmacology* 33(3): 237–242.

Standley, P.C. 1920–1926. *Trees and Shrubs of Mexico*. Washington, DC: Smithsonian Institution.

Standley, P. C. and J. A. Steyermark. 1952. *Flora of Guatemala*. Chicago: Field Museum of Natural History.

Steinmetz, E.F. 1957. *Codex vegetabilis*. Amsterdam: [s.n.].

Swaziland National Trust Commission, Swaziland's Flora Database. 2013. [Online database.] http://www.sntc.org.sz/flora (accessed 4 January 2013).

Tanira, M.O.M., A.M. Ageel, and M.S. Al-Said. 1989. A study of some Saudi medicinal plants used as diuretics in traditional medicine. *Fitoterapia* 60(5): 443–447.

Tesoriere, L., D. Butera, C. Gentile, and M.A. Livrea. 2007. Bioactive components of caper (*Capparis spinosa* L.) from Sicily and antioxidant effects in a red meat simulated gastric digestion. *Journal of Agricultural and Food Chemistry* 55(21): 8465–8471.

Uniyal, M.R. 1990. Utility of hitherto unknown medicinal plants traditionally used in Ladakh. *Journal of Research and Education in Indian Medicine* 9(2): 89–95.

Upadhya, L., S.S. Shukla, A. Agrawal, and G.P. Dubey. 1988. Changes in brain biogenic amines under influence of an indigenous drug, Geriforte, following immobilization stress. *Indian Journal of Experimental Biology* 26(11): 911–912.

Uphof, J.C.Th. 1968. *Dictionary of Economic Plants*. Lehre, Germany: Cramer.

Vasileva, B. 1969. *Plantes medicinales de Guinee*. Conakry, Guinea: [s.n.].

Vlietinck, A.J., L. van Hoof, J. Totte, A. Lasure, D. vanden Berghe, P.C. Rwangabo, and J. Mvukiyumwami. 1995. Screening of hundred Rwandese medicinal plants for antimicrobial and antiviral properties. *Journal of Ethnopharmacology* 46(1): 31–47.

Vyšniauskiene, R. and V. Ranceliene. 2008. Changes in the activity of antioxidant enzyme superoxide dismutase in *Crepis capillaris* plants after the impact of UV-B and ozone. Scientific Works of the Lithuanian Institute of Horticulture and Lithuanian University of Agriculture. *Sodininkyste Ir Daržininkyste* 27(2): 129–37.

Watt, J.M. and M.G. Breyer-Brandwijk. 1962. *The Medicinal and Poisonous Plants of Southern and Eastern Africa*. 2nd ed. London: Livingstone.

Wealth of India. 1985–1992. *The Wealth of India:* A Dictionary of Indian Raw Materials and Industrial Products. New Delhi: Publications and Information Directorate, Council of Scientific and Industrial Research.

Weniger, B., M. Rouzier, R. Daguilh, D. Henrys, J.H. Henrys, and R. Anton. 1986. Popular medicine of the Central Plateau of Haiti. 2. Ethnopharmacological inventory. *Journal of Ethnopharmacology* 17(1): 13–30.

Yadav, P., S. Sarkar, and D. Bhatnagar. 1997. Lipid peroxidation and antioxidant enzymes in erythrocytes and tissues in aged diabetic rats. *Indian Journal of Experimental Biology* 35(4): 389–392.

Yang, T., C. Wang, H. Liu, G. Chou, X. Cheng, and Z. Wang. 2010. A new antioxidant compound from *Capparis spinosa*. *Pharmaceutical Biology* 48: 589–594.

Yu, L., C.P. He, X.M. Zhang, N. Yu, and L.Q. Xie. 2011. Extraction and antioxidant activity of chemical compositions in *Capparis spinosa* L. *Harbin Shangye Daxue Xuebao, Ziran Kexueban* 27(4): 524–527.

14 Inflammation

INTRODUCTION

Inflammation underlies or is at least strongly associated with nearly all diseases and disorders, including those that are the subject of most or all of the chapters to follow. This statement, which now enjoys extensive agreement among physicians and scientists, was for many years not appreciated in modern times. The roots of the concept actually extend back many centuries and are highlighted in tables in this chapter (Tables 14.1 and 14.2) enumerating ancient and medieval uses of *Capparis* for treating inflammation in its many different forms that are not covered in similar tables in subsequent chapters.

In modern medical terminology, most types of inflammation carry the suffix *-itis*, with the first part of these words indicating the area that is inflamed. Some common examples include stomatitis (inflammation of the mouth), gastritis (of the stomach), mastitis (of the breast), encephalitis (of the brain), myocarditis (of the muscle of the heart), myositis (of muscles in general), neuritis (of the nerves), vasculitis (of blood vessels), arteritis (of arteries), phlebitis (of veins), pleuritis (of the pleura that envelop the lungs), arthritis (of the joints), osteomyelitis (of the bones), and so on. Although pneumonitis for inflammation of the lungs themselves is sometimes used, the anomalous term pneumonia is more common, whereas asthma describes a highly specific inflammation of the airways resulting in the characteristic syndrome. Although the common use of the word *inflammation* as a synonym for various types of infections by bacteria, fungi, and viruses is incorrect and improper, all of these infections (with a few specific exceptions) induce inflammation in its myriad forms.

Generally, inflammation can be thought of as a kind of rapid oxidation of tissues, a complex and specific set of responses that serve the purpose mainly of protecting the organism from numerous assaults by both biotic and abiotic stressors. The biotic stressors are primarily the aforementioned microbes and viruses, but also encompass attacks by insects, reptiles, and mammals, including human beings. Abiotic stressors are mainly those of temperature, overhydration or dehydration, wind, ultraviolet radiation and other forms of radiation, chemicals, and more. Both plants and animals are subject to all of these factors, and both plants and animals experience and undergo inflammation.

Higher animals respond with inflammation to such stressors in a protective way through the aegis of their immune systems, which include both innate immunity and adaptive immunity. Innate immunity is the more basic of the two types, involving the complement system in animals, and it is the basis of the immune response to these stressors of plants. Adaptive immunity is the more complex animalian cognitive network that identifies biotic stressors as different from the organism (self vs. not self) and, using the mobile type of white blood cells known as lymphocytes in a manner highly homologous to the manner in which the nervous system and brain use neurons, identifies such attackers and mounts immune responses to destroy such pernicious intrusions into its being by way of inflammation. Plants also utilize inflammation associated with their innate immunity to help repair damage to their botanical selves caused by biotic or abiotic assaults, although the physiology of botanical inflammation is of course much different from inflammation mounted by animals. In both the plant and animal kingdoms, inflammation is primarily geared to walling off and containing damage and then providing the increased movement of vital fluids to the problematic areas of organisms to effect repair.

TABLE 14.1
Medieval/Ancient Authors Cited in Table 14.2

Author # in Table 14.2	Biography	Works	Reference
1	23–79, Italy	[The Natural History]	Plinius (1967–1970)
2	ca. 40–90, Anazarbos (Turkey)	[The Materials of Medicine]	Dioscorides (1902)
3	Second century, Pergamum/Rome	[Complete Works]	Galen (1964–1965)
4	800–875, Persia	[The Paradise of Wisdom]	Al-Tabari (1928)
5	ca. 865–925, Persia	[The Comprehensive Book on Medicine]	Al-Razi (1955–1970)
6	980–1037, Persia	[The Canon of Medicine]	Ibn Sina (1877/1294)
7	d. 1066, Baghdad/Cairo/Antioch	[Tables of Health]	Ibn Butlan (1531)
8	Eleventh century, Italy	Flos medicinae scholae Salerni [Flower of Medicine of the School of Salerno]	Anonymous
9	1206–1280, Germany	[Book on Growing Things]	Albertus Magnus (1867)
10	d. 1248, Spain/Egypt/Syria	[Comprehensive Book of Simple Drugs and Foods]	Ibn al-Baytar (1992/1412)
11	Thirteenth century, Italy	Herbal	Rufinus de Rizardo (1946)
12	1498–1554, Germany	[Herbal on the Differences, Names and Properties of Herbs]	Hieronymus Bock (1964)
13	1522–1590, Germany	[New Complete Herbal]	Jacob Theodor Tabernaemontanus (1970)
14	1542–1596, Mexico	[Treasure of Medicines]	Gregorio López (1982)
15	d. 1599, Syria/Cairo	[Memorandum Book for Hearts and Comprehensive Book of Wonderful Marvels]	Al-Antaki (1356)
16	1545–1611/1612, England	Herball	John Gerard (1633)
17	1567–1650, England	[The Botanical Theater]	John Parkinson (1640)

Unfortunately, sometimes this vital and valuable natural function (i.e., inflammation) becomes hung up or stuck and, failing to effect definitive repair and resolution, becomes low grade, ineffectual, and chronic. The range of medical conditions that are caused by or closely associated with inflammation is astonishing. These conditions include resistance to the action of insulin for utilizing glucose and can result in the metabolic syndrome, which may lead to diabetes or obesity. These are cardinal examples of our new understanding of inflammation, which is based mainly on a much improved and expanded knowledge of the underlying biochemistry involving factors such as cytokines that help control and regulate inflammation. Other factors are 20 carbon fatty acids known as eicosanoids that include the subdivisions of prostaglandins and leukotrienes. Lipoxygenases are enzymes involved in the synthesis of leukotrienes, which bring about inflammatory cascades in both animals and plants (Paget 1880). This new knowledge also extends into the pathogenesis of atherosclerosis and hypertension and, even more poignantly, diseases of the brain such as schizophrenia, autism, senile dementia, and epilepsy (Kobow et al. 2012, Vezzani et al. 2012). Short-term pain in muscles during and after exertion is part of an inflammatory response that leads to a rebuilding and strengthening of muscle with increased tolerance of activity.

Characteristic of inflammatory responses in both animals and plants is a movement of fluids within the organism to the damaged part of itself to exact repair in response to pathogens and abiotic and biotic stressors. In animals, this involves the movement of blood cells, especially white blood cells such as leukocytes, monocytes, lymphocytes, and eosinophils. To effect this movement of fluid, changes occur in the circulatory systems, making blood vessels such as veins and capillaries more permeable. This "leakiness" is important to the process of inflammation, which facilitates destruction of external invading cells and repair of tissue following injury. Thus, inflammation is at once integral to the organism's biological survival strategy and yet, when chronic, indolent, and unresolved, is something like the "mother of all diseases," facilitating progression or initiation of numerous pathological processes.

Inflammation has been recognized in its many guises for millennia, and some sense of the role capers may play in mollifying this process in humans is reflected in Table 14.2 on historical uses of caper. The separate sections of the text that follow, such as those concerning hypertension, sunburn, epilepsy, diabetes, infections, cancer, and cardiovascular dyscrasias, are all at core specific forms or manifestations of inflammation and could technically all be contained in the present chapter. However, since they are treated in further chapters, this chapter deals essentially with inflammation in general or inflammation that is not subsumed under the topics that follow.

No treatment of inflammation would be complete without at least mentioning its famous four cardinal signs, which have been appreciated since antiquity. These are, in Latin, *calor* (heat), *rubor* (redness), *dolor* (pain), and *tumor* (swelling). Later, attributed to Galen, a fifth sign was noted, namely, immobility of the affected area.

The first sign, calor, may refer to localized heat, as in an inflamed extremity, or systemic heat, more reflective of general inflammation and known as fever. Although fever may be sometimes considered a phenomenon that might be separate from inflammation, for our purposes we appreciate fever as caused by inflammation, as the name inflammation suggests, consumed by flames. Therefore, fever, and the actions of drugs that reduce fever, known traditionally as febrifuges (agents that cause the fever to "flee") or in modern parlance as antipyretics, are considered in this chapter, as well as in a further chapter focusing on the twin signs of fever and pain.

Similarly, dolor in modern medicine may be classified at least some of the time as a complex neuropsychic phenomenon that might be separate from inflammation and treated by "pain specialists" or "dolorologists" using myriad methods of "pain management." However, as our knowledge of the pervasiveness and insidiousness of inflammation develops into a model inclusive of virtually all disease, we might accurately classify all pain as inflammation.

Redness, or rubor, is not generally considered a disease per se, although the specific example of sunburn may be so considered and is also given a separate chapter, particularly due to the close homology between the caper plants' evolution to withstand intense solar radiation and our own. In alchemy, the source of modern chemistry that was briefly discussed in the previous chapter on minerals, the *rubedo* constituted an important "yang" midpoint in the process of internal transformation, occupying the space between the undifferentiated *prima materia,* the *nigredo*, and the final stage of "purification," the white *albedo*. This model may also be seen as a relevant metaphor for our own understanding of inflammation as an intermediate stage of repair, burning out or clearing debris from the black nigredo before reinstating the fully differentiated (or undifferentiated) totally healed state of the albedo, which may appear as white scar tissue (fibrosis). The relative and culture-bound status of these associations, however, is evident in the general Western association of the state of ultimate entropy, death, with the color black (nigredo), while in the traditional Chinese culture, the color associated with death is white (albedo). Redness, though, rubedo or rubor, is in all cultures, to the best of our knowledge, associated with life, passion, vivacity, and even joy (the color of the heart and blood and reflected in the notion of a *sanguine* temperament), lending additional depth of understanding of inflammation as a process of life, which, however, may pass on to the ambivalent whiteness or regress to the undifferentiated blackness (as in gangrene, when the inflammatory process becomes completely stagnant and fails).

Swelling, *TUMOR*, may be associated with infection, as in abscess formation, cellulitis, and the like, and, as is most commonly perceived, with cancer. Tumescence may also be a part of normal physiology, reflected along with redness, in blushing. Cancer is taken up in detail in a separate chapter, although its roots are here in inflammation. Swelling (*kveling*) may also be understood intrapsychically, as in puffed up with pride or in self-importance. The first stage of reconstruction during inflammation of the subcutaneous tissue is known as "proud flesh."

The following section elaborates some of the historical uses of *Capparis* in treating a wide variety of diseases that, like most diseases, feature inflammation in their processes and highlights modern research in general relating to inflammation and *Capparis* both *in vitro* and *in vivo*. More specific aspects of these historical uses are taken in up in further chapters.

INFLAMMATION IN MEDIEVAL AND ANCIENT MEDICINE

Traditional uses of different parts of *Capparis* plants (Figures 14.1–14.6) in medicine have been documented over thousands of years and reflected in medieval and ancient writings by a number of prominent physicians in this period (although it was not described by them all as, for example, garlic and fig were; examples of this are the physician Ishaq Israeli from Kairouan [855–955] and Leonhart Fuchs, the German medical botanist [1501–1566] as found in the works of Israeli 1515; Fuchs 1549; and Lansky and Paavilainen 2011). Table 14.2 employs a scoring system based on the number of times an author discusses the use of the Mediterranean varieties of *Capparis* (probably mostly *C. aegyptia, C. spinosa* spp., *C. orientalis, C. sicula*) (Lev and Amar 2008, Rivera et al. 2003) for the purpose indicated. If the number of citations is regular, the reference is given in regular type and scored a 1. If the number of citations is regular, the author's number is given in regular type and scored a 1.

FIGURE 14.1 *Capparis aegyptia* was one of the sources for ancient and medieval capers and *Capparis* medications. Like in other *Capparis* spp., the buds, fruit, leaves, and young stems are all available simultaneously. (16.5.2011, near Kibbutz Ketura, Southern Arava, Israel: by Helena Paavilainen.)

748 THE SECOND BOOKE OF THE

The names.

Peruinkle is called in Greeke, bicauſe it bringeth foorth ſtalkes,which creepe like thoſe of the Vine, and *Daphnoides* by reaſon that the leaues are like thoſe of the Bay, as aforeſaide. *Pliny* calleth it *Vinca Peruinca*, and *Chamadaphne*: notwithſtanding there is another *Chamadaphne*, of which in his place. The ſame author likewiſe calleth it *Centunculus*: in high Dutch Ingrün: in lowe Dutch Uincoorde, maegden crupt: in French *Pucellage, Vauche & Peruauche*: in Italian *Provenca*: in Spaniſh *Peruinqua*: in ſhoppes *Clematis peruincla*: in Engliſh Peruinkle, Pervinkle, and Periwinkle.

The temperature.

Peruinkle is ſomething hot, but within the ſecond degree, ſomething drie and aſtringent.

The vertues.

A The leaues boiled in wine and drunken, ſtoppeth the laske and bloudie flixe.

B An handfull of the leaues ſtamped and the iuice giuen to drinke in red wine, ſtoppeth the laske and bloodie flixe, ſpetting of blood, which neuer faileth in any bodie, either man or woman: it likewiſe ſtoppeth the inordinate courſe of the monethly ſickneſſe.

Of Capers. Chap. 317.

The kindes.

There be two ſortes of Capers eſpecially, one with broad leaues ſharpe pointed: the other with rounder leaues. The Brabanders haue alſo another ſort, called *Capparis fabago*, or beane Capers

1 *Capparis folio acuto.* Sharpe leafed Capers.

2 *Capparis rotundiore folio.* Round leafed Capers.

The deſcription.

1 The Caper is a prickly ſhrubbe, the ſhootes or braunches whereof be full of ſharpe prickly thornes, trailing vpon the grounde if they be not ſupported or propped vp: whereupon doe growe leaues like thoſe of the Quince tree, but broader and ſharper pointed: among the which come foorth long ſlender footeſtalkes, whereon do grow rounde knops, which do

FIGURE 14.2 Two different *Capparis* varieties by John Gerard in his *Herball* in the beginning of the article on *Capparis*. Notice the roots. (Gerard, J. 1633. *The Herball or Generall Historie of Plantes*. London: Adam Islip Ioice Norton and Richard Whitakers, 748.)

FIGURE 14.3 Bark and root bark were the most important parts of *Capparis* for ancient and medieval medicine. An unusually well-developed *Capparis spinosa* trunk. (January 2010, Haifa, Israel: by Ephraim Lansky.)

FIGURE 14.4 Both the aerial parts and the roots of *Capparis decidua* are considered anti-inflammatory in traditional medicine. (24.3.2010, Rajasthan, India: by Ryan Brookes.)

If the number of citations is strong, the author's number is given in *italics* and scored as 2. If the number of citations is strongest, the author's number is given in **bold** and scored a 3. The final score in the rightmost column is the sum of these scores for the individual authors. In addition, the actual amount of citations for each particular part of the plant is given in parentheses. Authors' numbers correspond with the numbers in Table 14.1.

Table 14.3 reviews some of the salient anti-inflammatory activity discovered for *Capparis* fractions that are not discussed in any of the more specialized subdivisions of inflammation. The entries in the table include an important study describing antiallergic actions. A study that described an acute dermatitis precipitated by contact with *Capparis* aqueous extract used as a treatment and a study of both *C. spinosa* hydroalcoholic extract and pumpkin polycose that found immune-enhancing properties of the pumpkin polycose but not of the *C. spinosa* are also included. Immune response is of course not synonymous with inflammation, but the latter is often a characteristic of the former. For this reason, these rare studies focusing on the effect or lack of effect of *Capparis* extracts on immune function (or dysfunction) are also included in Table 14.3.

In modern ethnomedicine, the anti-inflammatory activity of *Capparis* spp. is much appreciated and beneficial. Table 14.4 lists only such anti-inflammatory uses that are not mentioned in more specific tables. Therefore, it should be compared in particular with Table 15.2, "Contemporary Ethnomedical Uses of *Capparis* spp. against Arthritis (Including Rheumatism and Gout)," but also with Tables 17.2, 19.2, 19.3, 21.2, and 22.2: on infections ("there is always inflammation with infection, but not always infection with inflammation"); on cancer, in which inflammation plays a crucial role; and on diabetes, pain, and fever.

TABLE 14.2
Use of the Different Parts of the *Capparis* Plant in Ancient and Medieval Sources

	All Parts (Unspecified)	Buds	Flower	Leaf	Twig/ Stalk	Fruit	Seed	Bark	Root	Bark of Root	Part Not Specified	Author #	Score
Historical Uses of *Capparis* spp., General													
Astringent (also taste)								(1)	(7)	(12)	(3)	*3*, 4, **5**, 6, *9*, *10*, 13, 17	13
Against phlegm/thick viscous humors			(1)			(1)			(3)	(2)	(2)	*5*, 10, 12, 15	5
Good for hot temperament/heated persons											(1)	10	1
Historical Uses of *Capparis* spp., Skin Problems													
Other discolorings of skin			(2)	(2)						(2)		17	1
Against pimples			(1)	(2)					(1)	(1)		14, 17	2
Against scurfy eruption on skin			(1)	(1)						(1)		17	1
Against ringworm											(1)	16	1
Historical Uses of *Capparis* spp., Unspecific Swellings and Tumors													
Dissolves swellings/tumors									(1)			5	1
Against hardenings/hard tumors/hard swellings				(3)					(6)	(4)		2, 3, *5*, 6, 10, 15, 16	9
Historical Uses of *Capparis* spp., Unspecified Ulcers, Sores, and Wounds													
Cleans wounds										(1)		14	1
Against bad/old/dirty ulcers/sores/wounds (cleaning/drying/healing)				(1)	(1)	(1)			(2)	(9)	(3)	2, 3, *5*, 6, 10, 12, 13, 16, 17	10
Against moist ulcers									(2)			10	1

(Continued)

TABLE 14.2 (*Continued*)
Use of the Different Parts of the *Capparis* Plant in Ancient and Medieval Sources

	All Parts (Unspecified)	Buds	Flower	Leaf	Twig/ Stalk	Fruit	Seed	Bark	Root	Bark of Root	Part Not Specified	Author #	Score
Historical Uses of *Capparis* spp., Neurological Problems, Joint Problems, and Tissue Problems													
Joint Problems													
Against (undefined) joint problems										(1)		15	1
Against gout									(1[a])	(1[b])	(1)	12, 17	2
Against sciatica						(2)	(1)		(1[c])	(4[d])	(4)	1, 2, 5, 6, 10, 14, 15, 16, 17	9
Against hip pains						(1)	(1)		(3[e])	(5[f])	(5)	*5*, 6, 10, *12*, 13, 16, 17	9
Breakages of Tissues													
Heals broken bones										(1)		15	1
Historical Uses of *Capparis* spp., Head and Its Organs													
Mouth, Including Teeth													
Against ulcers of mouth											(1)	1	1
Against toothache				(2)		(4)	(3)	(1[g])	(5)	(10[h])	(1)	1, 2, **5**, 6, 9, *10*, 12, 13, 14, 16, 17	14
Ears													
Against earache + against ear worms				(3)		(1)			(2)		(11)	1, 2, 3, 4, *5*, 6, 9, 10, 12, 13, 14, 17	13
Head and Neck													
Against headache										(1)		6	1

Removes phlegm/humors/moisture from head + cleanses the head					(2)	(2)	(5[i])	(6[j])		3, 6, *10*, 12, *13*, 14, 16, 17	10
Against scrofula/swellings/tumors of parotid glands			(10)	(2)	(2)		(8)	(7)	(1)	1, 2, 3, 4, *5*, 6, 9, *10*, 13, 15, 16, 17	14
Historical Uses of *Capparis* spp., Respiratory Organs											
Against asthma									(1)	6	1
Against pains of the sides									(1)	17	1
Historical Uses of *Capparis* spp., Digestive Organs, Including Liver and Spleen											
Digestive Organs											
Clears stomach/intestines of phlegm/ raw thick humor	(2)				(2)				(4)	5, 7, 9, 10, 13, 16, 17	7
Liver and Hepatic Problems											
Against liver pain									(1)	1	1
Against obstructions of liver	(2)			(1)	(2)				(6)	4, 5, 7, 8, 10, 13, 14, 16, 17	9
Cleans the liver				(1)	(2)					5, 10	2
Against hardness of liver									(2)	7, 11	2
Spleen											
Good for spleen		(1)		(1)			(1)	(3)	(4)	5, 6, 9, 10, 16	5
For spleen problems	(1)						(2[k])	(6[l])	(5)	1, 2, 5, 6, 9, 10, 11, 12, 13, 16, 17	11
Against obstructions of spleen	(2)			(1)	(2)		(1)	(5)	(6)	4, 5, 7, 8, 10, **13**, 14, 16, 17	11
Against spleen pain	(1)		(1)					(1)	(1)	5, 12, 13, 17	4

(Continued)

TABLE 14.2 (*Continued*)
Use of the Different Parts of the *Capparis* Plant in Ancient and Medieval Sources

	All Parts (Unspecified)	Buds	Flower	Leaf	Twig/ Stalk	Fruit	Seed	Bark	Root	Bark of Root	Part Not Specified	Author Number	Score
Prevents pains of spleen	(1[m])											1	1
Against hardness of spleen	(1[n])					(1)		(1)	(5[o])	(12[p])	(4)	2, 3, *5*, 6, 7, 9, 10, **13**, 16, 17	13
Against swellings/tumors of spleen				(1)		(2)			(1)	(3)	(2)	5, 10, 12, *13*	5
Makes spleen smaller						(1)	(3)			(1)	(1)	1, 8, 12, 13, 14	5
Cleans spleen/removes from it thick/ slow/viscous humors					(1)	(2)			(3[q])	(6[r])	(2)	3, 5, 6, 9, *10*, 13, 16, 17	9
Historical Uses of *Capparis* spp., Excretory Organs and Reproductive System													
Kidneys and Urinary System													
Diuretic			(1)			(4)	(1)			(1)	(1)	2, 5, 10, 12, 13, 14, 16	7
Against strangury + dysuria											(3)	8, 12	2
Excretory Organs													
Against hemorrhoids									(2)		(2)	5, 6, 10	3
Against worms											(5)	1, 6, 9	3

Historical Uses of *Capparis* spp., Fevers				
Calms fever		(1)	5	1
Against quartan malaria	(1)		16	1
Historical Uses of *Capparis* spp., Poisons and Antidotes				
Against poisons/theriac		(4)	4, 5, 6, 10	4

[a] Or root bark.
[b] Or root.
[c] Or root bark.
[d] Or root (1x).
[e] Or root bark (1x).
[f] Or root (1x).
[g] Or root bark.
[h] Or bark (1x).
[i] Or bark root (1x).
[j] Or root (1x).
[k] Or root bark (1x).
[l] Or root (1x).
[m] Plant.
[n] Plant.
[o] Or root bark (1x).
[p] Or root (1x).
[q] Or root bark (1x).
[r] Or root (1x).

FIGURE 14.5 The entire plant of *Capparis micracantha* var. *henryi*, especially the dried leaves and the root, have been used against asthma. (4.5.2010, Taiwan: by Ming-I Weng.)

FIGURE 14.6 Bark, root, and the entire plant of *Capparis sepiaria* have found use in ethnomedicine as anti-inflammatory agents, although modern research has not yet confirmed its anti-inflammatory effect. (Near Odzani River, northwest of Mutare, Zimbabwe, 21.10.2005: by Bart T. Wursten. In Hyde, M.A., B.T. Wursten, and P. Ballings. 2013. Flora of Zimbabwe: Species Information: Individual Images: *Capparis sepiaria*. http://www.zimbabweflora.co.zw/speciesdata/image-display.php?species_id=124450&image_id=2, retrieved 20 February 2013.)

TABLE 14.3
Anti-inflammatory and Immunological Properties of *Capparis* spp.

Capparis sp.	Plant Part	Pharmacological Action	Compounds Responsible (If Known)	References
decidua	EtOH ext. *decidua*	Ext. anti-inflammatory in carrageenan-induced ear edema assay in rat		Ageel et al. 1986
decidua	Fresh, powdered, and anatomical sections of stem w/different solvents	Anthelmintic, hepatoprotective, antidiabetic, hypolipidemic	Alkaloids, tannins, phenolic compound, flavonoids, steroids, glycosides, and carbohydrates	Garg et al. 2011
heyneana	Ethanolic extract	Transdermal anti-inflammatory and analgesic		Wang et al. 2009
spinosa		Formation of granular layer of scale epidermis on mouse tail, ↓ inhib. of mitosis of vaginal epithelial mitosis, ↓ total sustaining time and times of itch caused by dextran, inhibition of the auricle turgescence caused by xylene and permeability improvement of celiac capillary vessels caused by acetic acid, ↑ hemolysin antibody, for treating sclerosis and psoriasis, good transcutaneous absorbability, long acting	Alkaloids 1.2–2%	Li et al. 2009
spinosa		Allergic contact dermatitis secondary to topical applications		Angelini et al. 1991
spinosa	80% alcohol extract	Failed to increase weight of immune organs and phagocytic activity of macrophages (as did pumpkin polycose)		Bao et al. 2010
spinosa	Aqueous extract	↓ xylene-induced mouse ear edema (inflamm.) and ↓ acetic acid- and hot plate-induced writhing (analgesia)	Polysaccharides	Zhang et al. 2011
spinosa	Aqueous extract	Claims efficacy for chronic glomerulonephritis, IGA nephropathy, chronic interstitial nephritis, focal segmental glomerulus sclerosis		Ying 2008
spinosa	Aq. ext. *spinosa*	Ext. anti-inflammatory in carrageenan-induced ear edema assay in rat		Ageel et al. 1986

(Continued)

TABLE 14.3 (*Continued*)
Anti-inflammatory and Immunological Properties of *Capparis* spp.

Capparis sp.	Plant Part	Pharmacological Action	Compounds Responsible (If Known)	References
spinosa	Bud, cold, lyophilized ethanolic extract	14 mg/kg significantly inhibited antigen-induced bronchospasm in guinea pig and 1% solution inhibited histamine-induced skin erythema in human volunteers		Trombetta et al. 2005
spinosa	Fruit, aqueous extract	Anti-inflammatory *in vivo* (carrageenan-induced paw edema test)	Flavonoids, indoles, and phenolic acids by NMR and MS	Zhou et al. 2010
spinosa	Fruits	Potent inhibitor of NFκB	New flavonoids isoginkgetin, ginkgetin, and sakuranetin	Zhou et al. 2011
zeylanica	Leaves, methanol extract	Antiulcer against gastric ulcers caused by ethanol and indomethacin (>80%), ($p < 0.01$) ↓ gastric volume, free acidity and ulcer/compared to control		Sini et al. 2011

TABLE 14.4
Contemporary Ethnomedical Uses of *Capparis* spp. against Inflammation

Indication	Geography	*Capparis* sp.	Part Used	References
Anti-inflammatory	Saudi Arabia	*decidua*	Dried aerial parts	Al-Yahya 1986
Anti-inflammatory	India	*sepiaria*	Bark + root	Kakrani and Saluja 1994
Against asthma	India	*decidua*	Dried entire plant	Dhar et al. 1972
Against asthma	India	*decidua*	Dried root bark	Gaind and Juneja 1969
Against asthma	Indonesia	*micracantha*	Entire plant	Burkill et al. 1966
Against asthma	Thailand	*micracantha*	Dried leaf + root	Panthong et al. 1986
Against asthma	India	*spinosa*	Dried root	Jain and Puri 1984
Against allergies	Peru	*angulata*	Dried bark	Ramirez et al. 1988
Against carditis		*micracantha*		*Wealth of India* 1985–1992
Against earache	Oman	*spinosa*	Leaf	Ghazanfar and Al-Sabahi 1993
Against eruptions	Saudi Arabia	*decidua*	Dried aerial parts	Shah et al. 1989
Against eruptions	Sudan	*decidua*		Broun and Massey 1929
For swellings and eruptions	India	*grandis*	Dried bark + leaf	Sebastian and Bhandari 1984
Against reddish eruptions that cause itching around anus of children	India	*sepiaria*	Root	Alam and Anis 1987
Against sore eyes	Argentina	*tweediana*	Leaf	Filipoy 1994
Against gingivitis	Uganda	*corymbosa*	Dried root bark	Sawadogo et al. 1981
Against hemorrhoids	India	*horrida*		Uphof 1968

(*Continued*)

TABLE 14.4 (*Continued*)
Contemporary Ethnomedical Uses of *Capparis* spp. against Inflammation

Indication	Geography	*Capparis* sp.	Part Used	References
Against hemorrhoids	Iran	*spinosa*	Dried fruit	Miraldi et al. 2001
For hemorrhoids and boils as a counterirritant	India	*zeylanica*	Dried leaf	Sebastian and Bhandari 1984
Against piles	India	*zeylanica*	Dried leaf	Jain and Verma 1981
Against hepatitis	Rwanda	*tomentosa*	Dried leaf	Vlietinck et al.1995
Against neuritis	Peru	*cordata*		Ramirez et al. 1988
Against neuritis	Peru	*ovalifolia*	Fruit	Ramirez et al. 1988
Against ophthalmia	Rwanda	*tomentosa*	Dried leaf	Vlietinck et al.1995
Against lung complaints	Israel	*spinosa*	Ripe fruit	Dafni et al. 1984
Against pleurisy	India	*zeylanica*		Jain and Tarafder 1970
Against stomatitis	Venezuela	*jamaicensis*		Pittier 1926
Against swellings	Sudan	*decidua*		Broun and Massey 1929
Against swellings and eruptions	India	*grandis*	Dried bark + leaf	Sebastian and Bhandari 1984
Against swellings	India	*horrida*		Uphof 1968
Against swellings	Malaya	*micracantha*		Burkill et al. 1966
Against swellings	India	*sepiaria*	Dried entire plant	Sebastian and Bhandari 1984
Against swellings	India	*zeylanica*		Jain and Tarafder 1970
To reduce swellings on any part of the body	India	*zeylanica*	Fresh root	Sabnis and Bedi 1983
For swollen ankles	Venda	*tomentosa* (in an herbal mixture)	Dried root	Arnold and Gulumian 1984
Against toothache	India	*decidua*	Dried entire plant	Dhar et al. 1972
Against toothache	Saudi Arabia	*deserti*	Leaf	Al-Said 1993
Against toothache	Tanganyika	*elaeagnoides*	Fresh entire plant	Watt and Breyer-Brandwijk 1962
Against toothache	Jordan	*galeata*	Flowers	Al-Khalil 1995
Against toothache	Spain	*ovata*	Leaf + stem	Martínez-Lirola et al. 1996
Against toothache		*spinosa*		Duke 1993
Indication	**Geography**	***Capparis* sp.**	**Part Used**	**References**
Against ulcers	Saudi Arabia	*decidua*	Dried aerial parts	Shah et al. 1989
Against torpid ulcer	Spain	*ovata*	Leaf + stem	Martínez-Lirola et al. 1996
For ulcers (not specified whether internal or external)	India	*zeylanica*	Dried root bark	Jain and Verma 1981
Against wounds	Ethiopia	sp.	Dried root	Abebe 1986
To cure wounds	Italy	*spinosa*	Buds	De Feo et al. 1992
For wound healing	Spain	*spinosa*	Dried leaf + root	Rios et al. 1987

REFERENCES

Abebe, W. 1986. A survey of prescriptions used in traditional medicine in Gondar region, northwestern Ethiopia: general pharmaceutical practice. *Journal of Ethnopharmacology* 18(2): 147–165.

Ageel, A.M., N.S. Parmar, J.S. Mossa, M.A. Al-Yahya, M.S. Al-Said, et al. 1986. Anti-inflammatory activity of some Saudi Arabian medicinal plants. *Agents and Actions* 17(3–4): 383–384.

Alam, M.M. and M. Anis. 1987. Ethno-medicinal uses of plants growing in the Bulandshahr district of northern India. *Journal of Ethnopharmacology* 19(1): 85–88.

Al-Antaki, D.b.A. 1356 (A.H.). *Tadhkirah ula li-lbab wa-l-jami' li-l-'ajab al-'ujab* [Memorandum Book for Hearts and Comprehensive Book of Wonderful Marvels]. Misr, Egypt: Al-Matba'ah al-'Utmaniyyah.

Albertus Magnus. 1867. *De vegetabilibus libri VII: Historiae naturalis pars XVIII* [Book on Growing Things]. E.H.F. Meyer and K. Jessen, eds. Berlin: Reimeri.

Al-Khalil, S. 1995. A survey of plants used in Jordanian traditional medicine. *International Journal of Pharmacognosy* 33(4): 317–323.

Al-Razi, M.b.Z. 1955–1970. *Kitab al-Hawi fi al-Tibb* [The Comprehensive Book on Medicine]. Vol. 21: 1, pp. 197–207, 256–9. Hyderabad, India: Dairat al-Ma'arif al-'Utmaniyyah bi-ʿilmi'ah al-'Utmaniyyah.

Al-Said, M.S. 1993. Traditional medicinal plants of Saudi Arabia. *American Journal of Chinese Medicine* 21(3–4): 291–298.

Al-Tabari, A.b.R. 1928. *Firdaws al-hikmah fi al-tibb* [The Paradise of Wisdom]. M.Z. Al-Siddiqi, ed. Berlin: Aftab.

Al-Yahya, M.A. 1986. Phytochemical studies of the plants used in traditional medicine of Saudi Arabia. *Fitoterapia* 57(3): 179–182.

Angelini, G., G.A. Vena, R. Filotico, C. Foti, and M. Grandolfo. 1991. Allergic contact dermatitis from *Capparis spinosa* L. applied as wet compresses. *Contact Dermatitis* 24(5): 382–383.

Arnold, H.J. and M. Gulumian. 1984. Pharmacopoeia of traditional medicine in Venda. *Journal of Ethnopharmacology* 12(1): 35–74.

Bao, X., F. Li, H. Han, J. Zhu, X. Dai, Y. Wang, and J. Huang. 2010. Effects of pumpkin polycose and *Capparis spinosa* Linn polycose on immune function of mice. *Xinjiang Nongye Kexue* 47(3): 508–511.

Bock, H. 1964. *Kreütterbuch darin underscheidt Nammen und Würckung der Kreütter, standen* [Herbal in which are the Different Names and Properties of Herbs]. Strassburg, Austria: Josiam Rihel, 1577. Reprint Konrad Kolbl, Munich.

Broun, A.F. and R.E. Massey. 1929. *Flora of the Sudan: with a Conspectus of Groups of Plants and Artificial Key to Families*. London: [s.n.].

Burkill, I.H., W. Birtwistle, F.W. Foxworthy, J.B. Scrivenor, and J.G. Watson. 1966. *A Dictionary of the Economic Products of the Malay Peninsula*. Kuala Lumpur, Malaysia: Published on behalf of the governments of Malaysia and Singapore by the Ministry of Agriculture and cooperatives.

Dafni, A., Z. Yaniv, and D. Palevitch. 1984. Ethnobotanical survey of medicinal plants in northern Israel. *Journal of Ethnopharmacology* 10 (3): 295–310.

De Feo, V., R. Aquino, A. Menghini, E. Ramundo, and F. Senatore. 1992. Traditional phytotherapy in the Peninsula Sorrentina, Campania, Southern Italy. *Journal of Ethnopharmacology* 36(2): 113–125.

Dhar, D.N., R.P. Tewari, R.D. Tripathi, and A.P. Ahuja. 1972. Chemical examination of *Capparis decidua. Proceedings of the National Academy of Sciences, India: Section A* 42(1): 24–27.

Dioscorides, P. 1902. *Des Pedanios Dioscurides aus Anazarbos Arzneimittellehre in fünf Büchern*. J. Berendes, trans. and comm. Stuttgart, Germany: Enke.

Duke, J.A. 1993. Dr. Duke's Phytochemical and Ethnobotanical Databases. [Online database.] USDA–ARS–NGRL, Beltsville Agricultural Research Center, Beltsville, MD. http://www.ars-grin.gov/duke/ethnobot.html (accessed 12 May 2011).

Filipoy, A. 1994. Medicinal plants of the Pilaga of Central Chaco. *Journal of Ethnopharmacology* 44(3): 181–193.

Flos medicinae scholae Salerni [Flower of Medicine of the School of Salerno]. 1852–59. In G.E.T. Henschel, C. Daremberg, and S. de Renzi, eds., *Collectio Salernitana: Ossia documenti inediti, e trattati di medicina appartenenti alla scuola medica Salernitana*. Vol. 5. Napoli, Italy: Filiatre-Sebezio.

Fuchs, L. 1549. *De historia stirpium commentarii insignes* [Notable Commentaries on the History of Plants]. Lyon, France: Arnolletum. Electronic edition 1995, Bibliotheque nationale de France, Paris.

Gaind, K.N. and T.R. Juneja. 1969. Investigations on *Capparis decidua. Planta Medica* 17: 95–98.

Galen. Galenus, C.G. 1964–1965. *Opera omnia* [Complete Works]. C.G. Kuhn, ed. Hildesheim, Germany: Olms. [Facsimile reprint of Leipzig, Germany: C. Cnobloch, 1821–33 edition.]

Garg, P., D. Gandhi, P. Khatri, A. Pandey, and V. Jakhetia. 2011. Pharmacognostic and phytochemical evaluation of stem of *Capparis decidua* (Forsk) Edgew. *Indian Journal of Novel Drug Delivery* 3(1): 29–35.

Gerard, J. 1633. *The Herball or Generall Historie of Plantes*. London: Adam Islip Ioice Norton and Richard Whitakers.

Ghazanfar, S.A. and M.A. Al-Sabahi. 1993. Medicinal plants of Northern and Central Oman (Arabia). *Economic Botany* 47(1): 89–98.

Hyde, M.A., B.T. Wursten, and P. Ballings. 2013. Flora of Zimbabwe. [Online database.] http://www.zimbabweflora.co.zw (accessed 5 February 2013).

Ibn al-Baytar, a.M. 1992/1412. *Kitab al-jamiʿ li-mufradat al-adwyah w-al-aghdhyah* [Comprehensive Book of Simple Drugs and Foods]. Beirut, Lebanon: Dar al-kutub al-ʿilmiyah.

Ibn Butlan, M.b.H. 1531. *Tacuini sanitatis Elluchasem Elimithar: de sex rebus non naturalibus, earum naturis, operationibus, and rectificationibus, publico omnium usui, conservandae sanitatis, recens exarati* [Tables of Health]. Argentorati, France: Apud Ioannem Schottum.

Ibn Sina, H.a.A. 1877/1294. *Kitab al-Qanun fi-l-Tibb* [The Canon of Medicine]. Beirut, Lebanon: Dar Sadir.

Israeli, I. 1515. *Omnia opera Ysaac in hoc volumine contenta:cumquibusdam alijs opusculis* [Complete Works]. [Lyons, France: Jean de La Plate for Barthelemy Trot]. Electronic edition 1995 by Bibliotheque nationale de France (BnF), Paris.

Jain, S.K. and C.R. Tarafder. 1970. Medicinal plant-lore of the Santals. *Economic Botany* 24(3): 241–278.

Jain, S.P. and H.S. Puri. 1984. Ethnomedicinal plants of Jaunsar-Bawar Hills, Uttar Pradesh, India. *Journal of Ethnopharmacology* 12(2): 213–222.

Jain, S.P. and D.M. Verma. 1981. Medicinal plants in the folk-lore of North-East Haryana. *National Academy of Science Letters (India)* 4(7): 269–271.

Kakrani, H.K.N. and A.K. Saluja. 1994. Traditional treatment through herbs in Kutch district, Gujarat State, India. Part II. Analgesic, anti-inflammatory, antirheumatic, antiarthritic plants. *Fitoterapia* 65(5): 427–430.

Kobow, K., S. Auvin, F. Jensen, W. Löscher, I. Mody, H. Potschka, D. Prince, A. Sierra, M. Simonato, A. Pitkänen, A. Nehlig, and J.M. Rho. 2012. Finding a better drug for epilepsy: antiepileptogenesis targets. *Epilepsia* 53(11): 1868–1876.

Lansky, E.P. and H.M. Paavilainen. 2011. *Figs: The Genus* Ficus. Boca Raton, FL: CRC Press.

Lev, E. and Z. Amar. 2008. *Practical* Materia Medica *of the Medieval Eastern Mediterranean According to the Cairo Genizah*. Leiden, the Netherlands: Brill.

Li, F., Z. Song, Y. Hui et al. 2009. Method for producing *capparis spinosa* extract and *capparis spinosa* cream for treating sclerosis and psoriasis. Chinese Patent CN 2009-10301140.

López, G. 1982. *El tesoro de medicinas de Gregorio López 1542–1596* [Treasure of Medicines], F. Guerra, ed. and comm. Madrid: Instituto de cooperacion iberoamericana, Ediciones cultura hispanica.

Martínez-Lirola, M.J., M.R. Gonzalez-Tejero, and J. Molero-Mesa. 1996. Ethnobotanical resources in the Province of Almeria, Spain: Campos de Nijar. *Economic Botany* 50(1): 40–56.

Miraldi, E., S. Ferri, and V. Mostaghimi. 2001. Botanical drugs and preparations in the traditional medicine of West Azerbaijan (Iran). *Journal of Ethnopharmacology* 75(2–3): 77–87.

Paget, J. 1880. An address on elemental pathology. *British Medical Journal* 2(1034): 649–652.

Panthong, A., D. Kanjanapothi, and W.C. Taylor. 1986. Ethnobotanical review of medicinal plants from Thai traditional books. Part 1: Plants with antiinflammatory, antiasthmatic and antihypertensive properties. *Journal of Ethnopharmacology* 18(3): 213–228.

Parkinson, J. 1640. *The Theater of Plants, or, an Herball of a Large Extent*. London: Cotes. Available through Early English Books Online, http://eebo.chadwyck.com.

Pittier, H. 1926. *Manual de las plantas usuales de Venezuela*. Caracas, Venezuela: Litografia del comercio.

Plinius, C. (Pliny the Elder). 1967–1970. *C. Plini Secundi naturalis historiae, libri XXXVII* [The Natural History]. K.F.T. Mayhoff, ed. Stuttgart, Germany: Teubner.

Ramirez, V.R., L.J. Mostacero, A.E. Garcia, C.F. Mejia, P.F. Pelaez, C.D. Medina, and C.H. Miranda. 1988. *Vegetales empleados en medicina tradicional norperuana*. Trujillo, Peru: Banco Agrario del Peru and Universidad Nacional de Trujillo.

Rios, J.L., M.C. Recio, and A. Villar. 1987. Antimicrobial activity of selected plants employed in the Spanish Mediterranean area. *Journal of Ethnopharmacology* 21(2): 139–152.

Rivera, D., C. Inocencio, C. Obón, and F. Alcaraz. 2003. Review of food and medicinal uses of *Capparis* L. subgenus *Capparis* (Capparidaceae). *Economic Botany* 57(4): 515–534.

Rufinus de Rizardo. 1946. *Herbal of Rufinus*. L. Thorndike, ed. Chicago: University of Chicago Press.

Sabnis, S.D. and S.J. Bedi. 1983. Ethnobotanical studies in Dadra-Nagar Haveli and Daman. *Indian Journal of Forestry* 6(1): 65–69.

Sawadogo, M., A.M. Tessier, and P. Delaveau. 1981. Studies on the roots of *Capparis corymbosa* Lam. *Plantes Medicinales et Phytotherapie* 15: 234–239.

Sebastian, M.K. and M.M. Bhandari. 1984. Medico-ethno botany of Mount Abu, Rajasthan, India. *Journal of Ethnopharmacology* 12(2): 223–230.

Shah, A.H., M. Tariq, A.M. Ageel, and S. Qureshi. 1989. Cytological studies on some plants used in traditional Arab medicine. *Fitoterapia* 60(2): 171–173.

Sini, K.R., B.N. Sinha, and A. Rajasekaran. 2011. Protective effects of *Capparis zeylanica* Linn. leaf extract on gastric lesions in experimental animals. *Avicenna Journal of Medical Biotechnology (Tehran, Iran)* 3(1): 31–35.

Tabernaemontanus, J.T. 1970. *Neu vollkommen Kräuter-Buch, darinnen uber 3000 Kräuter, mit schönen und kunstlichen Figuren, auch deren Underscheid und Würckung, samt ihren Namen in mancherley Sprachen beschrieben* [New Complete Herbal]. 4th ed. C. Bauhinus and H. Bauhinus, eds. Munich: Kolbl. [Reprint of Basel, Switzerland: Konig, 1731.]

Trombetta, D., F. Occhiuto, D. Perri, et al. 2005. Antiallergic and antihistaminic effect of two extracts of *Capparis spinosa* L. flowering buds. *Phytotherapy Research* 19: 29–33.

Uphof, J.C.Th. 1968. *Dictionary of Economic Plants*. Lehre, Germany: Cramer.

Vezzani, A., S. Balosso, and T. Ravizza. 2012. Inflammation and epilepsy. *Handbook of Clinical Neurology* 107: 163–175.

Vlietinck, A.J., L. van Hoof, J. Totte, A. Lasure, D. vanden Berghe, P.C. Rwangabo, and J. Mvukiyumwami. 1995. Screening of hundred Rwandese medicinal plants for antimicrobial and antiviral properties. *Journal of Ethnopharmacology* 46(1): 31–47.

Wang, X., W. Chen, J. Xing, et al. 2009. Method for manufacturing antiinflammatory and analgesic cataplasma of *Capparis heyneana*. Chinese Patent CN 2009-10113221.

Watt, J.M. and M.G. Breyer-Brandwijk. 1962. *The Medicinal and Poisonous Plants of Southern and Eastern Africa*. 2nd ed. London: Livingstone.

Wealth of India 1985–1992. *The Wealth of India: A Dictionary of Indian Raw Materials and Industrial Products*. New Delhi: Publications and Information Directorate, Council of Scientific and Industrial Research.

Ying, X. 2008. Application of *Capparis spinosa* L. in preparing medicine for treating kidney diseases. Chinese Patent CN 2008-10059985.

Zhang, Y., H. Zhang, B. Han, and W. Chen. 2011. Extraction of polysaccharides in *Capparis spinosa* L. and anti-inflammatory and analgesic effects. *Shihezi Daxue Xuebao, Ziran Kexueban* 29(2): 205–209.

Zhou, H., R. Jian, J. Kang, X. Huang, Y. Li, C. Zhuang, F. Yang, L. Zhang, X. Fan, T. Wu, and X. Wu. 2010. Anti-inflammatory effects of caper (*Capparis spinosa* L.) fruit aqueous extract and the isolation of main phytochemicals. *Journal of Agricultural and Food Chemistry* 58(24): 12717–12721.

Zhou, H.F., C. Xie, R. Jian, J. Kang, Y. Li, C.L. Zhuang, F. Yang, L.L. Zhang, L. Lai, T. Wu, and X. Wu. 2011. Biflavonoids from Caper (*Capparis spinosa* L.) fruits and their effects in inhibiting NF-κB activation *Journal of Agricultural and Food Chemistry* 59(7): 3060–3065.

15 Rheumatism

The old-fashioned term *rheumatism*, treated by today's modern rheumatologists, encompasses diseases of the musculoskeletal system, including those of fibrous tissue (i.e., cartilage, ligaments, tendons) and of the joints proper (arthritis). While arthritides (plural of arthritis) gather most of the rheumatologists' attention, lesser-appreciated disorders and diseases of the fascia and other connective tissue supports of the musculature and skeleton also result in considerable human suffering. An example of this last group includes the vague condition known as fibromyalgia (i.e., usually long-term, chronic pain syndromes of the muscles and fascia, particularly in the extremities). Conventional pharmaceutical therapies for these conditions as well as for various arthritides related to primary dysfunction in the immune system are of only limited efficacy and considerable toxicity, resulting often in a general weakening of the defenses of the body and, in some cases, additional pathology.

The *Capparis* genus represents a great possibility and potential for treating rheumatological disorders. This is reflected in the treasure trove of mostly Chinese patents cited in Table 15.1 that deal with preparations of *C. spinosa* fruit for both internal and external remedies for arthritis and related disorders subsumed under the general field of rheumatology. The chemical mechanisms responsible for the antiarthritic or chondroprotective effect are a subject of active research. Isothiocyanates have been proposed as a likely possibility (Figure 15.1), while other groups have considered alkaloids, sterols, and polyphenols to be responsible. It is likely the mechanism relates to the complex medicinal chemistry of *Capparis* and is a collection of complex effects that include all of these compounds.

One question subsumes this complexity and looks deeper, namely, how evolution of *Capparis* parallels the important constraints on mammals that lead to rheumatism. Certainly, psychosomatic or, better, somatopsychic theorists (Andrew Weil comes to mind) have suggested theories for the psychic underlay of rheumatism. Sometimes these conditions, especially osteoarthritis, are secondary to physical trauma and overuse, while the autoimmune variants may involve intrapsychic issues related to stoic defenses and rejection of self (see *Archetypal Medicine* by the Jungian analyst A.J. Ziegler, 1983). Curiously, such self and the struggle between its acceptance and rejection have a key meaning in both psychology and immunology. In time, we may better appreciate the unity of, rather than the interaction between, these two great cognitive systems, the brain with its neurons and the immunological network with its lymphocytes, each with its own inner concept of self and not self. In some totally mysterious manner, the evolution of *Capparis* has endowed it with the biochemical infrastructure for inserting itself between the lymphocytes and the articulations between joints, as well as regarding the tendons and fascia that provide the secondary structure for the skeleton, and contain and guide the insertion, contraction, and relaxation of muscles. In all of these rheumatological scenarios, uncontained and imbalanced inflammation plays a key role in subverting their normal function.

In modern ethnomedicine, rheumatism (Figure 15.2) is one of the most popular indications for using the different *Capparis* spp. (Table 15.2). See also Tables 14.4 and 19.3 on the use of *Capparis* for inflammation and more general pain.

TABLE 15.1
Antirheumatic Properties of *Capparis* spp.

Capparis sp.	Plant Part	Pharmacological Action	Compounds Responsible (If Known)	References
acutifolia	Chloroform extract	Analgesic, anti-inflammatory, low-cost curative for neuropathological, inflammatory, cancerous, diabetic, rheumatic, traumatic, abdominal pains, laryngopharyngitis, toothache, and Freund's complete adjuvant (FCA) arthritis		Deng 2010
cartilaginea		Antiarthritic targeting needed costimulatory signal for full T-cell activation, ↓ synovial inflammation w/o disrupting cellular homeostasis, ↓ reduction in gene expression benefits bones, suggests a basis for radiological > w/ isothiocyanates in patients w/ rheumatoid arthritis, inhibitor comprising extracellular domain of human cytotoxic T-lymphocyte antigen 4 and Fc domain fragment of human IgGl, competing with CD28 for binding to CD80/CD86, w/ modulating second costimulatory signal for full T-cell activation	Isothiocyanates and derivatives	Balar and Nakum 2010
ovata	Stem bark	Anti-inflammatory	Catalposide, an iridoid	Villasenor 2007
spinosa	(By wt. parts) *C.* powder 1, baijiu powder 0.8–1.2, mixed, soaked with water, on afflicted part for 30–50 min after skin temp. of afflicted part rising, plus in equal parts by wt. *Carthamus tinctorius* 0.25, *Astragalus membranaceus* 0.25, *Manis pentadactyla* 0.25, *Achyranthes bidentata* 0.25, and borneol 0.25	Antiarthritic for external use, effective rate of 100% and curative rate of 95%		Yuan 2010

TABLE 15.1 (*Continued*)
Antirheumatic Properties of *Capparis* spp.

Capparis sp.	Plant Part	Pharmacological Action	Compounds Responsible (If Known)	References
spinosa	1:1 ethanol-water	Antiarthritic activity in adjuvant arthritic rat model	P-Hydroxy benzoic acid; 5-(hydroxymethyl) furfural; bis(5-formylfurfuryl) ether; daucosterol; α-D-fructofuranosides methyl; uracil; stachydrine	Feng et al. 2011
spinosa	Buds, lyophilized methanolic extract	↓ damage to human chondrocytes *in vitro* stim. by proinflammatory cytokine interleukin-1β (IL-1β): ↓ nitric oxide, glycosaminoglycans, prostaglandins, reactive oxygen species, note potential for repairing cartilage damage during inflammatory response		Panico et al. 2005
spinosa	Ethanolic extract	Dose and time dependent, significant ↓ fibroblast proliferation and expression of alpha2 (I) collagen mRNA and type I collagen protein in progressive systemic sclerosis but not in normal cells		Cao et al. 2008a
spinosa	Ext. to macroporous adsorptive resin, eluting with water, 20–30%, then 50–100% ethanol, active against rheumatoid arthritis			Wu et al. 2011
spinosa	Extract by refluxing, cold immersion or percolation, vacuum concentrating, degreasing with petroleum ether, and extg. with chloroform	For treating rheumatic arthritis, rheumatoid arthritis, and scapulohumeral periarthritis	β-Sitosterol, β-sitosterol-glucoside, and 1-hydro-3-acetonitrile-4-methoxy-indole	Li et al. 2008
spinosa	Fruit, defatting, contact with EtOAc and other solvents	Effective in treating "rheum-arthritis," rheumatoid arthritis, frozen shoulder, gout		Ajiaikebaier 2004

(*Continued*)

TABLE 15.1 (*Continued*)
Antirheumatic Properties of *Capparis* spp.

Capparis sp.	Plant Part	Pharmacological Action	Compounds Responsible (If Known)	References
spinosa	Fruit, ethanol ext.—ground to 10–30 mesh, soaking w/6- to 12-fold 90–95% ethanol aq. soln., extg. 1–3 times each for 0.5–2 h under refluxing, merging extg. liqs., filtering to obtain filtrate, vacuum concg. at 50–70°C and (-0.1) to (-0.06) MPa to relative density of 1.05–1.20 at 60°C, drying to obtain *C. spinosa* fruit ext., and grinding to 80–200 mesh—comprises (1) dispersing gel skeleton material, filler, and crosslinking agent with wetting agent at 55–60°C in water bath, swelling completely under stirring to obtain liq. mixt. I; (2) swelling thickening agent completely with 6- to 12-fold water, adding orderly pH regulator, cross-linking regulator, defoaming agent, penetration enhancer, and *C. spinosa* fruit ext. powder under stirring to obtain liq. mixt. II; and (3) mixing uniformly liq. mixt. I and II under stirring to obtain semisolid fluid mixt., spreading onto back-lining layer, standing for 12 h, and coating with a layer of protective film	Antiarthritic gel for external use		Yang et al. 2011
spinosa	Fruit, ethanolic or methanolic fraction with alkane defatting	Preparation for treating progressive systemic sclerosis w/ inhib. of fibroblast proliferation, expression of type I collagen and connective tissue growth factor (CTGF) mRNA and protein synthesis on patients, w/o effect on fibroblast proliferation, collagen synthesis, or CTGF expression on normal people		Cao et al. 2008b

TABLE 15.1 (*Continued*)
Antirheumatic Properties of *Capparis* spp.

Capparis sp.	Plant Part	Pharmacological Action	Compounds Responsible (If Known)	References
spinosa	Root, branch, leaf, flower and fruit	External prep. (e.g., plaster or medicinal wine) for treating arthritis, rheumatoid arthritis, gout, and scapulohumeral periarthritis; may be mixed with other herbs, incl. Flos tagetis erectae, *Saussurea medusa*, *Rhizoma zingiberis recens*, *Caulis* et *folium gaultheriae yunnanensis*, *Borneolum*, *Sabina chinensis*, *Ramulus mori*, *Fructus chaenomelis*, *Notopterygii rhizoma*, *Radix angelicae pubescentis*, *Ramulus cinnamomi*, *Rhizoma atractylodis*, *Radix tripterygii wilfordii*, *Caulis sinomenii*, *Caulis lonicerae*, and *Glycyrrhizae radix*		Ding 2007

FIGURE 15.1 The isothiocyanates in *Capparis cartilaginea* leaves have an antirheumatic effect. (16.5.2011, Kibbutz Ketura, Southern Arava, Israel: by Helena Paavilainen.)

FIGURE 15.2 Most of the parts of *Capparis spinosa*—roots, bark, fruit, leaves, and buds—have traditionally been used against rheumatism and other joint problems. Modern research supports the concept. (2.6.2010, near Mar Elias Monastery, Israel: by Helena Paavilainen.)

TABLE 15.2
Contemporary Ethnomedical Uses of *Capparis* spp. against Arthritis (Including Rheumatism and Gout)

Indication	Geography	*Capparis* sp.	Part Used	References
Against arthritis	Nigeria	*decidua*	Dried root	Iwu and Anyanwu 1982
Against arthritis	Sudan	*decidua*		Broun and Massey 1929
Against arthritis	Saudi Arabia	*decidua, spinosa*	Dried aerial parts	Ageel et al. 1986
Against arthritis	Saudi Arabia	*spinosa*	Dried bark	Agha 1984
Against arthritis	India	*zeylanica*	Dried leaf	Reddy et al. 1991
Against rheumatism	Peru	*cordata*		Ramirez et al. 1988
Against rheumatism	Saudi Arabia	*decidua*	Dried aerial parts	Ageel et al. 1986, Shah et al. 1989
Against rheumatism	India	*decidua*	Dried entire plant	Dhar et al. 1972
Against rheumatism	India	*decidua*	Dried root bark	Gaind and Juneja 1969
Antirheumatic	India	*deserti*	Leaf	Kakrani and Saluja 1994
Against rheumatism	Saudi Arabia	*deserti*	Leaf	Al-Said 1993
Against rheumatism	Egypt	*galeata*	Fresh fruit	Goodman and Hobbs 1988
Against rheumatism		*heyneana*		*Wealth of India* 1985–1992
Against rheumatism	Peru	*ovalifolia*	Fruit	Ramirez et al. 1988
Against rheumatism	Saudi Arabia	*spinosa*	Dried aerial parts	Ageel et al. 1986
Antirheumatismal	Morocco	*spinosa*	Fruit	Bellakhdar et al. 1991

TABLE 15.2 (*Continued*)
Contemporary Ethnomedical Uses of *Capparis* spp. against Arthritis (Including Rheumatism and Gout)

Indication	Geography	*Capparis* sp.	Part Used	References
Against rheumatism	Israel	*spinosa*	Fresh fruit + root	Dafni et al. 1984
Against rheumatism	Tunisia	*spinosa*	Dried leaf	Boukef et al. 1982
Against rheumatism	India	*spinosa*	Root bark	Abrol and Chopra 1962
Antirheumatic	Jordan	*spinosa*	Root bark	Al-Khalil 1995
Against rheumatism	India	*spinosa*	Dried root bark	Shirwaikar et al. 1996
Against rheumatism	India	*spinosa*	Fresh root bark	Singh et al. 1996
Against rheumatic pain	India	*zeylanica*	Dried leaf	Reddy et al. 1991
Against rheumatism	India	*zeylanica*		Jain and Tarafder 1970
Against gout	Saudi Arabia	*decidua*	Dried aerial parts	Ageel et al. 1986, Al-Yahya 1986, Shah et al. 1989
Against gout	India	*decidua*	Dried root bark	Gaind and Juneja 1969
Against gout	Saudi Arabia	*deserti*	Leaf	Al-Said 1993
Against gout	Saudi Arabia	*spinosa*	Dried aerial parts	Ageel et al. 1986
Against gout	Jordan	*spinosa*	Leaf	Al-Khalil 1995
Against arthralgia	Spain	*ovata* (in an herbal mixture)	Leaf + stem	Martínez-Lirola et al. 1996

REFERENCES

Abrol, B.K. and I.C. Chopra. 1962. Some vegetable drug resources of Ladakh (Little Tibet). Part I. *Current Science* 31: 324–326.

Ageel, A.M., N.S. Parmar, J.S. Mossa, M.A. Al-Yahya, M.S. Al-Said, et al. 1986. Anti-inflammatory activity of some Saudi Arabian medicinal plants. *Agents Actions* 17(3–4): 383–384.

Agha, Z.M. 1984. Medicinal uses of plants in Saudi Arabia. Personal communication with the author. Via NAPRALERT[SM]. Natural Products Alert Database. 1975. [Online database.] Program for Collaborative Research in the Pharmaceutical Sciences (PCRPS), College of Pharmacy, University of Illinois at Chicago. Chicago, IL. http://www.napralert.org (accessed June 24, 2012).

Ajiaikebaier, A., 2004. Method for extracting active arthritis resisting element from *Capparis* fruit. Chinese Patent CN 1541700.

Al-Khalil, S. 1995. A survey of plants used in Jordanian traditional medicine. *International Journal of Pharmacognosy* 33(4): 317–323.

Al-Said, M.S. 1993. Traditional medicinal plants of Saudi Arabia. *American Journal of Chinese Medicine* 21(3–4): 291–298.

Al-Yahya, M.A. 1986. Phytochemical studies of the plants used in traditional medicine of Saudi Arabia. *Fitoterapia* 57(3): 179–182.

Balar, C. and A. Nakum. 2010. Herbal ingredients for the treatment of arthritis. Indian Patent CODEN: INXXBQ.

Bellakhdar, J., R. Claisse, J. Fleurentin, and C. Younos. 1991. Repertory of standard herbal drugs in the Moroccan pharmacopoea. *Journal of Ethnopharmacology* 35(2): 123–143.

Boukef, K., H.R. Souissi, and G. Balansard. 1982. Contribution to the study on plants used in traditional medicine in Tunisia. *Plantes Medicinales Phytotherapie* 16(4): 260–279.

Broun, A.F. and R.E. Massey. 1929. *Flora of the Sudan: With a Conspectus of Groups of Plants and Artificial Key to Families*. London: [s.n.].

Cao, Y.L., X. Li, and M. Zheng. 2008a. [Effect of *Capparis spinosa* on fibroblast proliferation and type I collagen production in progressive systemic sclerosis]. *Zhongguo Zhong Yao Za Zhi* 33: 560–563.

Cao, Y., C. Zhou, J. Mo, L. Gan, and M. Zheng. 2008b. Method for preparing effective fraction of *Capparis spinosa* fruit, and its use in preparing medicinal preparation for treating progressive systemic sclerosis. Chinese Patent CN 101244099.

Dafni, A, Z. Yaniv, and D. Palevitch. 1984. Ethnobotanical survey of medicinal plants in northern Israel. *Journal of Ethnopharmacology* 10 (3): 295–310.

Deng, Y. 2010. Method for manufacturing extract of *Capparis acutifolia* for treating Freund's complete adjuvant (fca) arthritis. Chinese Patent CN 101780124.

Dhar, D.N., R.P. Tewari, R.D. Tripathi, and A.P. Ahuja. 1972. Chemical examination of *Capparis decidua. Proceedings of the National Academy of Science, India: Section A* 42(1): 24–27.

Ding, H., 2007. Novel use of *Capparis spinosa* L of Capparaceae family for treating arthritis by external application and pharmaceutical composition containing *Capparis spinosa* L. Chinese Patent CN 1958063. May 9.

Feng, X., J. Lu, H. Xin, L. Zhang, Y. Wang, and K. Tang. 2011. Anti-arthritic active fraction of *Capparis spinosa* L. fruits and its chemical constituents. *Yakugaku Zasshi* 131(3): 423–429.

Gaind, K.N. and T.R. Juneja. 1969. Investigations on *Capparis decidua. Planta Medica* 17: 95–98.

Goodman, S.M. and J.I. Hobbs. 1988. The ethnobotany of the Egyptian Eastern desert: a comparison of common plant usage between two culturally distinct Bedouin groups. *Journal of Ethnopharmacology* 23(1): 73–89.

Iwu, M.M. and B.N. Anyanwu. 1982. Phytotherapeutic profile of Nigerian herbs. 1. Anti-inflammatory and anti-arthritic agents. *Journal of Ethnopharmacology* 6(3): 263–274.

Jain, S.K. and C.R. Tarafder. 1970. Medicinal plant-lore of the Santals. *Economic Botany* 24(3): 241–278.

Kakrani, H.K.N. and A.K. Saluja. 1994. Traditional treatment through herbs in Kutch district, Gujarat State, India. Part II. Analgesic, anti-inflammatory, antirheumatic, antiarthritic plants. *Fitoterapia* 65(5): 427–430.

Li, J., L. Xiao, T. Wu, Y. Li, Y. Liu, et al. 2008. Preparation method and application of *Capparis spinosa* extract. Chinese Patent CN 101185663. May 28.

Martínez-Lirola, M.J., M.R. Gonzalez-Tejero, and J. Molero-Mesa. 1996. Ethnobotanical resources in the province of Almeria, Spain: Campos de Nijar. *Economic Botany* 50(1): 40–56.

Panico, A.M., V. Cardile, F. Garufi, et al. 2005. Protective effect of *Capparis spinosa* on chondrocytes. *Life Sciencs* 77: 2479–2488.

Ramirez, V.R., L.J. Mostacero, A.E. Garcia, C.F. Mejia, P.F. Pelaez, C.D. Medina, and C.H. Miranda. 1988. *Vegetales empleados en medicina tradicional norperuana.* Trujillo, Peru: Banco Agrario del Peru and Universidad Nacional de Trujillo.

Reddy, M.B., K.R. Reddy, and M.N. Reddy. 1991. Ethnobotany of Cuddapah district, Andhra Pradesh, India. *Int J Pharmacognosy* 29(4): 273–280.

Shah, A.H., M. Tariq, A.M. Ageel, and S. Qureshi. 1989. Cytological studies on some plants used in traditional Arab medicine. *Fitoterapia* 60(2): 171–3.

Shirwaikar, A., K.K. Sreenivasan, B.R. Krishnanand, and A.V. Kumar. 1996. Chemical investigation and antihepatotoxic activity of the root bark of *Capparis spinosa. Fitoterapia* 67(3): 200–204.

Singh, V., B.K. Kapahi, and T.N. Srivastava. 1996. Medicinal herbs of Ladakh especially used in home remedies. *Fitoterapia* 67(1): 38–48.

Villasenor, I.M. 2007. Bioactivities of iridoids. *Anti-inflammatory & Anti-allergy Agents in Medicinal Chemistry* 6(4): 307–314.

Wealth of India 1985–1992. *The Wealth of India: A Dictionary of Indian Raw Materials and Industrial Products.* New Delhi, India: Publications and Information Directorate, Council of Scientific and Industrial Research.

Wu, T., Y. Li, Q. Liu, H. Zhou, X. Huang, et al. 2011. Method for manufacturing refined extract from *Capparis spinosa*, and application of the extract. Chinese Patent CN 102091104.

Yang, W., D. Tuerxunjiang, J. Ma, C. Liu. 2011. Method for production of gel plaster of *Capparis* fruit extract and its antiarthritic application. Chinese Patent CN 2010-10612158.

Yuan, W. 2010. External-use traditional Chinese medicinal composition for treating rheumatic diseases. Chinese Patent CN 101766663.

Ziegler, A.J. 1983. *Archetypal Medicine*. Dallas: Spring Publications.

16 Lipid Dyscrasias

In general, biochemical compounds can be divided into two major categories: those that are "water loving," or hydrophilic, and those that are "water fearing," or hydrophobic. This refers to their solubility in and affinity to water. Water is, of course, the great substance of all life as we know it. It is the basis of blood, lymph, saliva, cerebrospinal fluid, the humors of the eyes, and tears of pain and joy. Digestive fluids, from the strong hydrochloric acid of the stomach to the bile and juices of the intestines, are largely composed of water. However, without the fatty layers comprising membranes of cells and their intracellular organelles, such as their nuclei and mitochondria, the varied and numerous aqueous-based fluids would run into each other and chaos would rule. The fats of the body therefore provide separation and containment of the different aqueous media of the body and channel, through tubular vessels such as arteries, veins, and special ducts, their transport from head to toe, from right to left, in clockwise and counterclockwise directions, and via all other possibilities for movement within the circumscribed structure of the body.

Thus, it should be evident that fats, also known as lipids, are vital for life, and without them, life as we know it could not exist. Special compounds that exist within the lipid layers and in lipids are suspended in the aqueous media, within both cells and subcellular particles, and are required for numerous important biochemical operations. One example is cholesterol, a fatty alcohol at home in lipid matrices but insoluble in water. Cholesterol is a precursor of the male and female sex hormones, the androgens and estrogens, and many other hormones, including vitamin D. Cholesterol is thus also essential for life. Lipids and lipid-soluble compounds may also form complexes with proteins that are water soluble, thus essentially introducing these proteins as lipoproteins into the lipid layers. Lipoproteins might be considered to have properties that are both hydrophobic, from the lipid component, and hydrophilic, from the protein component. Overall, the relations between aqueous and lipid phases are complex but, in a healthy organism, function together in a harmonious manner.

Oxidative processes that occur in lipid layers and are not adequately checked by the body's antioxidant defense mechanisms lead to disorders in lipid chemistry and ultimately inflammation. Inflammation in the walls of blood vessels may also descend, partly through proinflammatory mechanisms, to abnormal depositing of lipids, first as droplets and later taken up in specialized white blood cells, including macrophages (known as foam cells), to create, undesirable from the point of health, fatty blockages in aqueous fluid-transporting channels (e.g., blood vessels). The blockages consist of fat, cholesterol, and lipoproteins, often with a mineral core, generally calcium. This conception is oversimplified but gives an idea of the complexity, of the importance of lipids to healthy life, and of the possibility for the lipid balance to go awry and lead to numerous health problems and blockages, including many that can result in premature cessation, relative to the normal life cycle, of the biological existence and ongoing continuity of the organism. We refer to such distortions to the balance of lipids within aqueous media (such as blood) and their proclivity to result and create many diseases as lipid dyscrasias, dyscrasia implying fertile ground for the evolution of disease.

Thus, lipid dyscrasias, while leading to diseases (including those with potentially fatal outcomes such as myocardial infarctions and infarctions of neural tissue, namely stroke, transient ischemic attacks, and cerebrovascular accidents), may not be diseases strictly in their own right. However, in our time, we have learned to recognize these abnormal levels of blood lipids or lipoproteins as serious challenges to personal health, necessitating vast programs of preventive medication, largely with the class of drugs known as statins, for correcting these dyscrasias and preventing them from progressing to frank diseases and other life-threatening events. The success of these pharmaceutical programs for prolonging life are widely acknowledged and a great source of pride to the medical and pharmaceutical professions and, were it not for side effects associated with these therapies, seemingly beyond reproach. However, because the side effects may be untenable, there is a great demand for new therapies that are more benign, at least as effective, and in the public eye, natural.

The *Capparis* genus contains numerous chemical constructs, from water-soluble fibers to specific lipophilic compounds such as sterols, to hydrophilic compounds such as alkaloids and flavonoids that exert favorable influence over the lipid balance in the body and thus possess the potential for ameliorating lipid dyscrasias such as high serum cholesterol, low levels of high-density lipoprotein, and high levels of low-density lipoprotein. Table 16.1 illustrates some of these "lipid-corrective" actions of *Capparis* parts and their extracts that may ultimately reverse or reduce lipid dyscrasias and prevent disease (Figure 16.1). The relatively low toxicity and high safety of *Capparis* chemical compartments, both hydrophilic and hydrophobic, provide these edible components with a wide therapeutic window (i.e., a wide margin of safety) for use as therapeutic agents. Within this window lies their particular appeal and pharmaceutical potential for benefiting humankind.

TABLE 16.1
Lipid-Corrective Properties of *Capparis* spp.

Capparis sp.	Plant Part	Pharmacological Action	Compounds Responsible (If Known)	References
decidua	Dried plant matter, mixed with feed	Most pronounced hypocholesterolemic effect in rat through increased fecal excretion of cholesterol and bile acids compared to khejri beans (*Prsopsis cinceria*), peepalbanti (*Ficus religiosa*), barbanti (*Ficus bengalensis*), gullar (*Ficus glomerata*)	Fiber	Agarwal and Chauhan 1988
decidua	Fruit, shoot, ethanolic (50% in H_2O) extract	In rabbits fruit extract 500 mg/kg → signif. ↓ serum total cholesterol (61%), LDL cholesterol (71%), triglycerides (TG; 32%), phospholipids (25%); shoot extract: ↓ serum total cholesterol (48%), LDL cholesterol (57%), TG (38%), phospholipids (36%); aorta choles. ↓ by 44 and 28% in fruit and shoot extracts treatment, respectively; HDL/total cholesterol and atherogenic index was signif. ↓ in treated groups		Purohit and Vyas 2005
decidua	Unripe fruits (*ker*)	Daily feeding to hyperlipidemic adults over 3 mo significantly reduced total lipids, TG, and phospholipids		Goyal and Grewal 2003
decidua	Whole plant parts	Improved utilization of fats and total energy in weanling rat (feed efficiency ratio, FER; dry matter digestibility, DMD; and true protein digestibility, TPD)	Fiber	Agarwal and Chauhan 1989
spinosa	Aqueous extract	20 mg/kg: signif. ↓ plasma TG in normal rats 1 wk ($p < 0.05$) and 2 wk ($p < 0.01$) after daily p.o. dose, plasma cholesterol after 4 days ($p < 0.05$) and 1 wk ($p < 0.05$); diabetic rats signif. ↓ TG after 4 days, cholesterol ($p < 0.05$) and after 2 wk ($p < 0.01$) and body wt. after 4 days ($p < 0.05$)		Eddouks et al. 2005

FIGURE 16.1 (See color insert.) Aerial parts of *Capparis decidua*, including fruit, have a hypolipidemic effect. (Mauritania: by Sébastien Sant, in J.P. Peltier. 2006. Plant Biodiversity of South-Western Morocco, http://www.teline.fr, accessed 13 March 2013.)

We were unable to find any references to the use of *Capparis* for cholesterol or any other lipid dyscrasias, probably also because they are difficult to diagnose as they do not have clear symptoms. Table 24.2, "Contemporary Ethnomedical Uses of *Capparis* spp. against Hypertension or Cardiac Disorders," and Table 17.2, on diabetes, would be worth consulting despite their brevity.

REFERENCES

Agarwal, V. and B.M. Chauhan. 1988. A study on composition and hypolipidemic effect of dietary fibre from some plant foods. *Plant Foods for Human Nutrition* 38(2): 189–197.

Agarwal, V. and B.M. Chauhan. 1989. Effect of feeding some plant foods as source of dietary fibre on biological utilisation of diet in rats. *Plant Foods for Human Nutrition* 39(2): 161–167.

Eddouks, M., A. Lemhadri, and J.B. Michel. 2005. Hypolipidemic activity of aqueous extract of *Capparis spinosa* L. in normal and diabetic rats. *J Ethnopharmacology* 98: 345–350.

Goyal, R. and R.B. Grewal. 2003. The influence of teent (*Capparis decidua*) on human plasma triglycerides, total lipids and phospholipids. *Nutrition and Health* 17: 71–76.

Peltier, J.P. 2006. Plant Biodiversity of South-Western Morocco. http://www.teline.fr (accessed 13.3.2013).

Purohit A. and K.B. Vyas. 2005. Hypolipidaemic efficacy of *Capparis decidua* fruit and shoot extracts in cholesterol fed rabbits. *Indian Journal of Experimental Biology* 43(10): 863–866.

17 Diabetes Mellitus

The word *diabetes* is from the Greek, meaning approximately and alternatively, compass and sieve (Stedman 1976). The sieve is straightforward, as it relates to the polyuria, the excessive urination, which diabetes mellitus shares in common with diabetes insipidus, from inadequate output of pituitary antidiuretic hormone or hysterical causes. The *mellitus* of the subject of this chapter of course means sweet (Latin, sweetened with honey) and refers to a metabolic disease in which carbohydrate utilization is reduced and that of lipids and protein enhanced, resulting in excess glucose in blood, in later cases in urine, and if unchecked, by electrolyte loss, ketoacidosis, and coma. All diabetes mellitus stems from inadequate insulin production, whether from likely genetic deficiencies in (type 1 diabetes mellitus) juvenile-onset diabetes (unspecified, diabetes refers to diabetes mellitus) or from acquired insulin resistance (type 2 diabetes mellitus).

While both type 1 and type 2 are each relevant to the ensuing discussion and Table 17.1, type 2 is the subject of most of the research and the form that has continued to increase in frequency. Together with insulin resistance (a consequence of or closely related to inflammation) and the derivative metabolic syndrome, type 2 has reached epidemic proportions worldwide in both adult and pediatric groups and shows no signs as of this writing of abating. Metabolic syndrome results in abdominal obesity as well as cardiovascular pathology and cancer, or at least is related to these maladies on an acausal web (Kaptchuk 1983).

The experimental methods described in Table 17.1 relate mainly to animal models of diabetes in rats and mice. One classical way of inducing diabetes in rodents is with alloxan, a drug causing destruction of cells of the pancreatic islets of Langerhans, or with the alkylating agent used for metastatic islet cell cancers, which produces the same effect as alloxan. Both methods can be understood as producing models for type 1 disease.

TABLE 17.1
Antidiabetic Properties of *Capparis* spp.

Capparis sp.	Plant Part	Pharmacological Action	Compounds Responsible (If Known)	References
aphylla	Dried MeOH ext.	In streptozocin-induced diabetic rat, ↓ blood sugar 4.6-fold w/ single 50–100 mg oral dose, secondary to beta cell regeneration of diabetic-damaged pancreatic tissue		Dangi and Mishra 2011b
aphylla	Methanol extract (300 mg/kg body wt.) and active fraction (30 mg/kg body wt.) of stems	Significantly ($p < 0.01$) reduced blood glucose levels after 3 h in diabetic rats; ↓↓ glutathione level, superoxide dismutase (SOD), catalase (CAT), and glutathione peroxidase activity in liver, heart, and kidney; ↓ lipid peroxidation by scavenging free radicals		Dangi and Mishra 2011a

(Continued)

TABLE 17.1 (*Continued*)
Antidiabetic Properties of *Capparis* spp.

Capparis sp.	Plant Part	Pharmacological Action	Compounds Responsible (If Known)	References
decidua	Fruits	Improved diabetic parameters in streptozocin-induced diabetic mice	Alkaloids	Sharma et al. 2010
decidua	Powder in diet	In alloxan-induced diabetic rat tissue ↓ lipid peroxidation > intraperineal insulin, ↑ SOD activity in heart and kidneys secondary to ↑ dismutation of superoxide anions but ↓ SOD activity in the liver and kidney comparable to control rats, ↑ CAT activity while i.p. insulin did not, ↑ total and Se-dependent glutathione peroxidase (GSH-Px) in heart → ↓ H_2O_2 toxicity and ↓ oxidative stress in diabetes, ↓ GSH-R activity in kidney and heart → ↓ GSH in these tissues → adaptive response to neutralize superoxide anions → counteract oxidative stress in diabetes and ↑ glutathione S-transferase (GST) comp. to i.p. insulin, ↑ glucose-6-phosphate dehydrogenase (G6PDH) in kidney and heart → ↑ NADPH generation required for GSH-R activity and GSH production, regenerating antioxidant function that was ↓ in chronic diabetes		Yadav et al. 1997a
decidua	Powdered fruit	In aged diabetic rats, ↓ alloxan-induced lipid peroxidn. (LPO) in erythrocytes, kidney, and heart; ↑ erythrocyte SOD activity; ↓ kidney and heart SOD; may neutralize H_2O_2 toxicity by its increased decompn. by CAT; therefore ↓ alloxan-induced LPO and alters SOD and CAT enzymes to reduce oxidative stress		Yadav et al. 1997b
decidua	Stems, aq. and EtOH exts.	↓ fasting blood glucose 58.5%, 83.6% (aq. ext.) and 60.2%, 98.51% (EtOH ext.) at 250 and 500 mg/kg in alloxan-induced diabetic rats after 21 days or single oral doses		Rathee et al. 2010
masaikai	Seeds, EtOH ext.	↓ the ↑ in serum glucose after glucose load in rats	Ornithine, GABA, cystathionine, phosphoserine, 20 amino acids	Nakasugi et al. 2004

TABLE 17.1 (*Continued*)
Antidiabetic Properties of *Capparis* spp.

Capparis sp.	Plant Part	Pharmacological Action	Compounds Responsible (If Known)	References
moonii	Fruit, hydroethanolic extract	Insulinometic, improves glucose uptake	Chebulinic acid derivatives	Kanaujia et al. 2010
sepiaria	Leaf, ethanol extract	Reduced blood sugar *in vivo* at 100–300 mg/kg		Selvamani et al. 2008
sepiaria	Leaf, ethanol extract	Single oral dose from 100 to 300 mg/kg lowered blood sugar in streptozocin-induced diabetic rats 9–15%, 894 mg/kg acutely toxic (LD_{50})		Selvamani et al. 2008
sepiaria	Leaves, ethanolic extract	Maximum ↓ of blood glucose in streptozotocin-induced diabetic rat of maximum fall of plasma glucose level 12 h after single oral dose of 100, 200, and 300 mg/kg 9.40%, 13.57%, and 15.25%, resp., compared to 18.80% ↓ following single oral dose of glibenclamide 10 mg/kg, acute toxic effect of extr. 894.43 mg/kg		Selvamani et al. 2008
spinosa	Fruit, aqueous extract	Significant decrease in blood sugar in streptozocin-induced diabetic mice by single dose or daily doses for 20 days (20 mg/kg) ($p < 0.001$)		Eddouks et al. 2004

Acquired insulin resistance and type 2 disease emerge from problematic and stressful lifestyle, poor and overly processed diet, excessive sugar consumption, and other related factors, with medical sequelae such as diabetes mellitus, obesity, heart disease, and cancer and of brain diseases, dementia, autism, and schizophrenia, since oxidative tension → inflammation → insulin resistance → metabolic syndrome → diabetes and similar. In any event, diabetes is in the background if not actually manifested in the lives of every person, and not only in the 10% or so who are directly afflicted with the disease, requiring insulin injections in type 1 and oral hypoglycemic agents in type 2. It is among the type 2 agents that natural drugs may or may not yet make their greatest contribution, but certainly this is where most research has been directed and for which the natural compounds hold probably the greatest pharmaceutical potential.

Among other natural compounds and plants with similar actions beneficial to diabetic patients, *Capparis* scores very well. Table 17.1 summarizes this research in a number of species and includes direct effects not only on blood sugar but also on consequences of diabetes, especially on lipids and oxidative tension. Compounds in *Capparis* with potential antidiabetes or diabetes-benefiting properties occur in both the lipid and aqueous phases of the plants (Figure 17.1).

In modern ethnomedical literature, *Capparis* is mentioned only rarely as a possible therapy for diabetes (Figure 17.2, Table 17.2). As a strong antioxidant, however, it may be—and has been—used in the therapy of diabetic complications.

See also Table 13.2 for ethnomedical uses connected with the antioxidant potential of the different *Capparis* spp.

FIGURE 17.1 *Capparis sepiaria* leaves reduce blood sugar *in vivo*. (Near Odzani River, northwest of Mutare, Zimbabwe, 21.10.2005: by Bart T. Wursten. In Hyde, M.A., B.T. Wursten, and P. Ballings. 2013. Flora of Zimbabwe: Species Information: Individual Images: *Capparis sepiaria*. http://www.zimbabweflora.co.zw/speciesdata/image-display.php?species_id=124450&image_id=3 [accessed 20 February 2013].)

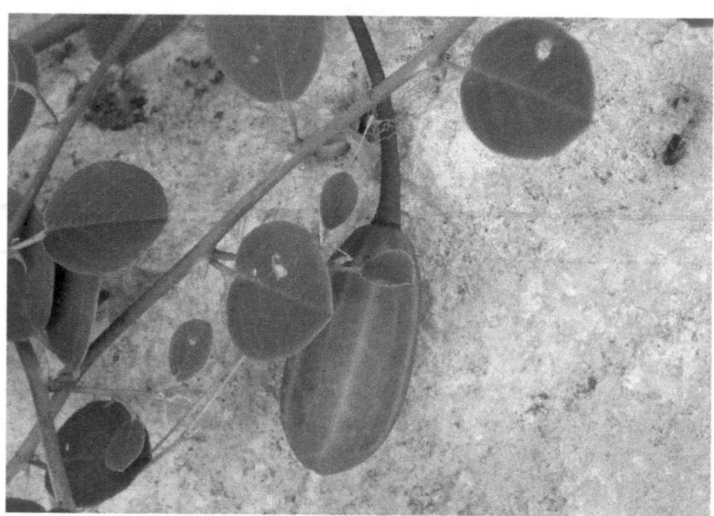

FIGURE 17.2 Dried fruit of *Capparis spinosa* is used in Israeli traditional medicine as an antidiabetic. Modern research supports the claim. Unripe fruit of *Capparis spinosa*. (25.7.2010, Jerusalem, Israel: by Helena Paavilainen.)

TABLE 17.2
Contemporary Ethnomedical Uses of *Capparis* spp. against Diabetes

Indication	Geography	*Capparis* sp.	Part Used	References
Claimed as antidiabetic	India	*decidua*	Dried fruit (pickled)	Yadav et al. 1997a
Against diabetes	Israel	*spinosa*	Dried fruit	Yaniv et al. 1987

REFERENCES

Dangi, K.S. and S.N. Mishra. 2011a. Antioxidative and β cell regeneration effect of *Capparis aphylla* stem extract in streptozotocin induced diabetic rat. *Biology and Medicine (Aligarh)* 3(3): 82–91.

Dangi, K.S. and S.N. Mishra. 2011b. *Capparis aphylla* extract compound as antidiabetic and antipathogenic agent. Indian Patent IN 2009DE00742 A.

Eddouks, M., A. Lemhadri, and J.B. Michel. 2004. Caraway and caper: potential anti-hyperglycaemic plants in diabetic rats. *Journal of Ethnopharmacology* 94: 143–148.

Hyde, M.A., B.T. Wursten, and P. Ballings. 2013. Flora of Zimbabwe. [Online database.] http://www.zimbabweflora.co.zw (accessed 5 February 2013).

Kanaujia, A., R. Duggar, S.T. Pannakal, et al. 2010. Insulinomimetic activity of two new gallotannins from the fruits of *Capparis moonii*. *Bioorganic & Medicinal Chemistry* 18: 3940–3945.

Kaptchuk, T.J. 1983. *The Web that Has No Weaver: Understanding Chinese Medicine*. New York: Congdon and Weed.

Nakasugi, T., T. Uekubo, and K. Komai. 2004. Carbohydrate absorption inhibitors containing *Capparis* (extracts) or amino acids, and foods containing the inhibitors. Japanese Patent JP 2004262904.

Rathee, S., O.P. Mogla, S. Sardana, M. Vats, and P. Rathee. 2010. Antidiabetic activity of *Capparis decidua* Forsk Edgew. *Journal of Pharmacy Research* 3(2): 231–234.

Selvamani, P., S. Latha, K. Elayaraja, P.S. Babu, J.K. Gupta et al. 2008. Antidiabetic activity of the ethanol extract of *Capparis sepiaria* L leaves. *Indian Journal of Pharmaceutical Sciences* 70(3): 378–380.

Sharma, B., R. Salunke, C. Balomajumder, et al. 2010. Anti-diabetic potential of alkaloid rich fraction from *Capparis decidua* on diabetic mice. *Journal of Ethnopharmacology* 127: 457–462.

Stedman, T.L. 1976. *Stedman's Medical Dictionary*. Baltimore: Williams and Wilkins.

Yadav, P., S. Sarkar, and D. Bhatnagar. 1997a. Action of *Capparis decidua* against alloxan-induced oxidative stress and diabetes in rat tissues. *Pharmacological Research: The Official Journal of the Italian Pharmacological Society* 36(3): 221–228.

Yadav, P., S. Sarkar, and D. Bhatnagar. 1997b. Lipid peroxidation and antioxidant enzymes in erythrocytes and tissues in aged diabetic rats. *Indian Journal of Experimental Biology* 35(4): 389–392.

Yaniv, Z., A. Dafni, J. Friedman, and D. Palevitch. 1987. Plants used for the treatment of diabetes in Israel. *Journal of Ethnopharmacology* 19(2): 145–151.

18 Sunburn

Much of *Capparis*'s evolution is of a desert plant, blessed with phytochemical defenses that have helped it to sustain the heat and brightness of the sun's energy. Simply and obviously, these powerful antisolar principles are available in the plants (Figure 18.1) and may be consumed (by us) once the products of this exceedingly adaptive and canny genus, such as its fruits, leaves, and buds, have been properly pickled in brine or vinegar or extracted with ethanol, supercritical carbon dioxide, water, or, it is hoped, some other nontoxic method.

The question of sunburn is both a clear-cut relationship between the sunburns suffered by plants and those suffered by humanity and the relationship between solar radiation and radiation developed by humans. As such, *Capparis* compounds possess great potential for use in skin salves, creams, milks, and ointments for mollifying, treating, and possibly preventing skin problems devolving from exposure to sunlight, most notably to its ultraviolet components. They also, though, may benefit radiation burns related to medical radiotherapy treatments and from industrial, military or, occupational exposure. The short Table 18.1 highlights three studies in *C. spinosa*, but the principles may be relevant for other species as well. Due to the excellent possibilities for advancing *Capparis* production to meet demand, with the aid of micropropagation and other evolving agronomical technology, safe *Capparis* agents for preventing and treating solar-acquired pathology and pathology from ionizing radiation may be practical.

We were not able to find any modern ethnopharmacological or historical data suggesting the use of *Capparis* spp. for sunburns.

FIGURE 18.1 *Capparis spinosa* buds (and leaves) show great potential for skin care and protection applications. (March 2011, Haifa, Israel: by Ephraim Lansky.)

TABLE 18.1
Potential Benefit of *Capparis* spp. for Photorepair and Photoprotection

Capparis sp.	Plant Part	Pharmacological Action	Compounds Responsible (If Known)	References
spinosa	Buds	Antioxidant activity *in vitro*, reduced post-UV skin erythema in human volunteers	Flavonols (kaempferol and quercetin derivatives) and hydroxycinnamic acids (caffeic acid, ferulic acid, p-cumaric acid, and cinnamic acid)	Bonina et al. 2002
spinosa	Compounded into topical agent	Significantly ↓ erythema following radiation therapy		Rizza et al. 2010
spinosa	Leaf extract	Dose-dependent ↑ melanogenesis in melanoma cells, potential as tanning agent	Quercetin	Matsuyama et al. 2009

REFERENCES

Bonina, F., C. Puglia, D. Ventura, R. Aquino, S. Tortora, et al. 2002. In vitro antioxidant and in vivo photoprotective effects of a lyophilized extract of *Capparis spinosa* L. buds. *Journal of Cosmetic Science* 53(6): 321–335.

Matsuyama, K., M. Villareal, A. El Omri, J. Han, M.E. Kchouk, et al. 2009. Effect of Tunisian *Capparis spinosa* L. extract on melanogenesis in B16 murine melanoma cells. *Journal of Natural Medicines* 63(4): 468–472.

Rizza, L., A. D'Agostino, A. Girlando, and C. Puglia. 2010. Evaluation of the effect of topical agents on radiation-induced skin disease by reflectance spectrophotometry. *Journal of Pharmacy and Pharmacology* 62(6): 779–785.

19 Pain and Fever

Pain and fever are two of the four cardinal symptoms of inflammation (the other two are redness and swelling), and each serves important preventive protective functions. Pain warns that borders within and of the body have been breached, and strong immunological or neurobehavioral actions will be mobilized. Fever is a way of intensifying immunological heat to help kill invading pathogens or to effect immunological repair of damage from abiotic causes.

From the plant's point of view, fever bears similitude to the heat of the sun, so compounds evolving to help *Capparis* adapt to severe xeric environments may help with ridding persons of fever, or so could go this line of thought. Similarly, pain is something at least *C. spinosa* puts forward with its very efficient system of thorns and its stoic persistence.

Nonetheless, pain can become chronic and require treatment, and pain is also associated with medical procedures, such as dental treatment or surgery. Headaches may be chronic, and pain may be associated with neoplastic or other inflammatory disease. All of these huge categories demand safe and effective pain medication; here also, *Capparis* may have something to offer as the animal studies indicated in Table 19.1 suggest. The studies were directed at both "pure pain" scenarios and pain

TABLE 19.1
Pain-Relieving Properties of *Capparis* spp.

Capparis sp.	Plant Part	Pharmacological Action	Compounds Responsible (If Known)	References
decidua	EtOH ext.	Anti-inflammatory in carrageenan ear edema test, antipyretic, not analgesic		Ageel et al. 1986
ovata	Bud, methanol extract	Antinociceptive in mice		Arslan and Bektas 2010
sepiaria	Etheric, methanolic, and aqueous extracts of roots	All exts. anti-inflammatory and analgesic		Chaudhari et al. 2004
spinosa	Aq. ext.	Anti-inflammatory in carrageenan ear edema test, not antipyretic, not analgesic		Ageel et al. 1986
spinosa	Polysaccharide-rich fraction by aq. ext., pptd. by alc., Sevage to deprotein, dioxogen to decolor, dialysis to enrich the polysaccharides to 36%	72% ↓ ear swelling in xylene-induced mouse inflam., 50% ↓ pain by hot plate test, 40% ↓ body contortions in writhing test		Zhang et al. 2011
zeylanica	Leaf, ethanolic and aqueous extracts	Dose-dependent, significant ($p < 0.05$) increases in pain threshold in tail-immersion test, (100–200 mg/kg) inhibition of writhing, significant ($p < 0.001$) inhibition of both phases of formalin pain test, water extract (200 mg/kg) significant ($p < 0.01$) reversal of yeast-induced fever	Alkaloids, flavonoids, saponins glycosides, terpenoids, tannins, proteins, carbohydrates	Ghule et al. 2007

associated with other symptoms of inflammation or of inflammation-induced disease (Figure 19.1). Such medication could be in the form of natural, over-the-counter pills against pain or in a concentrated form, for parenteral administration. Eating the pickled fruit and flowers may also benefit.

As febrifuges, *Capparis* has been used from ancient times (see Table 14.2). As for pain, safe and effective febrifuges (Figure 19.2) from natural sources could also provide a contribution to medicine; again, *Capparis* offers some potential worthy of further investigation.

The different parts of *Capparis* spp. are often recommended for fevers (particularly the root) (Table 19.2) and are popular in modern ethnomedicine, especially in treating pain (Table 19.3). See also Tables 14.4, 15.2, and 21.2 on contemporary treatments for inflammation, arthritis, and cancer, respectively.

FIGURE 19.1 The dried leaves of *Capparis tomentosa* have been used for malaria. (28.10.2008: by Linda Loffler via Swaziland's Flora Database, Swaziland National Trust Commission 2013, http://www.sntc.org.sz/flora/photo.asp?phid=2734)

FIGURE 19.2 Aerial parts of *Capparis sepiaria* are used in traditional medicine as febrifuges. Modern research has recognized the root extracts' analgesic and anti-inflammatory effects. (3.6.2010, India: by A. Lalithamba.)

TABLE 19.2
Contemporary Ethnomedical Uses of *Capparis* spp. as an Antipyretic

Indication	Geography	*Capparis* sp.	Part Used	References
Antipyretic	Thailand	*micracantha*	Dried root + stem	Mokkhasmit et al. 1971, Wasuwat 1967
Febrifuge	India	*sepiaria*	Dried entire plant	Sebastian and Bhandari 1984
Against fever	Egypt	*galeata*	Fresh fruit	Goodman and Hobbs 1988
Against intermittent fever	India	*decidua*	Dried entire plant	Dhar et al. 1972
Against typhoid fever	India	*sepiaria*	Fruit	Singh and Ali 1994
Against malaria		*decidua*		Duke 1993
Against malaria	Rwanda	*tomentosa*	Dried leaf	Vlietinck et al. 1995
Against malaria	South Africa	*tomentosa*	Fresh root bark	Watt and Breyer-Brandwijk 1962
Diaphoretic	India	*decidua*	Dried root bark	Gaind and Juneja 1969

TABLE 19.3
Contemporary Ethnomedical Uses of *Capparis* spp. as an Analgesic

Indication	Geography	*Capparis* sp.	Part Used	References
Analgesic	Saudi Arabia	*decidua*	Dried aerial parts	Al-Yahya 1986
Analgesic	India	*decidua*	Dried root bark	Gaind and Juneja 1969
Analgesic		*micracantha*		*Wealth of India* 1985–1992
Analgesic	India	*spinosa*	Dried root bark	Shirwaikar et al. 1996
Analgesic	India	*zeylanica*		Jain and Tarafder 1970
Against external pains	Israel	*spinosa*	Dried root	Dafni et al. 1984
For headache and general body pain	India	*assamica* (in an herbal mixture)	Dried leaf + root	Maikhuri and Gangwar 1993
Against neuralgia	India	*zeylanica*		Jain and Tarafder 1970
Against headache	Ghana	*erythrocarpos*		Ayensu 1978
Against headache	Egypt	*galeata*	Fresh fruit	Goodman and Hobbs 1988
Against headache	Tunisia	*spinosa*	Dried leaf	Boukef et al. 1982
Against headache	Venda	*tomentosa* (in an herbal mixture)	Dried root	Arnold and Gulumian 1984
Against headache	Venda	*tomentosa*	Dried root bark	Arnold and Gulumian 1984
Against backache	Tanganyika	*elaeagnoides*	Fresh entire plant	Watt and Breyer-Brandwijk 1962
Against toothache	India	*decidua*	Dried entire plant	Dhar et al. 1972
Against toothache	Saudi Arabia	*deserti*	Leaf	Al-Said 1993
Against toothache	Tanganyika	*elaeagnoides*	Fresh entire plant	Watt and Breyer-Brandwijk 1962
Against toothache	Jordan	*galeata*	Flowers	Al-Khalil 1995
Against toothache	Spain	*ovata*	Leaf + stem	Martínez-Lirola et al. 1996
Against earache	Oman	*spinosa*	Leaf	Ghazanfar and Al-Sabahi 1993
Against stomach pains	Kenya	*fascicularis*	Dried root	Johns et al. 1990
Against stomach pains	Java	*acuminata*		Burkill et al. 1966
Against abdominal pain	Haiti	*flexuosa*	Dried wood	Weniger et al. 1986
Against dysmenorrhea	India	*spinosa*	Fruit	Uniyal 1990
Against painful menstruation	Morocco	*spinosa*	Fruit	Bellakhdar et al. 1991

REFERENCES

Ageel, A M., N.S. Parmar, J.S. Mossa, M.A. Al-Yahya, M.S. Al-Said, et al. 1986. Anti-inflammatory activity of some Saudi Arabian medicinal plants. *Agents Actions* 17(3–4): 383–384.

Al-Khalil, S. 1995. A survey of plants used in Jordanian traditional medicine. *International Journal of Pharmacognosy* 33(4): 317–323.

Al-Said, M.S. 1993. Traditional medicinal plants of Saudi Arabia. *American Journal of Chinese Medicine* 21(3–4): 291–298.

Al-Yahya, M.A. 1986. Phytochemical studies of the plants used in traditional medicine of Saudi Arabia. *Fitoterapia* 57(3): 179–182.

Arnold, H.J. and M. Gulumian. 1984. Pharmacopoeia of traditional medicine in Venda. *Journal of Ethnopharmacology* 12 (1): 35–74.

Arslan, R. and N. Bektas. 2010. Antinociceptive effect of methanol extract of *Capparis ovata* in mice. *Pharmaceutical Biology* 48: 1185–1190.

Ayensu, E.S. 1978. *Medicinal Plants of West Africa.* Algonac, MI: Reference Publications.

Bellakhdar, J., R. Claisse, J. Fleurentin, and C. Younos. 1991. Repertory of standard herbal drugs in the Moroccan pharmacopoea. *Journal of Ethnopharmacology* 35(2): 123–143.

Boukef, K., H.R. Souissi, and G. Balansard. 1982. Contribution to the study on plants used in traditional medicine in Tunisia. *Plantes Medicinales et Phytotherapie*16(4): 260–279.

Burkill, I.H., W. Birtwistle, F.W. Foxworthy, J.B. Scrivenor, and J.G. Watson. 1966. *A Dictionary of the Economic Products of the Malay Peninsula.* Kuala Lumpur, Malaysia: Published on behalf of the governments of Malaysia and Singapore by the Ministry of Agriculture and cooperatives.

Chaudhari, S.R., M.J. Chavan, and R.S. Gaud. 2004. Phytochemical and pharmacological studies on the roots of *Capparis sepiaria. Indian Journal of Pharmaceutical Sciences* 66(4): 454–457.

Dafni, A., Z. Yaniv, and D. Palevitch. 1984. Ethnobotanical survey of medicinal plants in northern Israel. *Journal of Ethnopharmacology* 10(3): 295–310.

Dhar, D.N., R.P. Tewari, R.D. Tripathi, and A.P. Ahuja. 1972. Chemical examination of *Capparis decidua. Proceedings of the National Academy of Science, India: Section A* 42(1): 24–27.

Duke, J.A. 1993. Dr. Duke's Phytochemical and Ethnobotanical Databases. [Online database.] USDA–ARS–NGRL, Beltsville Agricultural Research Center, Beltsville, MD. http://www.ars-grin.gov/duke/ethnobot.html (accessed 12 May 2011).

Gaind, K.N. and T.R. Juneja. 1969. Investigations on *Capparis decidua. Planta Medica* 17: 95–98.

Ghazanfar, S.A. and M.A. Al-Sabahi. 1993. Medicinal plants of northern and central Oman (Arabia). *Economic Botany* 47(1): 89–98.

Ghule, B.V., G. Murugananthan, and P.G. Yeole. 2007. Analgesic and antipyretic effects of *Capparis zeylanica* leaves. *Fitoterapia* 78(5): 365–369.

Goodman, S.M. and J.I. Hobbs. 1988. The ethnobotany of the Egyptian Eastern desert: a comparison of common plant usage between two culturally distinct Bedouin groups. *Journal of Ethnopharmacology* 23(1): 73–89.

Jain, S.K. and C.R. Tarafder. 1970. Medicinal plant-lore of the Santals. *Economic Botany* 24(3): 241–278.

Johns, T., J.O. Kokwaro, and E.K. Kimanani. 1990. Herbal remedies of the Luo of Siaya District, Kenya: establishing quantitative criteria for consensus. *Economic Botany* 44(3): 369–381.

Maikhuri, R.K. and A.K. Gangwar. 1993. Ethnobiological notes on the Khasi and Garo tribes of Meghalaya, Northeast India. *Economic Botany* 47(4): 345–357.

Martínez-Lirola, M.J., M.R. Gonzalez-Tejero, and J. Molero-Mesa. 1996. Ethnobotanical resources in the Province of Almeria, Spain: Campos de Nijar. *Economic Botany* 50(1): 40–56.

Mokkhasmit, M., W. Ngarmwathana, K. Sawasdimongkol, and U. Permphiphat. 1971. Pharmacological evaluation of Thai medicinal plants. *Journal of the Medical Association of Thailand* 54(7): 490–504.

Sebastian, M.K. and M.M. Bhandari. 1984. Medico-ethno botany of Mount Abu, Rajasthan, India. *Journal of Ethnopharmacology* 12(2): 223–230.

Shirwaikar, A., K.K. Sreenivasan, B.R. Krishnanand, and A.V. Kumar. 1996. Chemical investigation and antihepatotoxic activity of the root bark of *Capparis spinosa. Fitoterapia* 67(3): 200–204.

Singh, V.K. and Z.A. Ali. 1994. Folk medicines in primary health care: common plants used for the treatment of fevers in India. *Fitoterapia* 65(1): 68–74.

Swaziland National Trust Commission. Swaziland's Flora Database. 2013. http://www.sntc.org.sz/flora (accessed 14 March 2013).

Uniyal, M.R. 1990. Utility of hitherto unknown medicinal plants traditionally used in Ladakh. *Journal of Research and Education in Indian Medicine* 9(2): 89–95.

Vlietinck, A.J., L. van Hoof, J. Totte, A. Lasure, D. vanden Berghe, P.C. Rwangabo, and J. Mvukiyumwami. 1995. Screening of hundred Rwandese medicinal plants for antimicrobial and antiviral properties. *Journal of Ethnopharmacology* 46(1): 31–47.

Wasuwat, S. 1967. *A List of Thai Medicinal Plants*, Bangkok, Thailand: Applied Scientific Research Corporation of Thailand (ASRCT).

Watt, J.M. and M.G. Breyer-Brandwijk. 1962. *The Medicinal and Poisonous Plants of Southern and Eastern Africa*. 2nd ed. London: Livingstone.

Wealth of India 1985–1992. *The Wealth of India: A Dictionary of Indian Raw Materials and Industrial Products*. New Delhi, India: Publications and Information Directorate, Council of Scientific and Industrial Research.

Weniger, B., M. Rouzier, R. Daguilh, D. Henrys, J.H. Henrys, and R. Anton. 1986. Popular medicine of the Central Plateau of Haiti. 2. Ethnopharmacological inventory. *Journal of Ethnopharmacology* 17(1): 13–30.

Zhang, Y., H. Zhang, B. Han, and W. Chen. 2011. Extraction of polysaccharides in *Capparis spinosa* L. and anti-inflammatory and analgesic effects. *Shihezi Daxue Xuebao, Ziran Kexueban* 29(2): 205–209.

20 Xerostomia

It is a bit of a stretch from the dryness of the desert to the dryness of the human mouth, but indeed, *Capparis* has enjoyed modern experimental employment as a remedy for xerostomia, specifically dryness of the oral environment. Several innovative mouthwashes, dentifrices, drinks, and similar items have been conceived and developed. This work is reflected in Tables 20.1 and 20.2 and Figure 20.1, which opens an intriguing market for oral health and dental care products from *Capparis*.

TABLE 20.1
Oral Hygiene Properties of *Capparis* spp.

Capparis sp.	Plant Part	Pharmacological Action	Compounds Responsible (If Known)	References
masaikai Levl.		Significant increase in oral moisture and oral conditions in 21 human volunteers as own controls		Kitada et al. 2009
sp.	Seeds	↓ heat and toxins, ↑ body fluid production, ↓ thirst, ↓ food stagnation		Yang 2009

TABLE 20.2
Contemporary Ethnomedical Uses of *Capparis* spp. for Oral Hygiene

Indication	Geography	*Capparis* sp.	Part Used	Reference
As a chew stick	Venezuela	*flexuosa*		Pittier 1926
Against aphthae	Venezuela	*jamaicensis*		Pittier 1926
As a gargle	Venezuela	*jamaicensis*		Pittier 1926
Against stomatitis	Venezuela	*jamaicensis*		Pittier 1926

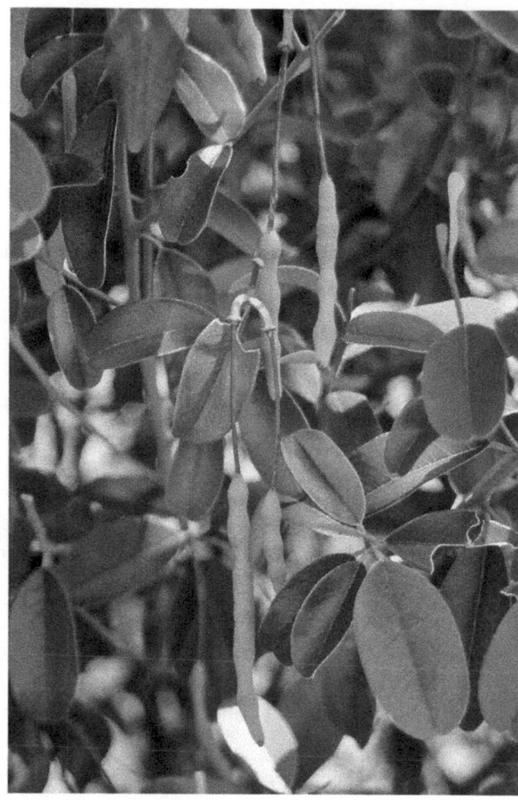

FIGURE 20.1 *Capparis jamaicensis* has been used in Venezuela against mouth problems. (29.12.2010, Sanibel, Florida, SCCF Native Plant Nursery.)

REFERENCES

Kitada, K., K. Shibuya, M. Ishikawa, et al. 2009. Enhancement of oral moisture using tablets containing extract of *Capparis masaikai* Levl. *Journal of Ethnopharmacology* 122: 363–366.

Pittier, H. 1926. *Manual de las plantas usuales de Venezuela*. Caracas, Venezuela: Litografia del comercio.

Yang, C. 2009. Semen Capparis used in health tea and drinking water packaging bottle filled with Semen Capparis. Chinese Patent CN 101416670.

21 Cancer

Cancer refers to a family of diseases characterized by growth of undifferentiated cells that form a "neoplastic clone," in the case of solid cancers, into palpable and imageable "growths" and tumors, and in the case of hematologic cancers, into populations of circulating cancer cells that are detectable only by microscopic analysis of the blood. The etiology of cancers appears to be multivaried and multifactorial but may include exposure to carcinogenic chemicals or radiation, derive from emotional or psychological stresses and trauma, and as already discussed in a nonspecific manner, from long-term chronic inflammation, which itself may be related to oxidative tension and other multiple causative factors.

Conventional treatments of cancer include surgery (to cut out the evil with a knife), radiation (to destroy it with ionizing rays), and chemotherapy (which uses alkylating agents and other cellular poisons to kill the cancer cells at a rate much higher than the same poisons kill healthy cells). Alternative therapies stress psychophysiological "cleansing," "purification," and detoxification and may include dietary restrictions and restructuring, physical exercise, and mental or psychological techniques. Medicinal plants, and medicinal foods that mainly include edible plants, provide a gentler form of treatment in most cases (although they may also include plants with high degrees of toxicity) but differ from simple regimens aimed at detoxification per se in that the medicinal botanicals presumably work because of the anticancer phytochemicals that they contain. Not necessarily, but sometimes also, poisons, these phytochemicals target processes in the cancer cells to prevent them from dividing, migrating, or dedifferentiating. The last refers to the mechanism by which normal cells become free of their normal restraints to stay differentiated with specific functions within the mammalian internal biochemical and cellular "community" and thus can be seen to "run amok" to form cancers. Normally, small cancers are said to form at the microscopic level in mammals all the time, but these are destroyed by the mammal's immune system, which conducts ongoing "surveillance" to recognize such neoplasms and then its destructive capacity kills these "foreign" bits of genetic matter that are at the heart of all cancers. Herbal medicines and medicinal foods may thus also work effectively by potentiating the immune system's capability for both surveillance and destruction, and this constitutes another type of botanical therapy.

Cancer cells also possess the ability to eliminate themselves, and botanical agents can work by stimulating these processes. As far as we know, such self-elimination takes one of two courses: apoptosis, which is something like a genetically available "self-destruct" button, and autophagy, by which the cancer cell undertakes to "eat itself" by self-digestion of its own internal cellular machinery and substance. Botanical agents might thus work to reduce or eliminate cancer by stimulating either apoptosis or autophagy.

Many botanical compounds, most famously flavonoids, can function as "differentiating agents" that reverse the downward spiral of cancer cells toward increasing differentiation and start them on a course back to their origins as normal cells. For example, a white blood cell gone bad loses its prime function of phagocytosis, the digestion of foreign particles, cancer cells, viruses, and bacterial or fungal invaders. The differentiating effect may thus be understood as a kind of cellular rehabilitation, returning the normal functionality to the cancer cells and making them revert to their healthy origins. Botanical substances, often at very low doses, much lower than those required to kill cells, either by promoting apoptosis or autophagy or by direct toxic effects that kill cells by inducing necrosis (which is different from apoptosis), are sometimes able to stop cellular invasion, which on the macroscopic level is known as metastasis. Since metastasis is the main killer of persons who succumb to their cancers, its inhibition, even if cellular death or differentiation is not achieved, is also an important and useful therapeutic goal when treating these diseases.

The anticancer effects of *Capparis*, like most of its physiological effects in mammals, draws from both its aqueous and lipid chemical compartments. These compounds, which include all of the aforementioned fatty acids, sterols, vitamins, alkaloids, flavonoids, volatile compounds, and proteins, act together in multiple ways, including the lowering of oxidative tension and inflammation that probably involve synergistic interactions. Presumably, the plants evolved such mechanisms in part to protect themselves from cancerous processes to which they themselves may be susceptible. Also, since Nature favors redundancy as well as multiple uses from single structures, both chemical and cellular, the anticancer effects may be related to phytochemicals in the plant with other well-recognized functions, sometimes against biotic pathogens such as viruses, bacteria, and fungi, and also functions related to regulation of the plant's own natural growth, reproduction, and energy metabolism.

Table 21.1 summarizes the few studies that have been conducted with *Capparis* fractions for anticancer effects (see also Figure 21.1). These include interference with the cell cycle of cancer cells, effectively stopping their ability for reproducing themselves, nonspecific interruption of their growth or proliferation, and promotion of apoptosis. Some workers have indicated that these mechanisms may be related to more precise processes, such as disturbing the transport of minerals within and without the cell, specifically by increasing intracellular concentrations of calcium through sabotaging the usual functions of "calcium channels" and ionic pumps. The work elaborated in these studies

TABLE 21.1
Anticancer Properties of *Capparis* spp.

Capparis sp.	Plant Part	Pharmacological Action	Compounds Responsible (If Known)	References
decidua (altoundob)	Stems, aq. and methanolic exts.			Ali et al. 2009
sikkimensis var. *formosana*	Roots	*In vitro* anticancer angst. ovarian (1A9), lung (A549), ileocecal (HCT-8), breast (MCF-7), nasopharyngeal (KB), vincristine-resistant (KB-VIN) human tumor cell lines, ED50 ≤ 4 μg/mL (mean GI50 15.1 μ*M*)	Cappamensin A (2H-1,4-benzoxazin-3(4H)-one, 6-methoxy-2-methyl-4-carbaldehyde)	Wu et al. 2003
spinosa		Induce apoptosis in SGC-7901 gastric carcinoma cells by ↑ intracellular Ca^{++}	Total saponins	Cui et al. 2008
spinosa		By MTT assay, ↓ growth of HepG2 human hepatocarcinoma cells, IC_{50} 471.53 mg/L, 75% apoptosis in high-dose group	Polysaccharides	Ji et al. 2008a
spinosa		Induce apoptosis in SGC-7901 human gastrocarcinoma cells by (1) obstructing transition from G1 to G2 of cell cycle, (2) ↑ intracellular Ca^{++}, (3) ↑ reactive oxygen species (ROS), (4) ↑ mitochondrial permeability, IC_{50} 127.5 μg/mL	Total alkaloids	Ji et al. 2009
spinosa		Apoptosis, IC_{50} 162.4 μg/mL in human liver carcinoma HepG-2 cells, blocked cycle S → G2, ↑ HepG-2 intracellular Ca^{2+} pos. correlation with drug dosage	Total alkaloids	Ji and Yu 2009

TABLE 21.1 (*Continued*)
Anticancer Properties of *Capparis* spp.

Capparis sp.	Plant Part	Pharmacological Action	Compounds Responsible (If Known)	References
spinosa		→ apoptosis in human hepatocarcinoma cell Line HepG-2 via ↑ ROS, intracellular Ca^{++}, caspase-9	Polar alkaloids	Yu et al. 2010
spinosa		Kills HepG2 cells, IC50 143 µg/mL, ↑ apoptosis, ↓ cell cycle S to G2, ↑ intracellular Ca^{++} (?) → apoptosis	Alkaloids	Yu et al. 2009
spinosa	Essential oil	In human hepatocarcinoma cell line HepG-2: dose-dependent ↓ growth w/IC_{50} 127.5 µg/mL, 44% apoptosis det. by fluorescent microscopy at 300 µg/mL, blocked cell cycle at G1, ↓ cells in S at 75 µg/mL and G2 at 150 µg/mL, ↓ mitochondrial membrane by MTT assay, ↓ potential w/shift of curve to left, ↑ intracellular Ca^{++} at medium and high doses		Ji et al. 2008b
spinosa	Fruit, total seed oil	Apoptosis in SGC-7901 human gastric carcinoma cells related to ↑ intracellular calcium and ↓ membrane potential of mitochondria	Seed oil	Wang et al. 2008
spinosa	Leaf extract	Upregulates melanogenesis in mouse B16 melanoma cells without cytotoxicity and by stimulating tyrosinase expression	Quercetin 1%	Matsuyama et al. 2009
spinosa	n-BuOH ext. of fruits	Inhibits proliferation of SKOV-3, Canpan-2, HepG-2, SGC-7901 cancer cell lines *in vitro*, IC_{50} is 46.160 µg/mL and 66.546 µg/mL for HepG-2 and SGC-7901, resp.	Alkaloids, flavones	Yang et al. 2009
spinosa	Seed oil	↓ proliferation of human gastric carcinoma cells SGC-7901 *in vitro*, apoptosis obsd. w/laser confocal microscope by staining with FITC-annexin V/PI after 24 h, IC_{50} 160.04 µg/mL, early, middle, and final phrases of apoptosis obsd. after 75, 150, and 300 µg/mL	Oil (fatty acids, sterols, tocols, carotenoids, etc.)	Ling et al. 2010a
spinosa	Seed oil	↓ growth and ↑ apoptosis in human hepatocarcinoma cell line HepG-2, by MTT and SRB assays and flow cytometry staining w/FITC-annexin V/PI, dose-dependent 14% apoptosis at 300 µg/mL	Oil (fatty acids, sterols, tocols, carotenoids, etc.)	Ling et al. 2010b

(*Continued*)

TABLE 21.1 (*Continued*)
Anticancer Properties of *Capparis* spp.

Capparis sp.	Plant Part	Pharmacological Action	Compounds Responsible (If Known)	References
spinosa	Seeds	Inhibited proliferation of hepatoma HepG2 cells, colon cancer HT29 cells, and breast cancer MCF-7 cells with an IC_{50} of about 1, 40, and 60 m*M*, respectively; inhibited HIV-1 reverse transcriptase with IC_{50} of 0.23 m*M*; inhibited mycelial growth in the fungus *Valsa mali*	Protein, monomeric	Lam and Ng 2009
spinosa	Total plant, saponin fraction	↓ prolif. HepG-2 cells, IC_{50} 46 µg/mL, apoptosis 67% if 50 µg/mL, ↓ cells in G1 and G2 of cell cycle after 24 h	Total alkaloids	Yu et al. 2008
spinosa	Whole plant, polysaccharides	In HepG2 human hepatocarcinoma cells, ↓ proliferation, calcium concn., mitochondrial membrane potential, Bcl-2 protein, ↑ apoptosis, Bax protein → controlling Bax/Bcl-2 in Ca^{2+} path	Polysaccharides	Ji et al. 2011

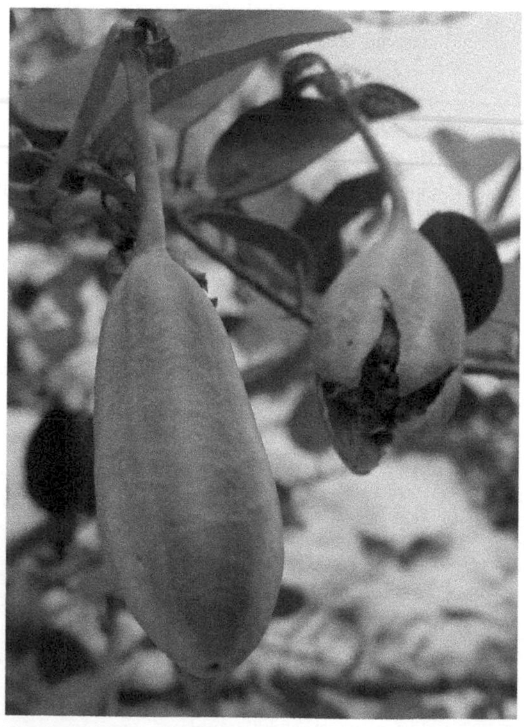

FIGURE 21.1 Seeds of *Capparis spinosa* show promise in anticancer research. (10.8.2011, Jerusalem, Israel: by Helena Paavilainen.)

TABLE 21.2
Contemporary Ethnomedical Uses of *Capparis* spp. against Neoplasias

Indication	Geography	*Capparis* sp.	Part Used	References
Against tumor	France	*rupestris*		Hartwell 1967–1971
Against tumor	Colombia	sp.		Hartwell 1967–1971
Against tumor	Nigeria	*thonningii*		Hartwell 1967–1971
Against tumor	Argentina	*tweediana*		Hartwell 1967–1971
Against cancer	Tanzania	*sepiaria*	Dried root bark	Kamuhabwa et al. 2000
Against cancer	Mozambique	*tomentosa*	Entire plant	Amico 1977

FIGURE 21.2 Only a few *Capparis* species are used in ethnomedicine for the treatment of cancers. *Capparis tomentosa* is one of them, with the whole plant used. (Gorongosa National Park, Mozambique, 13.7.2007: by Bart T. Wursten. In Hyde, M.A., B.T. Wursten, P. Ballings, and S. Dondeyne. 2013. Flora of Mozambique: Species Information: Individual Images: *Capparis tomentosa*. http://www.mozambiqueflora.com/speciesdata/image-display.php?species_id=124460&image_id=5 [accessed 21 February 2013].)

provides a solid basis for further pharmacognostical exploration of *Capparis* for its anticancer effects and as a possible source for new, better, and safer anticancer drugs and medicinal nutrients.

Cancer is given as an indication for the use of *Capparis* surprisingly seldom (Table 21.2, Figure 21.2), considering the anticancerous chemical compounds in the genus. See, however, Tables 14.4, 19.2, and 19.3, on modern ethnomedical use of *Capparis* spp. in the treatment of inflammations, fever, and pain, respectively.

REFERENCES

Ali, S.A., T.H. Al-Amin, A.H. Mohamed, and A.A. Gameel. 2009. Hepatoprotective activity of aqueous and methanolic extracts of *Capparis decidua* stems against carbon tetrachloride induced liver damage in rats. *Journal of Pharmacology and Toxicology* 4(4): 167–172.

Amico, A. 1977. Medicinal plants of southern Zambesia. *Fitoterapia* 48: 101–139.

Cui, R.T., L. Yu, W. Wang, K. Mo, and X. Zou. 2008. Preliminary study on apoptotic effect induced by total saponins in *Capparis spinosa* on SGC-7901. *Harbin Shangye Daxue Xuebao, Ziran Kexueban* 24(6): 652–656.

Hartwell, J.L. 1967–71. Plants used against cancer. A survey. *Lloydia* 30–34.

Hyde, M.A., B.T. Wursten, P. Ballings, and S. Dondeyne. 2013. Flora of Mozambique. [Online database.] http://www.mozambiqueflora.com (accessed 21 February 2013).

Ji, Y., F. Dong, S. Gao, and M. Yu. 2011. Study on *Capparis spinosa* L polysaccharide (CSPS) induced HepG2 apoptosis by controlling Ca^{2+} path. *Advanced Materials Research* 282: 203–208.

Ji, Y., F. Dong, S. Gao, and X. Zou. 2008a. Apoptosis induced by *Capparis spinosa* polysaccharide in human HepG2. *Zhongcaoyao* 39(9): 1364–1367.

Ji, Y.B., K. Mo, L. Yu, W. Wang, and R.T. Cui. 2009. Study on mechanism of total alkaloids in inducing apoptosis in SGC-7901 cells. *Harbin Shangye Daxue Xuebao, Ziran Kexueban* 25(1): 1–6.

Ji, Y.B., L. Yu, W. Wang, and X. Zou. 2008b. Effect on mitochondrial membrane potential and Ca^{2+} concentration in HepG-2 by CSEO. *Zhongguo Tianran Yaowu* 6(6): 474–478.

Ji, Y.B. and L. Yu. 2009. Analysis of relationship between apoptosis and change of Ca^{2+} in HepG-2 induced by CSA with laser scanning confocal technology. *Proceedings of SPIE 7280 (Photonics and Imaging in Biology and Medicine)*: 72800T/1–6.

Kamuhabwa, A., C. Nshimo, and P. de Witte. 2000. Cytotoxicity of some medicinal plant extracts used in Tanzanian traditional medicine. *Journal of Ethnopharmacology* 70(2): 143–149.

Lam, S.K. and T.B. Ng. 2009. A protein with antiproliferative, antifungal and HIV-1 reverse transcriptase inhibitory activities from caper (*Capparis spinosa*) seeds. *Phytomedicine* 16(5): 444–450.

Ling, N., Y. Ji, L. Yu, and X. Zou. 2010b. Inhibition of total oil from *Capparis spinosa* on proliferation of human hepatocarcinoma cell line HepG-2. *Harbin Shangye Daxue Xuebao, Ziran Kexueban* 26(3): 257–60, 264.

Ling, N., L. Yu, X. Zou, and Y. Ji. 2010a. Inhibition of total oil from *Capparis spinosa* on proliferation of human gastric carcinoma SGC-7901 cells. *Harbin Shangye Daxue Xuebao, Ziran Kexueban* 26(4): 393–397.

Matsuyama, K., M. Villareal, A. El Omri, J. Han, M.E. Kchouk, et al. 2009. Effect of Tunisian *Capparis spinosa* L. extract on melanogenesis in B16 murine melanoma cells. *Journal of Natural Medicines* 63(4): 468–472.

Wang, W., L. Yu, R.T. Cui, K. Mo, and X. Zou. 2008. Study on mechanism of total oil in *Capparis spinosa* inducing apoptosis in SGC-7901 cells. *Harbin Shangye Daxue Xuebao, Ziran Kexueban* 24(4): 396–399.

Wu, J.H., F.R. Chang, K.I. Hayashi, H. Shiraki, C.C. Liaw, Y. Nakanishi, K.F. Bastow, D. Yu, I.S. Chen, and K.H. Lee. 2003. Antitumor agents. Part 218: Cappamensin A, a new in vitro anticancer principle, from *Capparis sikkimensis*. *Bioorganic & Medicinal Chemistry Letters* 13(13): 2223–2225.

Yang, H.F., L. Yu, L. Pang, G.D. Liu, H. Li, and Y. Ji. 2009. Study on chemical constitutions of n-BuOH extract of *capparis spinosa* fruits and vitro antitumor activity. *Harbin Shangye Daxue Xuebao, Ziran Kexueban* 25(3): 264–267.

Yu, L., R. Cui, K. Mo, W. Wang, Y. Ji, and X. Zou. 2008. Preliminary study on apoptosis in HepG-2 of human hepatoma cell line induced by CSS. *Tianjin Zhongyiyao* 25(6): 509–511.

Yu, L., S. Gao, X. Zou, C. Ji, and Y. Ji. 2010. Effects of *Capparis spinosa* L. polar alkaloids on reactive oxygen species, Ca^{2+} and caspase-9,3 in HepG-2 cell. *Zhongguo Yao Cue Za Xhi* 45(23): 1827–1831.

Yu, L., K. Mo, W. Wang, R. Cui, X. Zou, and Y. Ji. 2009. Relationship between apoptosis and [Ca2+]I in HepG2 induced by *Capparis spinosa* alkaloid. *Zhongcaoyao* 40(2): 255–258.

22 Infections and Infestations

While the correctness of warfare as an appropriate metaphor for treating cancer may be arguable, for overcoming biotic invasion in the form of viruses, bacteria, and pathogenic fungi, the metaphor seems more appropriate. Modern medicine has developed a substantial armamentarium of both naturally occurring (as from fungi or bacteria) and synthetic agents that constitute the so-called antibiotics, antivirals, and antifungal drugs. However, the problems with these agents relates first to their potential toxicity and side effects to the host and second to the ability of the pathogens to develop resistance to the specific agents. When such invasion occurs on the community level rather than the individual level and involves higher organisms such as insects, arachnids, destructive fish, reptiles, and even lower mammals, the term *infestation* is used rather than infection, but for both infections and infestations, *Capparis* species may be a source of novel agents with minimal toxicity to humans that still manage to be toxic to the infective or infesting biotic invaders (Figure 22.1). The wide variation in these toxic effects, which undoubtedly developed in the plant to assist with its own defenses against many of the same pathogenic organisms, is depicted clearly in Table 22.1.

These effects again accrue from a great multiplicity of compounds within *Capparis*, but special attention in these actions is due to the alkaloids of *Capparis*, those compounds generally possessing the greatest toxicity in the mix. Although alkaloids occur in all parts of the plant, they are often most prominent in the roots. These preliminary findings also point to a role for *Capparis* as a source of novel antibiotic, antiviral, antifungal, antiprotozoal, insecticidal, and piscicidal agents, while

FIGURE 22.1 Root and bark of *Capparis spinosa* have the most alkaloids and thus the strongest anti-infective effect, but they are difficult to collect in great amounts. The plant in its natural surroundings and a closer view. (25.11.2011, French Hill, Jerusalem, Israel: by Helena Paavilainen.)

FIGURE 22.1 (*Continued*) Root and bark of *Capparis spinosa* have the most alkaloids and thus the strongest anti-infective effect, but they are difficult to collect in great amounts. The plant in its natural surroundings and a closer view. (25.11.2011, French Hill, Jerusalem, Israel: by Helena Paavilainen.)

TABLE 22.1
Anti-infective Properties of *Capparis* spp.

Capparis sp.	Plant Part	Pharmacological Action	Compounds Responsible (If Known)	References
aphylla	Dried MeOH ext.	↓ *Pseudomonas aeruginosa, Escherichia coli, Staphylococcus aureus, Candida albicans*		Dangi and Mishra 2011
decidua	Aq. and solvent exts.	Chloroform, acetone, methanol, and ether exts. ↓ *Lactobacilus* w/MIC 0.028–0.0625 μg/mL, aq. exts. have shown lowest MIC for *Klebsiella pneumoniae* and *Micrococcus luteus*, $CHCl_3$ ext. lowest MBC value for *Lactobacilus acidophilus*: 0.125 μg/mL, intermediate MBC against *Klebsiella pneumoniae* and *Escherichia coli* (0.25 μg/mL), in disk diffusion of growth inhibition zone diam. for MeOH ext. 36 mm in *Bacillus cereus*, 35 mm in *Klebsiella pneumoniae*, 36 mm in *Escherichia coli*, and 31 mm in *S. aureus* at 4 μg/disk, isolated pure compds. CS2, CS3, CS8, CDF1 → > ↓ > tetracycline, ampicillin, ciprofloxacin: ↓ zone diam. 22–40 mm at 4 μg/disk		Upadhyay et al. 2010
decidua	Stem and flower exts.	Dose- and time-dependent insecticidal and oviposition ↓ against pulse beetle *Bruchus chinensis*, LC_{50} ↑ w/↑ polarity of solvent (except aq.), e.g., 4–10 μg at 96 h	Triacontanol, 2-carboxy-1,1-dimethylpyrrolidine, 6-(1-hydroxy-non-3-enyl) tetrahydropyran-2-one	Upadhyay et al. 2006

TABLE 22.1 (*Continued*)
Anti-infective Properties of *Capparis* spp.

Capparis sp.	Plant Part	Pharmacological Action	Compounds Responsible (If Known)	References
moonii		Tuberculostatic		Shah 1962
sepiaria	Pet. ether and EtOH exts. (60–80°)	Best EtOH ext., significant, dose-dependent anthelmintic at 40 mg/mL, comp. to std. drug albendazole		Latha et al. 2009
sinaica	Hydroalcoholic extract	Antiviral activity against HSV *in vitro*		Soltan and Zaki 2009
sp.	Fruits	Antitubercular		Bundeally et al. 1962
spinosa	Aerial parts, acetone ext.	Up to 50% ↓ of larvicidal activity against the yellow fever mosquito, *Aedes aegypti*		Murthy and Rani 2009
spinosa	Aq. ext. of roots, as used in traditional southern Italian folk medicine	↓ growth of bacterium *Deinococcus radiophilus*	Heterocyclic compounds recovered from chloroformic ext.	Boga et al. 2011
spinosa	Aqueous extract	15 μg/mL medium → complete suppression of dermatophytes *Microsporum canis* and *Trichophyton violaceum in vitro*		Ali-Shtayeh and Abu Ghdeib 1999
spinosa	Buds, lyophilate, MeOH ext.	↓ HSV-2 replication in human peripheral blood mononuclear cells (PBMCs), ↓ extracellular virus release, ↑ production of IL-12, IFN-gamma, and TNF-alpha → ↑ immune surveillance by ↑ expression of peculiar proinflammatory cytokines→ putative treatment of HSV-2 infections in immunocompromised pts.	Flavonoids, incl. quercetin and kaempferol glycosides	Arena et al. 2008
spinosa	Dried leaves	Disrupted life cycle in snail parasiste *Bionimphalaria alexandrina* after 1-wk exposure		Mantawy et al. 2004
spinosa	Dried plant powder	Molluscidal effect correlated to ↓ of glucose-6-phosphatase (G-6Pase), fructose 1.6 diphosphatase (FDpase), phosphoenol pyruvate carboxykinase (PEPCK)		Aly Sanaa et al. 2004
spinosa	EtOAc fx. of MeOH ext.	Potently ↓ protozoan *Plasmodium falciparum*, IC_{50} 0.50 μg/mL, w/o ↓ MRC-5 cell line (CC_{50} > 30 μg/mL)		Abdel-Sattar et al. 2010
spinosa	Leaves, ultrasonic-assisted 75% EtOH ext. for 50 min	Antibacterial by agar plate	Flavonoids, alkaloids	Li et al. 2009

(*Continued*)

TABLE 22.1 (*Continued*)
Anti-infective Properties of *Capparis* spp.

Capparis sp.	Plant Part	Pharmacological Action	Compounds Responsible (If Known)	References
spinosa	Live plants	Killed > 50% of *Leishmania major* parasites in sand flies feeding on the plant		Schlein and Jacobson 1994
spinosa	MeOH ext.	↑ actions of chloramphenicol, neomycin, doxycycline, cephalexin, and nalidixic acid against both std. and resistant *Escherichia coli*		Darwish and Aburjai 2010
spinosa	Methanolic extract	Improved antibiotic activity against normal and resistant *E. coli in vitro*		Darwish and Aburjai 2010
spinosa	Petroleum ether, methanol, hexane, butanol, and aq. crude exts. of aerial parts of whole plant	Low to moderate ↓ of four species of bacteria and two fungi compared to antibiotics		Mahasneh et al. 1996
spinosa	Petroleum ether, methanol, hexane, butanol, and aq. crude exts. of aerial parts of whole plant	↓ *E. coli, P. aeruginosa, B. cerreus, S. aureus, C. albicans, A. flavus*, EtOH, butanol > EtOH >> aq., strongly ↓ gram-positive and gram-negative bacteria, ↓ fungi *C. albicans, A. flavus*		Mahasneh 2002
spinosa	Plant in situ	Agglutinates and kills parasite *Leishmania major* when ingested by sand flies, extract kills *L. major in vitro*	Lectins, toxins	Jacobson and Schlein 1999
spinosa	Seeds	Antiviral against human AIDS virus and antifungal	Protein	Lam and Ng 2009
stylosa	Root, MeOH ext.	Piscicidal to predatory fish *Channa punctatus*, LC_{50} 145, 118, 104, 89.33 ppm for 24, 48, 72, and 96 h, resp., speculative effect on respiration and energy metabolism		Ambedkar and Muniyan 2009
stylosa	Roots, ext. w/ dif. solvents in soxhlet	Potent antibacterial effects *in vitro* w/ acetone and MeOH exts., ethylene dichloride ext. weak inhibition, pet. ether and benzene exts. w/o effect		Mohan and Suganthi 1998

working their selective toxic effects within a wide therapeutic window and with a high margin of safety for humans.

The same wide variety of possible anti-infectious uses of *Capparis* spp. can be seen in modern ethnomedicine, with uses that range from an ordinary cold to malaria and leprosy (Table 22.2). Again, the roots and the bark are the most popular parts, but also leaves and fruit are mentioned (Figures 22.2 and 22.3).

TABLE 22.2
Contemporary Ethnomedical Uses of *Capparis* spp. against Infections

Indication	Geography	*Capparis* sp.	Part Used	References
Against infections	Tanzania	*erythrocarpos*	Dried root	Kisangau et al. 2009
Against leprosy	Nigeria	*tomentosa*	Dried bark, dried root	Nwude and Ebong 1980
Against leprosy	Rwanda	*tomentosa*	Dried leaf	Vlietinck et al.1995
Against leprosy	South Africa	*tomentosa*	Fresh root bark	Watt and Breyer-Brandwijk 1962
Against sleeping sickness	Tanzania	*elaeagnoides*	Dried root	Freiburghaus et al. 1996
Against sleeping sickness	Senegal	*tomentosa*	Bark	Kerharo 1974
Against typhoid fever	India	*sepiaria*	Fruit	Singh and Ali 1994
Against typhus	Ethiopia	sp.	Dried root	Abebe 1986
Against chicken pox	Paraguay	*retusa*	Dried root	Arenas 1987
Antisyphilitic	Argentina	*salicifolia*	Root	Filipoy 1994
Against hepatitis	Rwanda	*tomentosa*	Dried leaf	Vlietinck et al.1995
Against malaria		*decidua*		Duke 1993
Against malaria	Rwanda	*tomentosa*	Dried leaf	Vlietinck et al.1995
Against malaria	South Africa	*tomentosa*	Fresh root bark	Watt and Breyer-Brandwijk 1962
Against intermittent fever	India	*decidua*	Dried entire plant	Dhar et al. 1972
Against tuberculosis	Venda	*tomentosa* (in an herbal mixture)	Dried root	Arnold and Gulumian 1984
A pulmonary antihemorrhagic	Peru	*angulata*	Dried bark	Ramirez et al. 1988
Against gonorrhea	Somalia	*tomentosa*	Fresh root	Samuelsson et al. 1991
Against cholera	India	*zeylanica*	Dried leaf	Jain and Verma 1981
Against cholera	India	*zeylanica*		Jain and Tarafder 1970
For a cold remedy	Egypt	*galeata*	Fresh fruit	Goodman and Hobbs 1988
Against colds	Egypt	*spinosa*		Osborn 1968
Against colds	South Africa	*tomentosa*	Dried root	Dekker et al. 1987
Against cough	India	*decidua*	Dried entire plant	Dhar et al. 1972
Against cough	India	*decidua*	Dried root bark	Gaind and Juneja 1969
Against cough	Indochina	*micrantha*		Uphof 1968
Against cough	Israel	*spinosa*	Ripe fruit	Dafni et al. 1984
Against coughs	South Africa	*tomentosa*	Dried root	Dekker et al. 1987

(*Continued*)

TABLE 22.2 (*Continued*)
Contemporary Ethnomedical Uses of *Capparis* spp. against Infections

Indication	Geography	*Capparis* sp.	Part Used	References
Chronic cough	South Africa	*gueinzii*	Fresh root	Watt and Breyer-Brandwijk 1962
Antitussive	Argentina	*tweediana*	Leaf	Filipoy 1994
Expectorant	India	*spinosa*	Root bark	Abrol and Chopra 1962
Expectorant	Turkey	*spinosa*		Steinmetz 1957
Against bronchitis		*micracantha*		*Wealth of India* 1985–1992
Against pleurisy	India	*zeylanica*		Jain and Tarafder 1970
To expel worms from wounds	India	*zeylanica*	Fresh root	Sabnis and Bedi 1983
Anthelmintic	Haiti	*cynophallophora*		Liogier 1974
As an anthelmintic	India	*decidua*	Dried root bark	Gaind and Juneja 1969
Anthelmintic	India	*spinosa*		Uniyal 1990
Anthelmintic	India	*spinosa*	Root bark	Abrol and Chopra 1962
To eliminate worms	Venda	*tomentosa* (in an herbal mixture)	Dried root	Arnold and Gulumian 1984
To clean the nipples before nursing	Tanzania	*kirkii* (in an herbal mixture)	Dried root	Moller 1961
Against boils	Saudi Arabia	*decidua*	Dried aerial parts	Shah et al. 1989
Against boils	Sudan	*decidua*		Broun and Massey 1929
Against boils	India	*horrida*		Uphof 1968
For hemorrhoids and boils as a counterirritant	India	*zeylanica*	Dried leaf	Sebastian and Bhandari 1984
To cause furuncles to come to a head	Paraguay	*tweediana*	Fresh leaf	Arenas 1987

FIGURE 22.2 Modern research shows that both aerial and nonaerial parts of *Capparis spinosa* are effective against infections. (14.5.2010, Jerusalem, Israel: by Helena Paavilainen.)

FIGURE 22.3 Leaves, bark, and roots of *Capparis tomentosa* are considered effective against infections in ethnomedical practice. (Gorongosa National Park, Mozambique, 9.11.2007: by Bart T. Wursten. In Hyde, M.A., B.T. Wursten, P. Ballings, and S. Dondeyne. 2013. Flora of Mozambique: Species Information: Individual Images: *Capparis tomentosa*. http://www.mozambiqueflora.com/speciesdata/image-display.php?species_id=124460&image_id=6, accessed 21 February 2013.)

REFERENCES

Abdel-Sattar, E., L. Maes, and M.M. Salama. 2010. In vitro activities of plant extracts from Saudi Arabia against malaria, leishmaniasis, sleeping sickness and Chagas disease. *Phytotherapy Research* 24(9): 1322–1328.

Abebe, W. 1986. A survey of prescriptions used in traditional medicine in Gondar region, Northwestern Ethiopia: general pharmaceutical practice. *Journal of Ethnopharmacology* 18(2): 147–165.

Abrol, B.K. and I.C. Chopra. 1962. Some vegetable drug resources of Ladakh (Little Tibet). Part I. *Current Science* 31: 324–326.

Ali-Shtayeh, M.S. and S.I. Abu Ghdeib. 1999. Antifungal activity of plant extracts against dermatophytes. *Mycoses* 42(11–12): 665–672.

Aly Sanaa, A., F. Aly Hanan, N. Saba-el-Rigal, and E.M. Sammour. 2004. Induced changes in biochemical parameters of the molluscan tissues non-infected using two potent plants molluscicides. *Journal of the Egyptian Society of Parasitology* 34(2): 527–542.

Ambedkar, G. and M. Muniyan. 2009. Piscicidal activity of methanolic extract of *Capparis stylosa* on the freshwater fish *Channa punctatus* (Bloch). *Internet Journal of Toxicology* 6(1).

Arena, A., G. Bisignano, B. Pavone, A. Tomaino, F.P. Bonina, et al. 2008. Antiviral and immunomodulatory effect of a lyophilized extract of *Capparis spinosa* L. buds. *Phytotherapy Research* 22(3): 313–317.

Arenas, P. 1987. Medicine and magic among the Maka Indians of the Paraguayan Chaco. *Journal of Ethnopharmacology* 21(3): 279–295.

Arnold, H.J. and M. Gulumian. 1984. Pharmacopoeia of traditional medicine in Venda. *Journal of Ethnopharmacology* 12(1): 35–74.

Boga, C., L. Forlani, R. Calienni, T. Hindley, A. Hochkoeppler et al. 2011. On the antibacterial activity of roots of *Capparis spinosa* L. *Natural Product Research* 25(4): 417–421.

Broun, A.F. and R.E. Massey. 1929. *Flora of the Sudan: With a Conspectus of Groups of Plants and Artificial Key to Families*. London: [s.n.].

Bundeally, A.E., M.H. Shah, R.A. Bellare, and C.V. Deliwala. 1962. Antitubercular activity of *Capparis* fruits. *Journal of Scientific & Industrial Research (C)* 21C: 305–308.

Dafni, A., Z. Yaniv, and D. Palevitch. 1984. Ethnobotanical survey of medicinal plants in northern Israel. *Journal of Ethnopharmacology* 10(3): 295–310.

Dangi, K.S. and S.N. Mishra. 2011. *Capparis aphylla* extract compound as antidiabetic and antipathogenic agent. Indian Patent IN 2009DE00742 A.

Darwish, R.M. and T.A. Aburjai. 2010. Effect of ethnomedicinal plants used in folklore medicine in Jordan as antibiotic resistant inhibitors on *Escherichia coli*. *BMC Complementary and Alternative Medicine* 10:9.

Dekker, T.G., T.G. Fourie, E. Matthee, and F.O. Snyckers. 1987. An oxindole from the roots of *Capparis tomentosa*. *Phytochemistry* 26(6): 1845–1846.

Dhar, D.N., R.P. Tewari, R.D. Tripathi, and A.P. Ahuja. 1972. Chemical examination of *Capparis decidua*. *Proceedings of the National Academy of Science, India: Section A* 42(1): 24–27.

Duke, J.A. 1993. Dr. Duke's Phytochemical and Ethnobotanical Databases. [Online database.] USDA–ARS–NGRL, Beltsville Agricultural Research Center, Beltsville, MD. http://www.ars-grin.gov/duke/ethnobot.html (accessed 12 May 2011).

Filipoy, A. 1994. Medicinal plants of the Pilaga of Central Chaco. *J Ethnopharmacol* 44(3): 181–93.

Freiburghaus, F., E.N. Ogwal, M.H.H. Knunya, R. Kaminsky, and R. Brun. 1996. In vitro antitrypanosomal activity of African plants used in traditional medicine in Uganda to treat sleeping sickness. *Tropical Medicine & International Health* 1(6): 765–771.

Gaind, K.N. and T.R. Juneja. 1969. Investigations on *Capparis decidua*. *Planta Medica* 17: 95–98.

Goodman, S.M. and J.I. Hobbs. 1988. The ethnobotany of the Egyptian Eastern desert: a comparison of common plant usage between two culturally distinct Bedouin groups. *Journal of Ethnopharmacology* 23(1): 73–89.

Hyde, M.A., B.T. Wursten, P. Ballings, and S. Dondeyne. 2013. Flora of Mozambique. [Online database.] http://www.mozambiqueflora.com (accessed 21 February 2013).

Jacobson, R.L. and Y. Schlein. 1999. Lectins and toxins in the plant diet of *Phlebotomus papatasi* (Diptera: Psychodidae) can kill *Leishmania major* promastigotes in the sandfly and in culture. *Annals of Tropical Medicine and Parasitology* 93(4): 351–356.

Jain, S.K. and C.R. Tarafder. 1970. Medicinal plant-lore of the Santals. *Economic Botany* 24(3): 241–278.

Jain, S.P. and D.M. Verma. 1981. Medicinal plants in the folk-lore of North-East Haryana. *National Academy of Science Letters (India)* 4(7): 269–271.

Kerharo, J. 1974. Historic and ethnopharmacognosic review on the belief and traditional practices in the treatment of sleeping sickness in West Africa. *Bull Soc Med Afr Noire Lang Fr* 19: 400–410.

Kisangau, D.P., K.M. Hosea, H.V.M. Lyaruu, C.C. Joseph, Z.H. Mbwambo, P.J. Masimba, C.B. Gwandu, L.N. Bruno, K.P. Devkota, and N. Sewald. 2009. Screening of traditionally used Tanzanian medicinal plants for antifungal activity. *Pharmaceutical Biology* 47 (8): 708–716.

Lam, S.K. and T.B. Ng. 2009. A protein with antiproliferative, antifungal and HIV-1 reverse transcriptase inhibitory activities from caper (*Capparis spinosa*) seeds. *Phytomedicine* 16(5): 444–450.

Latha, S., P. Selvamani, T.K. Pal, and J.K. Gupta. 2009. Anthelmintic activity of *Heliotropium zeylanicum* and *Capparis sepiaria* L. *Asian Journal of Chemistry* 21 (7): 5780–5782.

Li, G., Z. Han, J. Li, X. Li, and S. Ababakri. 2009. Extraction and anti-microbial activities of total flavonoid from the leaves of *Capparis spinosa* L. *Shipin Gongye Keji* 30(1): 195–198.

Liogier, A.H. 1974. *Diccionario botanico de nombres vulgares de la española*. Santo Domingo, Dominican Republic: Impresora UNPHU.

Mahasneh, A.M. 2002. Screening of some indigenous Qatari medicinal plants for antimicrobial activity. *Phytotherapy Research* 16 (8): 751–753.

Mahasneh, A.M., J.A. Abbas, and A.A. El-Oqlah. 1996. Antimicrobial activity of extracts of herbal plants used in the traditional medicine of Bahrain. *Phytotherapy Research* 10(3): 251–253.

Mantawy, M.M., M.A. Hamed, E.M. Sammour, and M. Sanad. 2004. Influence of *Capparis spinosa* and *Acacia arabica* on certain biochemical haemolymph parameters of *Biomphalaria alexandrina*. *Journal of the Egyptian Society of Parasitology* 34(2): 659–677.

Mohan, R.T.S. and A.M. Suganthi. 1998. Antibacterial activity of the root extracts of *Capparis stylosa*. *Oriental Journal of Chemistry* 14(1): 137–138.

Moller, M.S.G. 1961. Custom, pregnancy and child rearing in Tanganyika. *Journal of Tropical Pediatrics and African Child Health* 7(3): 66–78.

Murthy, J.M. and P.U. Rani. 2009. Biological activity of certain botanical extracts as larvicides against the yellow fewer mosquito, *Aedes aegypti* L. *Journal of Biopesticides* 2(1): 72–76.

Nwude, N. and O.O. Ebong. 1980. Some plants used in the treatment of leprosy in Africa. *Leprosy Review* 51: 11–18.

Osborn, D.J. 1968. Notes on medicinal and other uses of plants in Egypt. *Economic Botany* 22(2): 165–177.

Ramirez, V.R., L.J. Mostacero, A.E. Garcia, C.F. Mejia, P.F. Pelaez, C.D. Medina, and C.H. Miranda. 1988. *Vegetales empleados en medicina tradicional norperuana*. Trujillo, Peru: Banco Agrario del Peru and Universidad Nacional de Trujillo.

Sabnis, S.D. and S.J. Bedi. 1983. Ethnobotanical studies in Dadra-Nagar Haveli and Daman. *Indian Journal of Forestry* 6(1): 65–69.

Samuelsson, G., M.H. Farah, P. Claeson, M. Hagos, M. Thulin, O. Hedberg, A.M. Warfa, A.O. Hassan, A.H. Elmi, A.D. Abdurahman, A.S. Elmi, Y.A. Abdi, and M.H. Alin. 1991. Inventory of plants used in traditional medicine in Somalia. I. Plants of the families Acanthaceae-Chenopodiaceae. *Journal of Ethnopharmacology* 35(1): 25–63.

Schlein, Y. and R.L. Jacobson. 1994. Mortality of *Leishmania major* in *Phlebotomus papatasi* caused by plant feeding of the sand flies. *American Journal of Tropical Medicine Hygiene* 50(1): 20–27.

Sebastian, M.K. and M.M. Bhandari. 1984. Medico-ethno botany of Mount Abu, Rajasthan, India. *Journal of Ethnopharmacology* 12 (2): 223–230.

Shah, H.H. 1962. Study on tuberculostatic activity of Rudanti (*Capparis moonii*). *Indian Journal of Medical Sciences* 16: 343–346.

Shah, A.H., M. Tariq, A.M. Ageel, and S. Qureshi. 1989. Cytological studies on some plants used in traditional Arab medicine. *Fitoterapia* 60(2): 171–173.

Singh, V.K. and Z.A. Ali. 1994. Folk medicines in primary health care: common plants used for the treatment of fevers in India. *Fitoterapia* 65(1): 68–74.

Soltan, M.M. and A.K. Zaki. 2009. Antiviral screening of forty-two Egyptian medicinal plants. *Journal of Ethnopharmacology* 126(1): 102–107.

Steinmetz, E.F. 1957. *Codex vegetabilis*. Amsterdam, the Netherlands: [s.n].

Uniyal, M.R. 1990. Utility of hitherto unknown medicinal plants traditionally used in Ladakh. *Journal of Research and Education in Indian Medicine* 9(2): 89–95.

Upadhyay, R.K., S. Ahmad, R. Tripathi, L. Rohtagi, and S.C. Jain. 2010. Screening of antimicrobial potential of extracts and pure compounds isolated from *Capparis decidua*. *Journal of Medicinal Plants Research* 4(6): 439–445.

Upadhyay, R.K., L. Rohatgi, M.K. Chaubey, and S.C. Jain. 2006. Ovipositional responses of the pulse beetle, *Bruchus chinensis* (Coleoptera: Bruchidae) to extracts and compounds of *Capparis decidua*. *Journal of Agricultural and Food Chemistry* 54(26): 9747–9751.

Uphof, J.C.Th. 1968. *Dictionary of Economic Plants*. Lehre, Germany: Cramer.

Vlietinck, A.J., L. van Hoof, J. Totte, A. Lasure, D. vanden Berghe, P.C. Rwangabo, and J. Mvukiyumwami. 1995. Screening of hundred Rwandese medicinal plants for antimicrobial and antiviral properties. *Journal of Ethnopharmacology* 46(1): 31–47.

Watt, J.M. and M.G. Breyer-Brandwijk. 1962. *The Medicinal and Poisonous Plants of Southern and Eastern Africa*. 2nd ed. London: Livingstone.

Wealth of India 1985–1992. *The Wealth of India: A Dictionary of Indian Raw Materials and Industrial Products*. New Delhi, India: Publications and Information Directorate, Council of Scientific and Industrial Research.

23 Lepsis

Lepsis is a Greek word meaning seizure in general and, like our own word for seizure, may refer to the seizure of property as well as a seizure affecting the brain caused by an irritable cerebral focus (such as from a tumor), a chronic proclivity (epilepsy), the effects of numerous drugs that may "lower the seizure threshold," fever, behavior and mental techniques that induce automaticity (Lansky and St. Louis 2006), strobe lights, and numerous other causes. Antiepileptic drugs are often effective but may also carry negative unwanted sequelae. Again, the old problem of the organism developing tolerance to these drugs also occurs. Thus, a great need exists in this sector for agents to help to control inflammation and oxidative stress that contribute to brain seizures and hence to curtail the seizures themselves, which are frequently terrifying to those witnessing the events if not to the patients themselves and which may result in permanent brain damage or even death.

Capparis compounds have been shown to exert a number of beneficial effects toward calming the central nervous system (Figure 23.1), including the prevention of seizures in animal models. These various findings are summarized in Table 23.1 and provide a basis for the use of *Capparis* species as a source of novel psychotropic and antiepileptic medications.

Its documented contemporary ethnomedical use for this purpose is rare (Table 23.2).

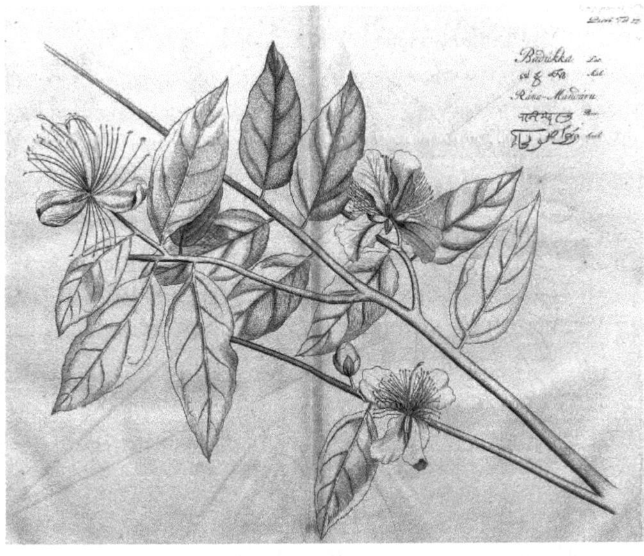

FIGURE 23.1 *Capparis baducca* has been used in Mexico as an antispasmodic. Modern research supports this use. (Illustration: Rheede van Draakestein, H.A. van, J. Casearius, J. Munnicks, A. Syen, J. Commelinus, and A. van Poot. 1686. *Hortus indicus Malabaricus, continens regni Malabarici apud Indos celeberrimi omnis generis plantae rariores, una cum floribus ... delin. ad vivum exhibitas.* Amsterdam: Someren et van Dyck. Vol. 6, t. 57. http://www.plantgenera.org/illustration.php?id_illustration=123019 2009. [Online collection of illustrations.] (accessed 5 February 2013).

TABLE 23.1
Antiepileptic Properties of *Capparis* spp.

Capparis sp.	Plant Part	Pharmacological Action	Compounds Responsible (If Known)	References
baducca	Dried leaves, 5% AcOH ext., neutralized w/Na_4OH, ext. again w/EtOH, and dried	LD_{30} 1.1 g/kg body wt. in mice (10% diln.), hypotensive, anticonvulsant, anti-inflammatory	20% alkaloids	Hashimoto 1967
decidua	Aerial parts including buds and fruits	Dose dependently decreased ($p < 0.05$) the number of animals with convulsions and increased convulsion latency ($p < 0.001$), sedative and anticonvulsant		Goyal et al. 2009
decidua	Stem, EtOH ext. (100 mg/kg p.o.)	↓ locomotor activity, → muscle relaxation, ↓ anxiety	Alkaloids	Garg et al. 2011

TABLE 23.2
Contemporary Ethnomedical Uses of *Capparis* spp. against Epilepsy

Indication	Geography	*Capparis* sp.	Part Used	References
Against epilepsy	South Africa	*oleoides*	Unripe fresh fruit	Watt and Breyer-Brandwijk 1962
Antispasmodic	Mexico	*baducca, flexuosa*		Standley 1920–1926

REFERENCES

Garg, P., K. Sachdeva, and I. Bhandari. 2011. Phytochemical and pharmacological evaluation of *Capparis decidua* (Forsk.) Edgew stem for central nervous system depressant activity. *Pharmacology Online* 2: 146–155.

Goyal, M., B.P. Nagori, and D. Sasmal. 2009. Sedative and anticonvulsant effects of an alcoholic extract of *Capparis decidua. Journal of Natural Medicines* 63: 375–379.

Hashimoto, Y. 1967. Alkaloidal extract from *Capparis baducca*. Japanese Patent JP 42010922 B4.

Lansky, E.P. and E.K. St. Louis. 2006. Transcendental meditation: a double-edged sword in epilepsy? *Epilepsy & Behavior* 9(3): 394–400.

Rheede van Draakestein, H.A. van, J. Casearius, J. Munnicks, A. Syen, J. Commelinus, and A. van Poot. 1686. *Hortus indicus Malabaricus, continens regni Malabarici apud Indos celeberrimi omnis generis plantae rariores, una cum floribus ... delin. ad vivum exhibitas*. Amsterdam: Someren et van Dyck. Vol. 6, t. 57. http://www.plantgenera.org/illustration.php?id_illustration=123019 2009. [Online Collection of Illustrations] (accessed 5.2.2013).

Standley, P.C. 1920–1926. *Trees and Shrubs of Mexico*. Washington, DC: Smithsonian Institution.

Watt, J.M. and M.G. Breyer-Brandwijk. 1962. *The Medicinal and Poisonous Plants of Southern and Eastern Africa*. 2nd ed. London: Livingstone.

24 Hypertension

In many ways, hypertension is a most modern disease, not known in ancient or even medieval times, in fact only known for little more than a century since the invention of the sphygmomanometer. Nevertheless, hypertension is a major health concern since it may lead to cerebral vascular or cardiovascular accidents that can claim and shorten life. The etiology of the disease is complex, but also involves inflammation and undoubtedly oxidative tension. Stress, with its effects on increasing sympathetic tone, including arterial smooth muscle constriction (and the associated increased intra-arterial pressure) also plays a prominent role. While the effect of *Capparis* fractions on this condition has not been extensively studied (Table 24.1), the value of other plant drugs (such as those obtained from olive leaves) and the one study described in this chapter lend hope for a possible beneficial use of *Capparis* drugs and species in the integrative and safe management of hypertension (i.e., high blood pressure).

There are few citations for the ethnomedical use of *Capparis* for hypertension, probably also because hypertension is difficult to diagnose without modern technology as it does not have any obvious symptoms (Table 24.2 and Figure 24.1).

TABLE 24.1
Hypotensive Properties of *Capparis* spp.

Capparis sp.	Plant Part	Pharmacological Action	Compounds Responsible (If Known)	References
aphylla	Aerial parts, crude extract	In anesthetized normotensive rats ↓ mean arterial pressure (MAP), partially ↓ by atropine (2 mg/kg), ↓ phenylephrine (1 μ*M*) and high K^+ (80 m*M*) precontractions in isolated rabbit aortic rings, EC_{50} 0.10 (0.07–0.15) and 1.22 mg/mL (1.00–1.50), resp., suggesting calcium channel blocking (CCB), pretreatment 0.1–1 mg/mL → rightward shift in Ca^{++} conc. response curves, sim. to verapamil, in isolated rat aorta → partial endothelium-dependent L-NAME/atropine-sensitive vasodilation, in guinea-pig atria, ↓ rate and force of spontaneous atrial contractions, w/EC_{50} of 1.35 (1.01–1.79) and 1.60 mg/mL (1.18–2.17), resp., unchanged by atropine (1 μ*M*), highlighting vasodilator and cardiac depressant effects with potential for antihypertensive treatment		Jabbar and Hassan 1993

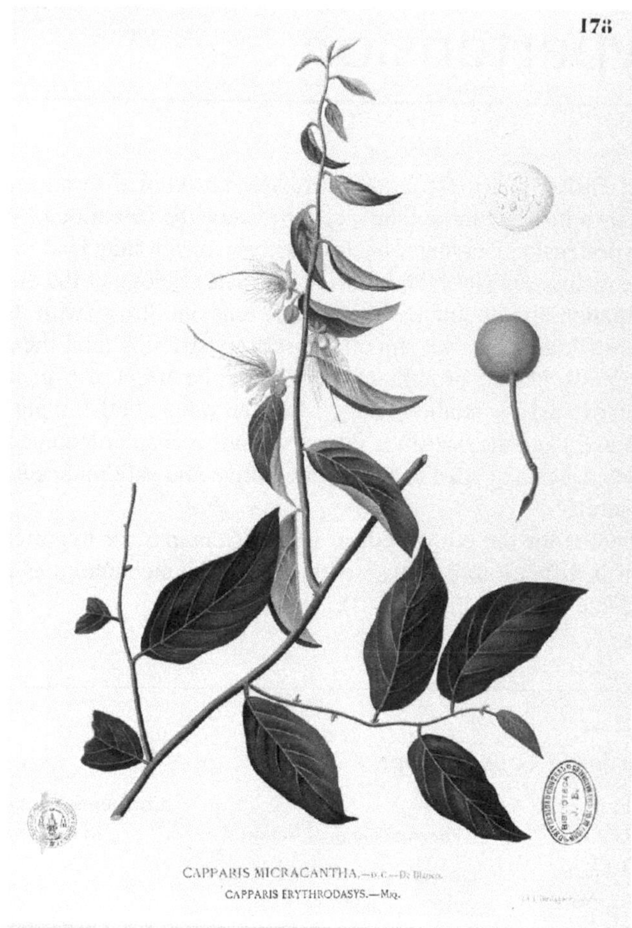

FIGURE 24.1 *Capparis micracantha* is used in ethnomedicine for the treatment of carditis. (Blanco, F.M., I. Mercado, A. Llanos, A. Naves, and C. Fernandez-Villar. 1880–83. *Flora de Filipinas [...] Gran edicion [...] [Atlas I]*. Manila, the Philippines: Plana. Via Wikimedia Commons, http://commons.wikimedia.org/wiki/File%3ACapparis_micracantha_Blanco1.178-original.png.)

TABLE 24.2
Contemporary Ethnomedical Uses of *Capparis* spp. against Hypertension or Cardiac Disorders

Indication	Geography	*Capparis* sp.	Part Used	References
To regulate blood pressure	Peru	*angulata*	Dried bark	Ramirez et al. 1988
Against dropsy	Mexico	*baducca*		Standley 1920–1926
Against dropsy	Haiti	*cynophallophora, gonaivensis*		Pierre-Noël and Brutus 1960, Liogier 1974
Against dropsy	Mexico, Haiti	*flexuosa*		Liogier 1974, Martínez 1933, Standley 1920–1926
Against arteriosclerosis	Saudi Arabia	*deserti*	Leaf	Al-Said 1993
Against carditis		*micracantha*		*Wealth of India* 1985–1992

REFERENCES

Al-Said, M.S. 1993. Traditional medicinal plants of Saudi Arabia. *American Journal of Chinese Medicine* 21(3–4): 291–298.

Blanco, F.M., I. Mercado, A. Llanos, A. Naves, and C. Fernandez-Villar. 1880–83. *Flora de Filipinas [...] Gran edicion [...] [Atlas I]*. Manila, the Philippines: Plana.

Jabbar, S.A. and G.A. Hassan. 1993. Blood pressure lowering effect of the extract of aerial parts of *Capparis aphylla* is mediated through endothelium-dependent and independent mechanisms. *Clinical and Experimental Hypertension* 33(7): 470–477.

Liogier, A.H. 1974. *Diccionario botanico de nombres vulgares de la española*. Santo Domingo, Dominican Republic: Impresora UNPHU.

Martínez, M. 1933. *Las plantas medicinales de México*. Mexico City: Ediciones Botas.

Pierre-Noël, A.V. and T.C. Brutus. 1960. *Les Plantes et les légumes d'Haïti qui guérissent: mille et une recettes pratiques*. Port au Prince, Haiti: Impr. de l'Etat.

Ramirez, V.R., L.J. Mostacero, A.E. Garcia, C.F. Mejia, P.F. Pelaez, C,D. Medina, and C.H. Miranda. 1988. *Vegetales empleados en medicina tradicional norperuana*. Trujillo, Peru: Banco Agrario del Peru and Universidad Nacional de Trujillo.

Standley, P.C. 1920–1926. *Trees and Shrubs of Mexico*. Washington, DC: Smithsonian Institution.

Wealth of India 1985–1992. *The Wealth of India: A Dictionary of Indian Raw Materials and Industrial Products*. New Delhi, India: Publications and Information Directorate, Council of Scientific and Industrial Research.

Section III

Miscellany

25 Reviews

This brings us nearly to the end of our survey of the chemistry and medical uses of the plants and their extracts in the *Capparis* genus. Although, to the best of our knowledge, ours is the first book-length treatment of this subject, the importance of *Capparis* as a source of valuable phytochemicals with medicinal applications has certainly not gone unnoticed, either in the annals of antiquity or in the modern literature of research. We were inspired and frequently benefited from several excellent article-size reviews previously published on the pharmacology of *Capparis*. Table 25.1 summarizes these reviews, which provide excellent sources to individuals seriously interested in the genus who wish to plan their own future courses of study and research.

TABLE 25.1
Reviews of *Capparis* spp.

Capparis sp.	Plant Part	Pharmacological Action	Compounds Responsible (If Known)	References
decidua	Climbing, thorny shrub, densely branched, spinous shrub or tree, up to 6 m in height	Carminative, tonic, emmenagogue, aphrodisiac, alexipharmic, appetizer, antirheumatic, benefits lumbago, hiccough, cough, asthma	Alkaloids, phenols, sterols, glycosides	Singh et al. 2011
decidua	Xerophytic, from dry deserts to cooler mountain terrains as shrubs, trees, or creepers; threatened status in their resp. niches	Bark used for coughs, asthma, inflammation; roots for fever, buds in treatment of boils, in Unani, leaves as appetizer, in cardiac troubles, alveolaris, pyorrhea, leaves and fruit in biliousness; root bark anthelmintic, purgative, pharmacologically: hypercholesterolemic, anti-inflammatory, analgesic, antidiabetic, antimicrobial, antiplaque, antihypertensive, antihelmintic, purgative; female flowers as vegetable, fruits for pickles	Alkaloids, terpenoids, glycosides, fatty acids; buds and fruits: proteins, carbohydrate, minerals, and vitamins	Joseph and Jini 2011
sepiaria, spinosa, tomentosa, zeylanica	Distributed in tropical and subtropical India	Usage in traditional medicine, e.g., Ayurvedha, Siddha, Unani, rasayana, adaptogen to ↑ nonspecific resistance, antioxidant, immunostimulant		Rajesh et al. 2009
sp.		Botany, properties, commercialization, processing, extraction, chemical constituents		Sozzi and Vicente 2006

(Continued)

TABLE 25.1 (*Continued*)
Reviews of *Capparis* spp.

Capparis sp.	Plant Part	Pharmacological Action	Compounds Responsible (If Known)	References
spinosa		Anti-inflammatory, odynolysis, antifungal, hepatoprotective effect, hypoglycemic activity, antioxidation, antihyperlipemia, anticoagulated blood, smooth muscle stimulation, antistress reaction, improves memory; clinical therapy of arthrolithiasis, rheumarthritis, dermatosis	Saccharides and glycosides, flavonoids, alkaloids, terpenoids and volatile oils, fatty acids, and steroids	Yang et al. 2008
spinosa		Antimicrobial and antifungal, anti-inflammatory, antihepatotoxic, antioxidant, hypolipidemic, hypoglycemic, diuretic, treating gout	Volatile (isothiocyanate, thymol, etc.), indole, flavonoid, sitosterol, glucosinolate, novel compounds	Ji et al. 2006
spinosa		Preclin. anti-inflammatory, analgesic, antifungal, hepatoprotective, hypoglycemic, antioxidant, antihyperlipemic, anticoagulant, smooth muscle stimulant, antistress, memory improvement; clin. for arthrolithiasis, rheumarthritis, dermatosis	Saccharides, glycosides, flavonoids, alkaloids, terpenoids, volatile oils, fatty acids, steroids	Yang et al. 2008
spinosa	Distribution	Antihepatotoxic, antidiabetic, hypolipemic, antioxidant, anti-inflammatory, antiallergic	Volatile oils, alkaloids, thioglycosides, terpenes, flavonoids, others	Ao et al. 2007
spinosa	From China NW arid region, Sinkiang desert, fruit, epigean parts	Rheumatic disease, ↓ phlegm, spasm, pain, blood glucose, hypertension		Xie et al. 2007
spinosa	Fruits		Thin-layer chromatography spot of betaine as quality mark, moisture < 11%	Yang et al. 2011
spinosa	Roots		Lipids, carbohydrates	Yili et al. 2006
spinosa	Seeds, subspecies segregation between *C. spinosa* subsp. *rupestris* (as a homogeneous group) and *C. spinosa* subsp. *spinosa* (heterogeneous group)		Oil (ca. 30%), protein (26%)	Tlili et al. 2011

TABLE 25.1 (*Continued*)
Reviews of *Capparis* spp.

Capparis sp.	Plant Part	Pharmacological Action	Compounds Responsible (If Known)	References
spinosa	Winter-deciduous medical shrubs, drought resistant, windbreak, sand fixation, root, stem, leaf	Rapid propagation; biol. characteristic, seed germination, tissue culture		Li et al. 2008
spinosa	Xingiang region, roots, stem, leaves, fruits	↓ blood glucose, fat, ↓ inflammation, oxidn., protects liver		Geng et al. 2007
spinosa, zeylanica, decidua, sepiaria, incanescens, sikkimensis, grandis, moonii, rotundifolia, longispina		Glucocapparin but not its isothiocyanate aglycon highly antibacterial, anti-inflammatory, analgesic, antimicrobial, anthelmintic, hepatoprotective in many spp.	Fatty acids, glucosinolates (glucocapparin) in all spp., triterpenoids (α-amyrin), sterols, β-carotene, saponins in most spp., flavonoids, glycosides, alkaloids in some spp.	Satyanarayana et al. 2008
spp.		Hypoglycemic, anti-inflammatory, paregoric, antibiotic, antitumor	Glucosinolates, alkaloids, flavonoids, terpenoids, org. acids	Yu et al. 2007
spp.		Personal care products for skin conditions		Fox 2010
spp.			Glucosinolates, alkaloids and flavonoids	Zhao et al. 2007
zeylanica		"Rasayana" drug in Ayurveda, antioxidant, antipyretic, analgesic, anti-inflammatory, antimicrobial, immunostimulant	Fatty acids, flavonoids, tannins, alkaloids, E-octadec-7-en-5-ynoic acid, saponins, glycosides, terpenoids, p-hydroxybenzoic, syringic, vanillic, ferulic, and p-coumanic acids	Lather et al. 2010
zeylanica	Bark of root commonly known as Waghata or Vyaghranakhi, thorny, climbing, woody shrub distributed throughout the greater part of India	In indigenous medicine as bitter, cooling, sedative, stomachic, antihydrotic	Alkaloids, xanthoproteins, cysteine, oils, saponins, tannins, glycosides	Rathod 2011

REFERENCES

Ao, M., Y. Gao, and L. Yu. 2007. Advances in studies on constituents and their pharmacological activities of *Capparis spinosa*. *Zhongcaoyao* 38(3): 463–467.

Fox, C. 2010. From caper and prickly pear for dermatitis to dual phase mouthwash: literature review. *Cosmetics and Toiletries* 125(7): 20, 22–26.

Geng, D.S., J.J. Wu, and J.J. Liang. 2007. Chemical constituents and pharmacological activity of *Capparis spinosa* Linn. *Jiefangjun Yaoxue Xuebao* 23(5): 369–371.

Ji, Y.B., S.D. Guo, and C.F. Ji. 2006. Progress of study on *Capparis spinosa*. *Harbin Shangye Daxue Xuebao, Ziran Kexueban* 22(1): 5–10.

Joseph, B. and D. Jini. 2011. A medicinal potency of *Capparis decidua*—a harsh terrain plant. *Research Journal of Phytochemistry* 5(1): 1–13.

Lather, A., A.K. Chaudhary, V. Gupta, P. Bansal, and R. Bansal. 2010. Phytochemistry and pharmacological activities of *Capparis zeylanica*: an overview. *International Journal of Research in Ayurveda and Pharmacy* 1 (2): 384–389.

Li, M., W. Liu, L. Gan, Y. Wang, and L. Yu. 2008. Progress on the botanical characteristic of *Capparis spinosa* L. *Xiandai Shengwuyixue Jinzhan* 8(11): 2194–2197, 2178.

Rajesh, P., P. Selvamani, S. Latha, A. Saraswathy, and V. Rajesh Kannan. 2009. A review on chemical and medicobiological applications of Capparidaceae family. *Pharmacognosy Reviews* 3(6): 378–387.

Rathod, V.S. 2011. Ethnopharmacognostical studies of *Capparis zeylanica* (Linn.): a potential psychoactive plant. *Journal of Pharmacy Research* 4(3): 910–911.

Satyanarayana, T., A.A. Mathews, and P. Vijetha. 2008. Phytochemical and pharmacological review of some Indian *Capparis* species. *Pharmacognosy Reviews* 2(4): Suppl. 36–45.

Singh, P., G. Mishra, S. Sangeeta, J.K.K. Shruti, and R.L. Khosa. 2011. Traditional uses, phytochemistry and pharmacological properties of *Capparis decidua*: an overview. *Pharmacia Lettre* 3(2): 71–82.

Sozzi, G.O. and A.R. Vicente. 2006. Capers and caperberries. In K.V. Peter, ed., *Handbook of Herbs and Spices*. Cambridge, UK: Woodhead, 3: 230–256.

Tlili, N., E. Saadaoui, F. Sakouhi, W. Elfalleh, M. El Gazzah, S. Triki, and A. Khaldi. 2011. Morphology and chemical composition of Tunisian caper seeds: variability and population profiling. *African Journal of Biotechnology* 10(11): 2112–2118.

Xie, L.Q., S.Y. Xue, X. Yan, and C. Tian. 2007. Chemical constituents and pharmacological study of Uygur herb medicine *Capparis spinosa* L. *Xibei Yaoxue Zazhi* 22(4): C3–C4.

Yang, T., Y. Liu, C. Wang, and Z. Wang. 2008. Advances on investigation of chemical constituents, pharmacological activities and clinical applications of *Capparis spinosa*. *Zhongguo Zhongyao Zazhi* 33(21): 2453–2458.

Yang, W., S. Abudou, Y. Chen, J. He, and M. Manerhaba. 2011. Quality standards for fruits of *Capparis spinosa* L. *Shizhen Guoyi Guoyao* 22(1): 133–134.

Yili, A., T. Wu, B.T. Sagdullaev, H.A. Aisa, N.T. Ul'chenko, A.I. Glushenkova, and R.K. Rakhmanberdyeva. 2006. Lipids and carbohydrates from *Capparis spinosa* roots. *Chemistry of Natural Compounds* 42(1): 100–101.

Yu, Y., Z.H. Wu, and L.J. Wu. 2007. Progress on chemical constituents and pharmacological activities of plants of *Capparis*. *Shenyang Yaoke Daxue Xuebao* 24(2): 123–128.

Zhao, G., Z. Yin, and J. Dong. 2007. Progress of chemical constituents and pharmacological activities of plants of *Capparis*. *Zhongnan Yaoxue* 5(3): 250–254.

26 Propagation

Although *Capparis* genus in general and especially *C. spinosa* in particular evolved in the wild in numerous niches over millennia, most recently the genus and the species have received another push through technological progress driven by economic incentives in the realm of artificial propagation. In short, these methods have allowed for nodal cuttings of caper plants to be grown *in vitro* in tissue culture, with the possibility of upscaling to large industrial production. Once these young shoots or roots are established in such artificial milieux, they might be cultivated further to some form of artificial maturity, after which the completely artificially grown product might be diverted to chemical extraction or other methods by which the pharmacological richness of this biomass can be concentrated and purified to produce drugs, or even foodstuffs. More commonly, once the young plants are established *in vitro*, they are then transferred to plantations to continue their growth to maturity as they might in the wild, by normal processes of rooting, flowering, fruiting, and the like. Then, from these mature plants, regular methods of harvesting and postharvesting treatment can be instituted to obtain the value parts of the plant for commerce (Figure 26.1). Although this field is only in its infancy, substantial inroads have already been laid and are summarized in Table 26.1. In the future, pharmaceutical pioneers seeking to cultivate *Capparis* species for industrial applications may benefit greatly from these preliminary studies.

FIGURE 26.1 Young cotyledons of *Capparis sola*. (By C. E. Timothy Painc.)

TABLE 26.1
Propagation Science for the Genus *Capparis*: 2012

Capparis sp.	Finding	References
cartilaginea	Germinated without treatment, compared to 10 other species from same semiarid area requiring acid scarification + growth hormone	Raole et al. 2010
decidua	Method for micropropagation of mature trees w/multiple shoots of nodal explants on Murashige and Skoog (MS) medium + 0.1 mg L^{-1} NAA + 5.0 mg L^{-1} BAP + additives (50 mg L^{-1} ascorbic acid and 25 mg L^{-1} each of adenine sulfate, L-arginine, and citric acid) at 28 ± 2°C, 12 h/day photoperiod w/35–40 μmol $m^{-2}s^{-1}$ photon flux density, shoots multiplied by (1) subculture of nodal shoot segments onto MS + 0.1 mg L^{-1} IAA + 1.0 mg L^{-1} BAP + additives and (2) repeated transfer of original explant onto MS + 0.1 mg L^{-1} IAA + 2.5 mg L^{-1} BAP + additives, at 3-wk intervals → 60–70% shoots rooted when pulse treated w/100 mg L^{-1} IBA in half-strength MS liq. medium 4 h, then transferred onto hormone-free half-strength agar-gelled MS basal salt medium, incubated in dark at 33 ± 2°C	Deora and Shekhawat 1995
spinosa	Highest germination of 62% when seeds pretreated w/H_2SO_4 40 min, then 400 ppm gibberellic acid soaking (2 h), from interventions, including hot water treatment, scarification, stratification, concd. acids (H_2SO_4, HNO_3, and HCl), gibberellic acid, potassium nitrate, EtOH, acetone, and gamma-ray irradn. to dried seeds in germination chambers 20–28 days at 25 ± 2°C under continuous light (20 h) photoperiod following treatments	Bhoyar et al. 2010
spinosa	*In vitro* propagation subsp. *rupestris* w/single nodal cuttings; multiple shoots on MS medium supplemented w/1 mg/L zeatin, optimized by subculturing shoot segments w/2–3 nodes/6 wk; proliferation maintained during nine subcultures > 20 new shoots/explants, 92% shoot rooting response after 4 h pulse treatment in darkness w/100 mg/L indole acetic acid, followed by culture on solid half-strength MS basal medium → potential for large-scale multiplication	Chalak and Elbitar 2006
spinosa	Tissue culture optimum explants rinsing for 8 h and 0.1% $HgCl_2$ for 12 min, medium MS + 6-BA 0.6 mg/L + NAA 0.1 mg/L for primary and second culture, medium MS + 6-BA 0.6 mg/L + 2,4-D 1.0 mg/L for proliferation, for rooting MS + IBA 0.8 mg/L + 300 mg/L activated carbon → rapid proliferation of seeds for large-scale seedlings	Ma et al. 2010
spinosa	Scarification through seed coat rupture → ↑ germination percentage, *in vitro* obtained seedlings → micropropagation protocol for caper, usual percentage of germination of caper seeds low	Germana and Chiancone 2009
spinosa	Seeds germinated *in vitro* 6 wk, seedlings cut into pieces, propagated on medium w/4 mg/L kinetin	Horshati and Jambor-Benczur 2006
spinosa	Soaking in H_2SO_4 15–30 min → ~ 40% seed germination, contact w/100 ppm gibberilin in water → ↑germination to ~ 80%	Orphanos 1983
spinosa	For germination, optimum presoaking with sul. ac. 70 min, optimum treatment with gibberilin 700 mg/L, ($p < 0.05$)	Sun and Ma 2010
spinosa	*In vitro* thermotherapy combined w/shoot tip tissue culture → 89–93% elim. of caper latent virus (CapLV) by RT-PCR and improved survival of shoot tips 60% (Salina) and 90% (Pantelleria)	Tomassoli et al. 2008
spinosa	*In vitro* propagation w/simple nodal cuttings, supplemented w/1 mg/L zeatin, shoot multiplication optimized by subculturing shoot segments w/2–3 nodes q 6 wk → mean rate > 20 new shoots/explant, 92% rooting response after 4-h pulse treatment in darkness w/100 mg/L IAA soln., followed by culture on solid half-strength MS basal medium →→ potential large-scale multiplication	Chalak and Elbitar 2006

REFERENCES

Bhoyar, M.S., G.P. Mishra, R. Singh, and S.B. Singh. 2010. Effects of various dormancy breaking treatments on the germination of wild caper (*Capparis spinosa*) seeds from the cold arid desert of trans-Himalayas. *Indian Journal of Agricultural Sciences* 80: 621–625.

Chalak, L. and A. Elbitar. 2006. Micropropagation of *Capparis spinosa* L. subsp. *rupestris* Sibth. & Sm. by nodal cuttings. *Indian Journal of Biotechnology* 5: 555–558.

Deora, N.S. and N.S. Shekhawat. 1995. Micropropagation of *Capparis decidua* (Forsk.) Edgew.—a tree of arid horticulture. *Plant Cell Reports* 15(3–4): 278–281.

Germana, M.A. and B. Chiancone. 2009. In vitro germination and seedling development of caper (*Capparis spinosa* L.) mature seeds. *Acta Horticulturae* 839 (Issue *Proceedings of the First International Symposium on Biotechnology of Fruit Species 2008*): 181–186.

Horshati, E. and E. Jambor-Benczur. 2006. In vitro propagation of *Capparis spinosa*. *Acta Horticulturae* 725 (Issue *Proceedings of the Fifth International Symposium on In Vitro Culture and Horticultural Breeding 2004*): 151–154.

Ma, S., T. Lu, A. Zhang, Y. Wang, and L. Zhou. 2010. Tissue culture and rapid propagation of seeds of Uyghur traditional herbal. *Zhongyaocai* 33: 1833–1836.

Orphanos, P.I. 1983. Germination of caper (*Capparis spinosa* L.) seeds. *Journal of Horticultural Science* 58: 267–270.

Raole, V.M., A.G. Joshi, S.K. Garge, and R.J. Desai. 2010. Seed germination of selected taxa from Kachchh desert, India. *Notulae Scientia Biologicae* 2: 41–45.

Sun, Q. and M. Ma, M. 2010. Effects of sulphuric acid and gibberellins on seed germination of *Capparis spinosa* seeds. *Shihezi Daxue Xuebao, Ziran Kexueban* 28: 144–146.

Tomassoli, L., G. Di Lernia, A. Tiberini, G. Chiota, E. Catenaro, and M. Barba. 2008. In vitro thermotherapy and shoot-tip culture to eliminate caper latent virus in *Capparis spinosa*. *Phytopathologia Mediterranea* 47: 115–121.

27 Fermentation

The table that follows is a review of the literature relating to fermentation of caper buds and fruits, the latter also known as caperberries or capsules. Although the caper leaves can also be subjected to the same process with similar results, these studies are related to either the buds or the (caper) berries.

The process is always related to lactic acid fermentation, which is accomplished primarily through selected lactic acid producing bacilli, also known as lactobacilli. NaCl may be used up to 10% during fermentation, and higher during storage. NaCl above 7% retards fermentation. NaOAc (sodium acetate, sodium ethanoate) may be used to buffer the solution and so increase the strength of fermentation. The possibility for contamination of the ferments with human, other mammalian, or avian faeces is an ever-present danger that must be heeded. Strict cleanliness of facilities and workers must be heeded when engaging in this activity. Still, though, fermentation can ensue in water with no salt whatsoever, with the possibility for preservation with salt still available further down the line.

There are pleasures in fermentation, but it is not for the squeamish. The results and possibilities are infinite, and the range of their pleasures astonishing.

TABLE 27.1
Fermentation Methods for *Capparis* spp.

***Capparis* sp., plant part(s)**	**Method, Results**	**References**
ovata, buds	Small (<8 mm diam) and big (8–13 mm diam), fermented in control, citric acid, lactic acid, yoghurt added, starter culture, *Lactobacillus plantarum* inoculated brines: small higher dry material, crude protein, crude fiber, energy than big, raw buds: high K, P, Ca, Mg, Na, control highest Al, B, Ca, Cu, Fe, Mg, Ni, P, Sr, V, Zn, *L. plantarum* inoculated: highest Al, B, Cr, Cu, Fe, Mg, Mn, Ni, Sr, Zn	Arslan and Ozcan 2007
ovata, buds	Citric acid and lactic acid added brine types best in 10% brines also including yogurt and starter culture (*Lactobacillus plantarum*) inoculation types from physiochemical, microbiological and sensory assessments, → pH 2.14–2.60, acid 1.2–1.7%, *L. plantarum* inoculation or yogurt addn. → ↓ % acid, taste in small and large buds +, but small buds harder ($P < 0.01$)	Arslan et al. 2008
ovata, *spinosa*, buds	When fermented in 5% and 10% brines, then subsequently stored in brines of 2% or 6% but with acetic acid, citric acid, sugar, or tarragon ext. added, pasteurized or not at 80°C, with sensory evaluation at intervals up to 360 days, *C. ovata* buds fermented in 5% brine and stored in 2% brine + 1% acetic acid, 6% brine + 1% acetic acid, or 6% brine + citric acid judged best, *C. ovata* buds retained bright yellowish color during storage, for all samples, firmness gradually ↓ during storage, *C. spinosa* ↓ > *C. ovata*, bud texture significantly protected by pasteurization, sugar, and tarragon ext.	Ozcan 2001
ovata, *spinosa*, fruits	0.7–1.9 cm in diam. collected and brined, crude oil, fiber, pH, starch, Mn, Zn of *C. spinosa* > of *C. ovata*, with less oleic and linoleic acids, crude protein, oil, fiber and energy, reducing sugars, starch, total carotenoids, ether-sol. ext., hardness, K, P, Cu, Mn, Zn, palmitic and oleic acids of the fermented products < raw fruits, major fatty acids in both species and products linoleic, oleic, linolenic and palmitic, *C. spinosa* fruits more suitable for use due to more nutrients	Ozcan 1999a

(*Continued*)

TABLE 27.1
Fermentation Methods for *Capparis* spp.

Capparis sp., plant part(s)	Method, Results	References
ovata, *spinosa*, fruits	0.7–1.9 cm diam. collected in June 1996, pickled 1 mo in 5% and 10% brine, lactic acid bacteria (LAB) ↑ during fermn. in 5% brine, but not after 15 days in 10% brine, in both caper species, best for LAB activity 5–10% NaCl, best for product color, flavor, acidity, pH, and LAB activity in brine 20–25 days fermentation, all maintained w/15% old or fresh brine during storage, color esp. of *C. ovata*, stored with old brine darker than of fresh berries	Ozcan 1999b
sp., fruits	Home fermented contaminated w/17 enterococci isolates: *Enterococcus faecium* (9 isolates), *E. fecalis* (4), *E. avium* (3), some w/α- and/or β-activity, broad antibiotic resistance → CAUTION: CLEANLINESS!	Perez-Pulido et al. 2006
sp., fruits	Spontaneous lactic acid fermentation following immersion in water, isolates by gene sequencing: *Lactobacillus plantarum* (37 isolates), *L. paraplantarum* (1 isolate), *L. pentosus* (5 isolates), *L. brevis* (9 isolates), *L. fermentum* (6 isolates), *Pediococcus pentosaceus* (14 isolates), *P. acidilactici* (1 isolate), *Enterococcus faecium* (2 isolates)	Perez-Pulido et al. 2005a
spp., fruits	Fresh caperberry fruits 0.6–1.8 cm diam. collected from Konya (Selcuklu), Turkey fermented in 8 % brines at 30°C for 45 days utilized in paste of crushed fermented fruits, virgin olive oil, spices, yogurt, with total minerals: Ca 2341.6 mg/kg, K 13583, P 321.4, Na 41257, Cu 23.7, Fe 17.4, Mn 32.4 and Zn 173.1	Ozcan and Duman 2010
spinosa	In caper fermentation: function f that (Y = f(X1, X2, ..., Xn)) between magnitude Y, i.e. ↓ of pH (called response), and variables Xi, i.e., brine, lactic acid, citric acid, "lactic ferment factors"; brine, lactic acid and citric acid → significant effect on ↓of pH, but not by lactic ferment; interactions between brine and lactic acid, between brine and lactic ferment, between lactic acid with citric acid and between lactic acid with lactic ferment → significant effects on ↓ of pH ($p < 0.0001$), fermented at ~30°C, between June and July in Marocapres-Fez lab	Douieb et al. 2010
spinosa, bud	Homofermentative metab. key for successful starter cultures	Barba and Diaz 1995
spinosa, buds	Pickling → ↓ N, P, Ca, Mg, K, acid-detergent fiber, thiamin, riboflavin, rutin, amino acids, ↑ Na, Fe, S	Nosti Vega and De Castro Ramos 1987
spinosa, buds	Fermented in 10% NaCl of brine for 2 mo at 30°C yielded *Lactobacillus* spp. *L. plantarum, L. casei, L. fermentum* and *L. brevis* dtnd. by plasmid, DNA and protein profiling	Ozkalp et al. 2009
spinosa, fruit	58 lactobacilli from fermented caper fruit: *Lactobacillus plantarum, L. paraplantarum, L. pentosus, L. brevis, L. fermentum* with active leucine aminopeptidase, acid phosphatase, β-galactosidase, β-glucosidase. Many strains able to degrade raffinose and stachyose. Phytase activity and bile salt hydrolase in certain strains of *L. plantarum* only → caper lactobacilli: metabolic diversity, functional potential: industrial caper fermentation	Perez-Pulido et al. 2007
spinosa, fruits	Ferment in 0, 4, 7 and 10% NaCl with 7 and 10% → delay and NaOAc brine buffering → stronger → →10% NaCl suits long term storage, lactic acid bacteria key	Sanchez et al. 1992
spinosa, fruits	*Lactobacillus plantarum* (43 isolates), *L. brevis* (9 isolates) and *L. fermentum* (6 isolates) from spontaneous fermentation in water resistant to vancomycin and teicoplanin (MIC > 16 μg/mL), ampicillin, erythromycin, chloramphenicol, gentamicin, streptomycin, quinupristin/dalfopristin, resistance to ciprofloxacin (MIC > 2 μg/mL) in all isolates of *L. brevis* and *L. fermentum* and in most isolates of *L. plantarum*	Perez-Pulido et al. 2005b

TABLE 27.1
Fermentation Methods for *Capparis* spp.

Capparis sp., plant part(s)	Method, Results	References
spinosa, leaf	Fungi of 35 species of 19 genera of fungi isolated, most during May, least during July, *Ulocladium* (4 species) most prominent genus, *Alternaria alternata, Aspergillus flavus, A. niger, Penicillium chrysogenum* isolated during all seasons	Bokhary et al. 2000

REFERENCES

Arslan, D. and M.M. Ozcan. 2007. Effect of some organic acids, yoghurt, starter culture and bud sizes on the chemical properties of pickled caper buds. *Journal of Food Science and Technology* 44: 66–69.

Arslan, D., A. Unver, and M. Ozcan. 2008. Determination of optimum fermentation quality of capers (*Capparis ovata* Desf. var. *canescens*) in different brine conditions. *Journal of Food Processing and Preservation* 32: 219–30.

Barba, J.L.R. and R.J. Diaz. 1995. Capers and olives. Strain selection criteria for controlled fermentation and quality. In G. Novel and J.F. Le Querler, eds., *Bacteries Lactiques, Actes du Colloque LACTIC 94, Caen, Fr., Sept. 7–9, 1994:* 153–61.

Bokhary, H.A., S. Al-Sohaibany, Q.H. Al-Sadoon, and S. Parvez. 2000. Fungi associated with *Calotropis procera* and *Capparis spinosa* leaves. *Journal of King Saud University, Science* 12: 11–23.

Douieb, H., M. Benlemlih, and F. Errachidi. 2010. Improvement of the lactic acid fermentation of capers (*Capparis spinosa* L) through an experimental factorial design. *Grasas y Aceites (Sevilla, Spain)* 61: 398–403.

Nosti Vega, M. and R. De Castro Ramos. 1987. Effect of pickling on the composition of capers. *Grasas y Aceites (Sevilla, Spain)* 38: 173–5.

Ozcan, M. 1999a. The physical and chemical properties and fatty acid compositions of raw and brined caper berries (*Capparis* spp.). *Journal of Agriculture and Forestry* 23(3): 771–6.

Ozcan, M. 1999b. Pickling and storage of caperberries (*Capparis* spp.). *Zeitschrift für Lebensmittel-Untersuchung und -Forschung A: Food Research and Technology* 208: 379–82.

Ozcan, M. 2001. A research note. Pickling caper flower buds. *Journal of Food Quality* 24: 261–9.

Ozcan, M.M. and E. Duman. 2010. Sensory evaluation, microbiological, chemical properties and mineral contents of pickling caperberries (*Capparis* spp.) paste. *Asian Journal of Chemistry* 22: 6600–6604.

Ozkalp, B., M. Onur Aladag, Z. Ogel, M. Ozcan, and B. Celik. 2009. Determination of some metalic antimicrobial activities and plasmid and DNA profiles of Lactobacillus strains isolated from fermented caper pickle. *World Applied Sciences Journal* 6(3): 347–54.

Perez-Pulido, R., H. Abriouel, N. Ben Omar, R. Lucas, M. Martinez-Canamero, and A. Galvez. 2006. Safety and potential risks of enterococci isolated from traditional fermented capers. *Food and Chemical Toxicology* 44: 2070–77.

Perez-Pulido, R., N. Ben Omar, H. Abriouel, R. Lucas Lopez, M. Martinez-Canamero, J.-P. Guyot, and A. Galvez. 2007. Characterization of lactobacilli isolated from caper berry fermentations. *Journal of Applied Microbiology* 102: 583–90.

Perez-Pulido, R., N. Ben Omar, H. Abriouel, R.L. Lopez, M. Martinez Canamero, and A. Galvez. 2005a. Microbiological study of lactic acid fermentation of caper berries by molecular and culture-dependent methods. *Applied and Environmental Microbiology* 71: 7872–9.

Perez-Pulido, R., N. Ben Omar, R. Lucas, H. Abriouel, M. Martinez Canamero, and A. Galvez. 2005b. Resistance to antimicrobial agents in lactobacilli isolated from caper fermentations. *Antonie van Leeuwenhoek* 88: 277–81.

Sanchez, A.H., A. De Castro, and L. Rejano. 1992. Controlled fermentation of caperberries. *Journal of Food Science* 57(3): 675–8.

28 Recipes

Please do not be discouraged by the lack of quantitation to which you may be accustomed. These recipes need your own intuitive guidance concerning amounts based on your experience and common sense. We welcome your feedback! Thank you.

Caper Sauce. Pound pre-fermented caperberries into a pulp. Add any or all of olive oil, yogurt, crushed garlic, squeezed and grated lemon, sesame tahini, ground rosemary, hot chili sauce. Blend together well with a fork. May be used as an accoutrement to grilled fish or tofu.

Caper Pasta Sauce. Cube fresh organic tomatoes and simmer alone for several hours. Add pickled caper buds and diced caperberries to the tomatoes along with crushed garlic, parsley, sage, rosemary, thyme, hyssop, oregano. Allow to simmer for another hour, add salt or umeboshi plum sauce to taste.

Caper Nut Loaf. Combine chopped, fresh, raw, organic pine nuts, cashews, Brazil nuts, filberts, walnuts, almonds, pistachios. Add to the mix a little black cumin seeds (*Nigella sativa*), sesame tahini and raw organic pristine egg yolk. Wheat germ and a bit of organic cooked grains such as buckwheat or brown rice can be used to thicken the mix. If moisture is needed, try a bit of rice wine (sake). Keep mixing and adding ingredients until the desired consistency is attained. Mix in a few choice, petite fermented or raw caper buds. Spoon carefully into a glass baking dish and cover with the lid or with aluminum foil if none. Bake around 200°C or 350°F for 20–25 minutes. Any of the above ingredients can be left out if not available or undesired. Serve with a Caesar salad or similar.

Caper Lamb. Obtain a fresh, young neck of lamb. Wash and dry and marinate in red wine or red wine vinegar, to which has been added sprigs of fresh rosemary, sage or thyme for 24 hours. Remove from the marinade and simmer on a small fire with a heat diffuser for six hours. Add pickled or fermented caper berries and/or capers to the pot throughout the cooking as desired.

Caper Chicken. Obtain a nice organic chicken. Wash well with water and remove any remaining feathers. Dry, and rub the skin with soft pickled caperberries, leaving remaining caperberry material in the cavity. Add chopped carrots and parsley roots and greens to the cavity as well and bake covered for an hour or more. Serve with red wine.

Caper Tofu. Slice the tofu lengthwise into three flat sheets. Marinate briefly in soy sauce and sake, then place on a baking sheet. Press fermented caper buds gently into the tofu-flesh. Sprinkle with a *za'atar* herb mix (hyssop and garden herbs) and bake at 300°F (180°C) for 25 minutes. Goes with quinoa.

Caper Tempeh. Obtain some "coconut water" (or cream if desired), lemon grass, fresh lemon juice, garlic, ginger and cayenne pepper to go with the tempeh. Chop the tempeh into cubes and stew with the other ingredients including caper buds or chopped caperberries if no bud available for an hour, adding soy sauce (tamari, shoyu) according to personal preference during the cooking. This one requires organic short grain brown rice pressured and cooked for one hour as a side dish.

Caper Fish. Wash and dry the fish filet (salmon is nice but also many others will do), dip in shoyu soy sauce, place on baking sheet and dot with caper buds. Bake at a medium heat for half an hour or until done (not burned!). Serve with lemon wedges.

Caper Fish Balls. Procure ground live carp or other fish as available from the fishmonger or raise it and grind it yourself! Mix with egg yolk (a little), chopped parsley, ground black pepper, wheat germ or whole spelt organic matzo meal. Mix the caper buds into the mix, press into balls and place in boiling water for 20 minutes. Serve on a bed of arugula with horseradish.

Caper-Pasto. Mix the pickled caper buds and/or sliced caperberries with black olives, chopped tomatoes, sliced cucumbers, iceberg or other lettuce and chunks of feta cheese. Toss with virgin olive oil and lemon as desired.

Caper-Shrooms. Cut into large pieces the trimmed and washed, preferably wild and properly identified mushrooms. Cultivated shitakes or portobellos also work. Stew with a little chopped parsley and capers (buds or berries) in dry red wine for 15 minutes. Salt or ume to taste.

Caper Chutney. Into a small mixing bowl add chopped pickled caperberries, caper buds, diced onion and garlic, hot red cayenne pepper, fresh ginger, coriander leaves and lemon juice. Honey can be added sparingly according to taste. A welcome accompaniment for everything, from falafel on up.

Caper-Touile. Cut washed and dried organic zucchinis lengthwise into quarters and chop further into cubes. Do the same with organic tomatoes and leeks. Combine and simmer for a couple hours at least with peeled garlic cloves and capers, buds preferred. For this one, fresh sweet corn is best by its side.

Caper Caesar. Break a well-separated egg yolk into a large mixing bowl. Add crushed garlic, ground black pepper, pickled caper buds, and the traditional anchovies in olive oil, extra olive oil to taste, fresh squeezed lemon juice, and touch of lemon zest. Mix well, add washed and dried romaine lettuce leaves and toss. Additional black pepper for grinding at the table is desirable.

Basic Brine Pickle. Place fresh, washed caper buds or berries in a sterile glass jar containing a 9 percent salt solution. Keep covered in a cool dark place for two to six weeks. Can be rinsed in fresh water before use if you are salt sensitive. Store the jar in the refrigerator and exercise maximum cleanliness precautions when dispensing the pickles.

Basic Vinegar Pickle. Just put fresh-picked caper buds and/or berries into a clean glass jar and cover with apple cider or balsamic wine vinegar. You can eat them in a week or two, but they will last nicely on the shelf with no refrigeration for a year or more. Try adding whole peeled garlic cloves to the mix as a variation. No problem to mix buds and caperberries in the same jar. They get along fine.

Caper Pizza. Lay out your favorite dough, sauce and toppings. Include caper buds and/or sliced buds on the top. Of course, best with anchovies!

29 Breaking Advances in Medical Capparology

TABLE 29.1
Advances in Medical Capparology in 2012–2013

Capparis species and part(s)	Finding	References
decidua, stems	Ethanolic extract → ↓ proliferation of human A549 lung cancer cells	Rathee et al. 2012
ovata	↓ seizures and oxidative brain injury in pentylentetrazol induced epileptic rats	Nazıroğlu et al. 2013
ovata, buds, fruits	Methanol extracts → anti-inflammatory and antithrombotic effects *in vivo*	Bektas et al. 2012
sinaica	EtOAc and 25% MeOH in EtOAc extracts inhibit bird flu virus	Ibrahim et al. 2013
spinosa	A Se capparis polysaccharide complex (SeCSPS) created, dose dependently inhibiting human gastric cancer cell SGC-7901 proliferation	Ji et al. 2012a
spinosa, bud, fruit	Food allergy	Alcántara et al 2013
spinosa, fruit	Polysaccharide effective against H22 mouse hepatoma cells	Ji et al. 2012b
thonningii, roots	Improved memory deficit and oxidative brain damage in scopolamine treated mice by Morris water maze test and enzyme studies → → potential for Alzheimer's disease	Ishola et al. 2013

REFERENCES

Alcántara, M., M. Morales, and J. Carnés. 2013. Food allergy to caper (*Capparis spinosa*). *J Investig Allergol Clin Immunol* 23: 67–69.

Bektas, N., R. Arslan, F. Goger, N. Kirimer, and Y. Ozturk. 2012. Investigation for anti-inflammatory and anti-thrombotic activities of methanol extract of *Capparis ovata* buds and fruits. *J Ethnopharmacol* 142: 48–52.

Ibrahim, A.K., A.I. Youssef, A.S. Arafa, and S.A. Ahmed. 2013. Anti-H5N1 virus flavonoids from *Capparis sinaica* Veill. *Nat Prod Res* 2013 May 7. [Epub ahead of print.]

Ishola, I.O., O.O. Adeyemi, E.O. Agbaje, S. Tota, and R. Shukla. 2013. *Combretum mucronatum* and *Capparis thonningii* prevent scopolamine-induced memory deficit in mice. *Pharm Biol* 51: 49–57.

Ji, Y.B., F. Dong, L. Lang, L.W. Zhang, J. Miao, Z.F. Liu, L.N. Jin, and Y. Hao. 2012a. Optimization of synthesis, characterization and cytotoxic activity of Seleno-*Capparis spionosa* L. polysaccharide. *Int J Mol Sci* 13: 17275–89.

Ji, Y.B., F. Dong, D.B. Ma, J. Miao, L.N. Jin, Z.F. Liu, and L.W. Zhang. 2012b. Optimizing the extraction of anti-tumor polysaccharides from the fruit of *Capparis spionosa* L. by response surface methodology. *Molecules* 17:7323–35.

Nazıroğlu, M., M.B. Akay, Ö. Çelik, M.İ. Yıldırım, E. Balcı, and V.A. Yürekli. 2013. *Capparis ovata* modulates brain oxidative toxicity and epileptic seizures in pentylentetrazol-induced epileptic rats. *Neurochem Res* 38: 780–8.

Rathee, P., D. Rathee, D. Rathee, and S. Rathee. 2012. *In-vitro* cytotoxic activity of β-sitosterol triacontenate isolated from *Capparis decidua* (Forsk.) Edgew. *Asian Pac J Trop Med* 5: 225–30.

30 Centers of Capparology

AFRICA

Egypt

Chemistry of Medicinal Plants Department, National Research Centre (NRC), Dokki, Cairo, Egypt. mahasoltan@netscape.net

Department of Medicinal Chemistry, National Research Center, Dokki, Cairo, Egypt.

Department of Pharmacognosy, Faculty of Pharmacy, Suez Canal University, 41522, Ismailia, Egypt

Phytochemistry Department, National Research Centre, El Bohoth Street, 12311, Dokki, Egypt.

Phytochemistry and Plant Systematics Department, National Research Centre, Dokki-12311, Cairo, Egypt. sharafali@hotmail.com

Morocco

Laboratory of Endocrinian Physiology, FSTE Boutalamine and Pharmacology, EDDOUKS, UFR PNPE, BP 21, Errachidia 52000, Morocco. m.eddouks@caramail.com

Nigeria

Department of Pharmacology, Faculty of Basic Medical Sciences, College of Medicine, University of Lagos, Idi-araba, Nigeria

Tunisia

Laboratoire de Biochimie, Département de Biologie, Faculté des Sciences de Tunis, Université Tunis El-Manar, Tunis 2092, Tunisia. Nizar.Tlili@fst.rnu.tn

ASIA

Bangladesh

Phytochemistry Research Laboratory, Department of Pharmacy, University of Rajshahi, Rajshahi 6205, Bangladesh.

China

College of Chemistry and Biological Science, Ili Normal University, Yining, Xinjiang, People's Republic of China 835000

College of Chemistry & Chemical Engineering, Xinjiang University, Urumqi, People's Republic of China 830046

College of Pharmaceutical Sciences, Xinjiang Agricultural University, Urumqi 830052, China. xjmshj@sina.com

Center of Research and Development Center of Life Sciences and Environmental Sciences, Harbin University of Commerce, Harbin, People's Republic of China 150076
Department of Biochemistry, Faculty of Medicine, The Chinese University of Hong Kong, Shatin, New Territories, Hong Kong, China. lamszekwan@yahoo.com.hk
Department of Dermatology, Second Affiliated Hospital, School of Medicine, Zhejiang University, 88 Jiefang Road, 310009 Hangzhou, China. CYL123321@sohu.com
Institute of Biophysics, Chinese Academy of Sciences, Beijing, People's Republic of China 100101
Institute of Chinese Materia Medica, Shanghai University of Traditional Chinese Medicine, Shanghai, China.
Institute of Materia Medica, Peking Union Medical College and Chinese Academy of Medical Sciences, Beijing 100050, People's Republic of China.
Institute of Medicinal Plant Development, Chinese Academy of Medical Sciences and Peking Union Medical College, Beijing, People's Republic of China.
Institute of Radiation Medicine, Academy of Military Medical Sciences, Beijing, People's Republic of China 100850
Key Laboratory of Bioactive Substances and Resources Utilization of Chinese Herbal Medicine, Institute of Materia Medica, Ministry of Education, Peking Union Medical College and Chinese Academy of Medical Sciences, Beijing, People's Republic of China 100050
Key Laboratory of Standardization of Chinese Medicines of the Ministry of Education, Shanghai, People's Republic of China.
National Key Biotechnology Laboratory for Tropical Crops, CATAS, Hainan, People's Republic of China.
National Laboratory of Biomacromolecules, Institute of Biophysics, Chinese Academy of Sciences, 15 Datun Road, Chaoyang District, Beijing 100101, People's Republic of China.
Plant Biotechnology Research Center, School of Agriculture and Biology, Fudan-SJTU-Nottingham Plant Biotechnology R&D Center, Shanghai Jiao Tong University, China.
Research Center on Life Sciences and Environmental Sciences, Harbin University of Commerce, Harbin 150076, China. jyb@hrbcu.edu.cn
School of Life Science, Tarim University, Alar, People's Republic of China 843300
School of Traditional Chinese Materia Medica, Shenyang Pharmaceutical University, Shenyang, People's Republic of China 110016
Second Affiliated Hospital, College of Medicine, Zhejiang University, Hangzhou 310009, China. CYL123321@sohu.com
Shanghai Institute of Pharmaceutical Industry, 1320 West Beijing Road, Shanghai 200040, People's Republic of China.
State Key Laboratory of Natural and Biomimetic Drugs, Peking University Health Science Center, Beijing, People's Republic of China 100083
State Key Laboratory of Systematic and Evolutionary Botany, Institute of Botany, Chinese Academy of Sciences, No. 20 Lin xin cun, Xiangshan, Beijing 100093, China.
Xinjiang Key Laboratory of Grassland Resources and Ecology, College of Grassland and Environment Sciences, Xinjiang Agricultural University, Urümqi 830052, China.
Xinjiang Medical University, Urumqi, Xinjiang Province, People's Republic of China 830000
Xinjiang Production and Construction Corps Key Laboratory of Protection and Utilization of Biological Resources in Tarim Basin, Tarim University, Alar, Xinjiang Province, People's Republic of China 843300

India

Department of Biotechnology, Alagappa University, Karaikudi, Tamil Nadu, India.
Department of Foods and Nutrition, CCS Haryana Agricultural University, Hisar-125004, India.

Department of Pharmaceutical Chemistry, Grace College of Pharmacy, Palakkad, India.

Department of Pharmaceutical Technology, Bharathidasan Institute of Technology, Anna University, Tiruchirappalli-620 024, India.

Department of Quality Assurance, PDM College of Pharmacy, Bahadurgarh-124507, India.

Department of Zoology, DDU Gorakhpur University, Gorakhpur 273009, India

Department of Zoology, JNV University, Jodhpur 342 001, India. purohit1411@yahoo.com Ranbaxy Research Labs, R&D-III, Gurgaon, Haryana, India. anil.kanaujia@ranbaxy.com

Faculty of Pharmacy, Jamia Hamdard (Hamdard University), New Delhi, India 110062

Hindu College of Pharmacy, Sonepat, Haryana, India

Institute of Pharmaceutical Education and Research, Borgaon, Wardha, MS, India. dr_yeole@rediffmail.com

Molecular Endocrinology Laboratory, Department of Biotechnology, Indian Institute of Technology Roorkee, Roorkee 247 667, Uttarakhand, India.

New Drug Discovery, R&D-III, Udyog Vihar Industrial Area Research, Ranbaxy Research Labs, Gurgaon, Haryana, India

Pharmacognosy and Phytochemistry Laboratory, Faculty of Pharmaceutical Sciences, Guru Jambheshwar University of Science and Technology, Hisar, India 125001

Pharmacy Department, Faculty of Technology and Engineering, M.S. University of Baroda, Gujarat, India.

Pharmacy Wing, Lachoo Memorial College of Science and Technology, Jodhpur 342008, India. manojgoyal620@yahoo.com

Plant Molecular Biology Research Unit, Department of Plant Biology and Plant Biotechnology, St. Xavier's College (Autonomous), Palayamkottai, Tamil Nadu, India 627 002

School of Pharmaceutical Sciences, Shobhit University, NH-58, Modipuram, Meerut, 250110, UP, India.

Iran

Department of Chemistry, Science and Research Branch, Islamic Azad University, Tehran, Iran

Department of Medicinal Chemistry, School of Pharmacy, Ahvaz Jundishapur University of Medical Sciences, Ahvaz, Iran.

Israel

Department of Parasitology, The Kuvin Center for the Study of Infectious and Tropical Diseases, The Hebrew University-Hadassah Medical School, 91120, Jerusalem, Israel. jacobsr@cc.huji.ac.il

Laboratory of Applied Metabolomics and Pharmacognosy, Institute of Evolution, University of Haifa, Haifa, Israel punisyn@gmail.com

Japan

Department of Preventive Dentistry, Kagoshima University Graduate School of Medical and Dental Sciences, 8-35-1 Sakuragaoka, Kagoshima 890-8544, Japan.

Graduate School of Life and Environmental Sciences, University of Tsukuba, 1-1-1 Tennodai, Tsukuba, Ibaraki 305-8572, Japan.

National Food Research Institute, Tsukuba, Japan 305-8642

School of Pharmaceutical Sciences, University of Shizuoka, 52-1 Yada, Shizuoka 422-8526, Japan.

Jordan

Chemistry Department, University of Jordan, Amman, 11943, Jordan

Department of Pharmaceutics and Pharmaceutical Technology, Faculty of Pharmacy, University of Jordan, Amman, Jordan. rulad@ju.edu.jo

Pakistan

Department of Pharmacognosy, Research Institute of Pharmaceutical Sciences, University of Karachi, Karachi-75270, Pakistan. ahirzia@gmail.com

Department of Zoology and Fisheries, University of Agriculture, Faisalabad 38040, Pakistan. zafaruaf@yahoo.com

Institute of Chemistry, University of Sindh, Jamshoro, Pakistan

Natural Products Research Unit, Department of Biological and Biomedical Sciences, Aga Khan University Medical College, Karachi, Pakistan.

Saudi Arabia

Department of Natural Products, Faculty of Pharmacy, King Abdulaziz University, Jeddah 21589, Saudi Arabia. abdelsattar@yahoo.com

Turkey

Department of Chemistry, College of Education, Ataturk University, Erzurum, Turkey. 25240

Department of Field Crops, Faculty of Agriculture, Dicle University, Diyarbakir-21280, Turkey.

Department of Food Engineering, Faculty of Agriculture, University of Selçuk, Konya, Turkey. mozcan@selcuk.edu.tr

Department of Horticulture, Faculty of Agriculture, Kahramanmaras Sutcu Imam University, Turkey. gulat@ksu.edu.tr

Department of Pharmacognosy, Faculty of Pharmacy, Hacettepe University, 06100 Ankara, Turkey.

Department of Pharmacology, Faculty of Pharmacy, Anadolu University, Eskisehir, Turkey. nurcanbektas@anadolu.edu.tr

Kafkas University, Artvin Faculty of Forestry, 08000 Artvin, Turkey. zaferolmez@yahoo.com

Uzbekistan

Inst. Khim. Rastit. Veshchestv, AN RUz, Tashkent, Uzbekistan

S. Yu. Yunusov Institute of Chemistry of Plant Substance, Academy of Sciences of Uzbekistan, Tashkent, Uzbekistan 100170

EUROPE

Germany

Institute for Lipid Research, Federal Research Center for Nutrition and Food, 48006 Münster, Germany. matthaus@uni-muenster.de

Greece

Section of Botany, Department of Biology, University of Athens, Athens 157 81, Greece

Italy

Department Farmaco-Biologico, School of Pharmacy, University of Messina, Contrada Annunziata, 98168 Messina, Italy. trombett@pharma.unime.it

Department of Life and Environmental Science and Dipartimento di Scienze Biomediche, Università degli Studi di Cagliari, Via Ospedale 72, 09124 Cagliari, Italy

Department of Organic Chemistry, A. Mangini, Alma Mater Studiorum-Università di Bologna, Bologna, Italy

Department of Pharmaceutical Sciences, Faculty of Pharmacy, University of Catania, V. le A. Doria 6, 95125 Catania, Italy. panico@unict.it

Department of Pharmaceutical Sciences, Faculty of Pharmacy, Nutritional and Health Sciences, University of Calabria, Arcavacata di Rende (CS), Italy I-87036

Department of Pharmaceutical Sciences, University of Calabria, Arcavacata di Rende, Italy. filomena.conforti@unical.it

Dipartimento di Chimica Organica e Biologica, Facolta di Scienze, Contrada Papardo, Universita di Messina, Messina, Italy 98166

Department of Surgical Science, Unit of Microbiology, University of Messina, Italy.

Dipartimento Farmacochimico Tossicologico e Biologico, Università di Palermo, Via M. Cipolla 74, 90128 Palermo, Italy. mal96@unipa.it

Dipartimento di Scienze e Tecnologie Biomediche, Cattedra di Chimica Biologica, Universita di Cagliari, Cittadella Universitaria, 09042 Monserrato, Italy. sanjust@unica.it

Istituto del CNR di Chimica Biomolecolare, Via P. Gaifami 18, 95126 Catania, Italy. laura.siracusa@icb.cnr.it

Pharmaco-Biological Department, School of Pharmacy, University of Messina, SS. Annunziata, 98168 Messina, Italy. germamp@pharma.unime.it

Spain

Area de Microbiología, Departamento de Ciencias de la Salud, Facultad de Ciencias Experimentales, Universidad de Jaén, Jaén, Spain.

Food Biotechnology Department, Instituto de la Grasa (C.S.I.C.), Seville, Spain 41012

Switzerland

Nestlé Research Center, NESTEC Ltd., Lausanne, Switzerland 1000

UK

School of Pharmacy and Biomedical Science, University of Portsmouth, Portsmouth, UK PO1 2DZ

NORTH AMERICA

USA

Center for New Crops & Plant Products, Purdue University. Department of Horticulture and Landscape Architecture, 625 Agriculture Mall Drive, West Lafayette, IN 47907-2010 Fax: 765-494-0391

Department of Basic Sciences and Craniofacial Biology, New York University College of Dentistry, 345 East 24th Street, New York, NY 10010, USA. ec46@nyu.edu

Missouri Botanical Garden, P.O. Box 299, St. Louis, Missouri 63166

Natural Products Laboratory, School of Pharmacy, University of North Carolina, 27599, Chapel Hill, NC, USA

SOUTH AMERICA

Brazil

Departamento de Quimica, Universidade Estadual de Maringa, Maringa, PR, Brazil 87020-900

Epilogue

The steadfastness of the genus *Capparis*, its adaptability and innovation, might well, in an evolutionary sense, be the story of our own. Faced with adversity throughout the niches of planet Earth, *Capparis* responded with diversity. This has been evident in its physical being, from the sheer breadth of its fabulous phytochemistry to the robust nature of its physical form: tough leaves, resourceful roots, sharp thorns, tough little buds, even tougher fruits, and opalescent flowers. The success of *Capparis* at a casual glance may seem to have been capricious, but its staying power and ability to adapt has been unassailable.

As "outside observers" to this internal story of a species in another kingdom, we have also seen caper make its impact on our own adventurous and curious psyches. We, like its many pollinating friends in the animal world, are struck by these plants' beauty, sobered by their defenses, and curiously drawn to their riches. We sense something tasty in this plant and possibly something of even greater value. For beyond the culinary wealth of capers for our hungry palates lies the medical potential of capers for our damaged health.

In a nutshell, this potential derives from a simple principle at the heart of the modern science of pharmacognosy, namely, the panoply of phytochemicals that plants evolve for their own botanical needs—defense, attracting pollinators, and advancing their migration throughout the world (and perhaps beyond) into new niches and new species—these very same phytochemicals are useful to us (and other animals) for healing and correcting of our own imbalances and diseases. What *Capparis* evolved chemically over countless millennia for its own purposes, we, as humans, may borrow for our own purposes, namely, the healing of our beings, minds, bodies, and even emotions. If this seems overdone or too far reaching, consider the value of this amazing genus for curing our infections, ridding our abodes and food sources of infestation, soothing our nerves, curbing our seizures, melting our cancers, cleansing our hearts and circulatory systems, correcting our sugar metabolism, balancing our weight, mitigating our pain, and cooling our fevers and inflammation. All of these lofty objectives are clearly, as our text has shown, found within the province of *Capparis*.

From the common caper of commerce known to most of us, *C. spinosa*, to the much esteemed medicinal article of Ayurveda, *C. decidua*, and dozens of species in between, *Capparis* shows its great medical potential from its pervasive yet humble footholds. From the quietest peripheries of the natural world, *Capparis* sheds great light and offers much hope to our troubled souls and our diseased bodies. Clearly, *Capparis* is a genus worthy of our attention, our respect, and our pursuit.

Index

Note: Numbers in *italic* indicate figures.

A

B

U

V

W

X

Y

Z

Milton Keynes UK
Ingram Content Group UK Ltd.
UKHW050453071024
449327UK00015B/359